COMPUTER
INTEGRATED
MANUFACTURING

An Introduction
with Case Studies

To my sons,
Gergely and Richard

COMPUTER INTEGRATED MANUFACTURING

An Introduction
with Case Studies

PAUL G. RANKY

Associate Professor
Industrial Technology Institute
The University of Michigan, Ann Arbor
(Formerly of Trent Polytechnic,
Nottingham, UK)

Prentice/Hall PHI International

Englewood Cliffs, NJ   London   Mexico   New Delhi
Rio de Janeiro   Singapore   Sydney   Tokyo
Toronto   Wellington

Library of Congress Cataloging in Publication Data

Ránky, Paul G., 1951–
Computer integrated manufacturing.

Includes bibliographies and index.
1. Computer integrated manufacturing systems.
2. CAD/CAM systems.   3. Flexible manufacturing
systems.   I. Title.
TS155.6.R36   1985   670.42'7   85-19199
ISBN 0-13-165655-4

British Library Cataloguing in Publication Data

Ránky, Paul G.
Computer integrated manufacturing: an
introduction (with case studies).
1. Computer integrated manufacturing systems
I. Title
670.42'7   TS155.6

ISBN 0-13-165655-4

ISBN 0-13-165655-4

Prentice-Hall, Inc., *Englewood Cliffs, New Jersey*
Prentice-Hall International, UK, Ltd, *London*
Prentice-Hall of Australia Pty Ltd, *Sydney*
Prentice-Hall Canada, Inc., *Toronto*
Prentice-Hall Hispanoamericana, S.A., *Mexico*
Prentice-Hall of India Private Ltd, *New Delhi*
Prentice-Hall of Japan, Inc., *Tokyo*
Prentice-Hall of Southeast Asia Pte Ltd, *Singapore*
Editora Prentice-Hall do Brasil Ltda, *Rio de Janeiro*
Whitehall Books Ltd, *Wellington, New Zealand*

Printed in Great Britain by
A. Wheaton & Co. Ltd, Exeter

Contents

viii   Contents

Foreword

It is widely known and agreed that we are encountering a worldwide upheaval in the way manufacturing processes are conducted. In marked contrast to former times, when repetitive automation worked side-by-side with human craftsmanship, the new wave of manufacturing process technology combines the ideas of flexibility, adaptability and modularity to produce better products in new ways. It will eventually touch all sectors of manufacturing: first in the durable goods, electronics, transportation and aerospace industries where its presence is already strong; later in building, clothing, mining and other industries as limited only by our imaginations and courage to take risks:

We now stand on the threshold of the *Factory of the Future*, which differs from the *Factory of the Past* essentially by the capabilities brought forth by modern digital computing. The early stages of this metamorphosis, dating back to the early 1960's, were characterized by NC machining, CNC machining, robotics, group technology and flexible manufacturing systems (FMS). These tools gave us islands of automation, usually separated by gulfs of inefficiencies. All these modern tools, however, were a prolog to the *Factory of the Future*, whose essence is the integration of these various machines so that they may work harmoniously in the overall process of manufacturing. With the power of VLSI microcomputers literally at our fingertips, real-time process control and reprogrammability of the tools of manufacturing enable us to explore new possibilities of product variety, life cycle and customization that were previously impossible.

The present important book follows closely the publication of Professor Ránky's two earlier books on FMS and Robotics (Paul G. Ránky: *The Design and Operation of FMS*, IFS (Publications) Ltd. and North-Holland, 1983; and Paul G. Ránky and C.Y. Ho: *Robot Modelling, Control and Applications with Software*, IFS (Publications) Ltd. and Springer Verlag, 1985.)

This book successfully makes the complex milieu of computer integrated manufacturing accessible to the student and the professional alike. It regards CIM as the interrelationship among four elements: the business data processing system, CAD, CAM and flexible manufacturing. By charting out the details of these elements and their mutual connections, the reader is provided both with a map describing the process of CIM, and a convenient means of analyzing the important features of the process. Thus

CIM-A*

the reader is led in a natural way to understanding the four elements mentioned above, as well as operational control utilizing dynamic scheduling, capacity planning and control, batchsize analysis, alternative routing using expert rule-base systems, and system balancing. Also a practical approach to CIM plant design layout is provided. The book is rich with useful case studies which are worked out in detail to facilitate deeper insight into the principles Professor Ránky discusses.

The book is at once comprehensive and understandable, pedagogical and practical. It is well illustrated, with many examples of equipment and implementations from around the world reflecting Professor Ránky's international activity in the CIM field. It will be useful both as a textbook and as a reference tool. In this book, Professor Ránky sets a high standard for others on the subject that will surely follow.

Philip H. Francis
Industrial Technology Institute
Ann Arbor, Michigan

Author's preface

One might be shocked at the enormous speed of development in computer technology and its application to different manufacturing processes, but one should also remember that humans are required and will always be required for designing, implementing, maintaining and redesigning such systems. The development of CIM and all of its subsystems is a continuous ('never ending') process providing exciting future employment for those who are willing and able to explore this new industrial revolution.

The aim of this text is to give a systematic introduction to some important parts of this rapidly developing technology with software examples and practical case studies to help the reader not only to understand, but also to gather experience in some important areas of CIM by studying simulation results.

To focus on the main objectives and system development tools of CIM we start the discussion with the core modules of such systems, including distributed data processing systems, computer networks, Database Management Systems (DBMS) and expert systems.

After this introduction the subject is discussed in four major parts, these being:

1. The business data processing system.
2. Integrated Computer Aided Design (CAD).
3. Computer Aided Manufacturing (CAM) systems.
4. Flexible Manufacturing Systems (FMS).

The reason for discussing CIM like this is logical and simple: the business data processing system provides the financial, organizational and data processing foundation, integrated CAD and CAM offer computer assistance in design, analysis, NC/CNC part programming, robot programming, etc. and FMS systems (meaning not only machining, but assembly, welding, testing and other processes) are there to execute the established plans and schedules in a flexible way.

It must be emphasized that although CIM must be tailored to each individual business and/or organization a major part of it is common from the system design point of view to many industries. In this text an attempt is made to explain and illustrate this important point with examples and case studies.

Following this 'general systems approach' throughout the book the word 'manufacturing' is used to refer to a number of different and often integrated processes including machining, inspection, testing, welding, assembly, painting, packaging, etc. In other words 'manufacturing' in this book means not only 'machining' but rather 'computer assisted fabricating' with all necessary related activities and processes.

The book relies on a series of case studies offering useful examples and computer simulation results. They do not give a full list of all CIM activities, but illustrate and explain some important concepts and can be used following a 'teach-yourself' method as well as in the classroom to understand the following fields:

- Database Management Systems.
- Computer Aided Design and computer graphics, including two dimensional spline interpolation and two and three dimensional graphics manipulation using interactive systems and high level graphics languages.
- Numerically controlled and CNC machine programming, including the manual programming of CNC machines and a COMPACT II macro library.
- Industrial robot programming and off-line robot program generation.
- Integrated CAD/CAM systems, computer graphics and design applications.
- Flexible Manufacturing Systems including a COMAO, a KTM, a Cincinnati-Milacron and a Yamazaki FMS.
- FMS simulation and operation software including loading sequencing, dynamic scheduling, capacity planning and control, batch size analysis and manufacturing system balancing with special interest towards flexible, robotized assembly.

The main emphasis in the whole text is put on the understanding of principles and relationships between different sub-systems of CIM by a series of self documented case studies and a brief introduction of the topic they relate to.

The software case studies are written as much as possible in a self explanatory way. However it is not our aim to replace programming manuals or handbooks, thus for the inexperienced reader the availability of Pascal, Compact II, Unimation's VAL, McAuto's Unigraphics and other language reference manuals as well as the use of business system, CAD, CAM and FMS software and related machinery is advisable in order to gain as much 'hands-on' experience as possible.

Note that further details on the case studies and the FMS Software Library used in this text can be gained directly from the author, address:

ITI, The University of Michigan,
P.O. Box 1485, Ann Arbor,
Michigan 48106, USA
Tel. (313) 764-6775

Software relating to FMS is also available from the following two system houses and software publishers and distributors:

MALVA Ltd, Real-time Software Specialists,
70 Ashchurch Drive, Wollaton,
Nottingham NG8 2RA, United Kingdom
Tel. (0602) 284593
Contact: Mr John G. Crouchley, BSc., Director

or from:

COMPORGAN System House,
H-1022 Budapest, Beg u. 3-5,
Hungary
Tel. 351-335
Telex: H-226708 coorg h
Contact: Mr K. Pogany, MSc., Director

Acknowledgements

There are many friends, colleagues and students as well as companies and institutions to whom I would like to express my thanks for their contribution and/or for sending me photographs of their systems, or machines.

Among the many let me mention first of all Dr Philip Francis at ITI in Ann Arbor for his kind foreword and the interesting talks we had during the past few years on CIM.

I would like to express my thanks to Mr Ian Taylor at Cincinnati-Milacron for his interesting comments on FMS, Mr Peter T. Rayson at Trent Polytechnic for his comments and useful assistance regarding the McAuto case studies.

Thanks are also expressed to Mr Jim Corlett and my editor Mr Glen Murray for reading the text, Mr Malcolm Roberts and Dr Peter Johnson for their comments regarding the business data processing system of CIM and my friends and previous colleagues at Plessey Office Systems Ltd, Nottingham, for providing access to some outstanding computer and robot systems and implementing some of my ideas and designs regarding flexible automation and robotized systems.

I would also like to mention here some of my students, namely Mr Colin Godson, Mr Paul Smith, Mr Kevin Walker, Mr David Johnson, Mr Mark Duddles, Mr John Burgess and my BSc, MSc and Double Honours degree classes at Trent Polytechnic, Nottingham, to whom I have lectured FMS, robotics, CIM and related material during the past five years, for helping me in testing some of the ideas, algorithms and programs included in this text.

Last but not least let me express my warmest thanks to my family and my wife Marti for their love, full support and hard work without which this project couldn't have been completed.

Paul G. Ránky
May 1985

CHAPTER ONE

Introduction to Computer Integrated Manufacture (C I M)

1.1 What is Computer Integrated Manufacture and why is it important to understand it?

Money is tight, worldwide competition is tough, demand and technology are changing at an exponentially growing rate, thus companies need to react to changes much faster and in a more flexible way than in the past.

But how can one see all different but important aspects of a complex business? What is required to be able to make correct decisions at a "superfast" speed? How can one optimize information flow, manufacturing and all related processes at the same time? How can one increase design and/or manufacturing productivity by 2 to 10 times and decrease production costs at the same time by 10 to 50%?

To summarize, how can one take the best possible decisions at the right time?

There is no perfect solution in terms of computing and human decision-making skills cannot be ignored either, but:

* if the decision makers in the organization have up-to-date information on the requested processes and data,

* if there is the possibility of controlling and analyzing large amount of business and technical data by computers,

* if design and production engineering are integrated, in

other words if the designers can get guidance and feedback from databases, knowledge bases of expert systems,
from the shop-floor and from other levels (e.g. production engineering) of the organization, on how to solve
certain design tasks to suit the manufacturing facilities best),

* if manufacturing in its broad sense (i.e. not only
machining, but also inspection, test, assembly, etc.)
is done mainly by Flexible Manufacturing Systems
producing the necessary workparts on order rather than
for stock,

then the company will have a better chance for survival
in the age of CIM.

Computer Integrated Manufacture (CIM) is concerned with
providing computer assistance, control and high level integrated automation at all levels of the manufacturing industries,
by linking islands of automation into a distributed processing
system. The technology applied in CIM makes intensive use of
distributed computer networks and data processing techniques,
Artificial Intelligence and Data Base Management Systems
(Figure 1.1).

Dr Bunce provides the CAM-I definition of CIM in [1.9]
as follows: "A series of interrelated activities and operations involving the design, materials selection, planning,
production, quality assurance, management and marketing of
discrete consumer and durable goods... CIM is the deliberate
integration of automated systems into processes of producing a
product...CIM can be considered as the logical organization of
individual engineering, production and marketing/support
functions into a computer integrated system".

Computer Integrated Manufacturing, in this respect, cov-
ers all activities related to the manufacturing business, including:

* Evaluating and developing different product strategies.

* Analyzing markets and generating forecasts.

* Analyzing product/market characteristics and generating
concepts of possible manufacturing systems (i.e. FMS
cells and FMS systems).

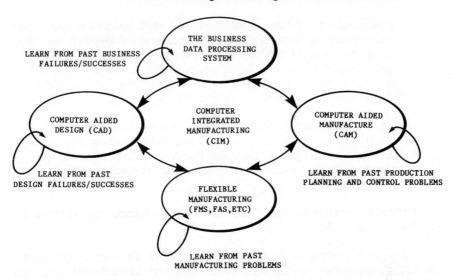

Figure 1.1 A Computer Integrated Manufacturing (CIM) system
concept indicating the importance of integrated data
flow, distributed data base management and communication as well as expert "learning and decision
making systems" adding some level of "intelligence"
to each software module and machinery of the total
system.

* Designing and analyzing components for machining, inspection, assembly and all other processes relating to the
nature of the component and/or product, (i.e. welding,
cutting, laser manufacturing, presswork, painting, etc.).

* Evaluating and/or determining batch sizes, manufacturing
capacity, scheduling and control strategies relating to
the design and fabrication processes involved in the
particular product.

* Analysis and feedback of certain selected parameters
relating to the manufacturing processes, evaluation of
status reports from the DNC (Direct Numerical Control)
system (source data monitoring and machine function
monitoring in real-time).

* Analyzing system disturbances and economic factors of the
total system.

It is quite obvious from this list that the aim of CIM is to let the advanced information processing technology penetrate into all areas of the manufacturing industry in order to:

* Make the total process more productive and efficient.

* Increase product reliability.

* Decrease the cost of production and maintenance relating both to the manufacturing system as well as to the product.

* Reduce the number of hazardous jobs and increase the involvement of well educated and able humans in the manufacturing activity and design ([1.1],[1.2],[1.3],[1.6]).

In comparison with FMS, CIM is mainly concerned with the information processing tasks at all levels of the factory and its management, whereas FMS provides the essential computer controlled manufacturing tools and systems for CIM to execute the computer generated plans and schedules that take account of a total system rather than just one cell or shop. One could also say that CIM integrates "FMS islands" with the overall computer network, the business system, with different design and and manufacturing databases of the company and allows the optimization of the data flow and eventually all activities at a much broader level than FMS ([1.1],[1.4],[1.4],[1.5]).

Figure 1.2 illustrates the CIM concept by showing some of the most important modules of the business system, (CAD Computer Aided Design), CAM (Computer Aided Manufacture) and FMS. It also indicates the most important data paths and the macro level relationship between the indicated modules and subsystems. Note that this figure will appear again in the book and each time different parts will be discussed in more detail. To offer in-depth knowledge in many cases source code level and/or application level case studies are shown and some of the more important details discussed.

1.2   The most important skills required when designing, implementing and working with CIM

What is the necessary knowledge engineers and managers should have when designing, implementing and/or working with CIM?

Since computers are involved in the increasingly fast and intelligent decision making process at all levels of the manufacturing industry, more skilled engineers and managers are required who have both a broader view of the business, the design and manufacturing facility, and a good understanding of the processes involved. The CIM engineer is working with computers, thus he or she must be familiar with the way software projects are designed, coded, tested, maintained and documented. Some of the most important activities and the required skills are summarized in Figures 1.3 and 2.1 (discussed in Chapter 2).

These Figures underline the important fact that working with CIM means that the software produced and used should be developed, tested, documented and installed in a similar way as any other product.

The most important areas of knowledge the "CIM engineer" should have include the ability to describe, define and analyze different computer integrated models as a system, by specifying its components and their functional relationships.

It is almost impossible to outline the areas of the required engineering knowledge without knowing the actual processes involved, but he must have a thorough understanding of the technical aspects of his own job as well as other areas.

The knowledge of computers and interfaces used in intelligent machines, the ability to develop several different logical structures for the same specified problem, to outline, design and implement medium and large size software projects in the areas of database processing, computer graphics, real-time control, digital network communication and applications software design are also very important.

The CIM engineer must have the ability to perform an economic and time analysis, based on known facts and information which is not available, but can be obtained by simulation.

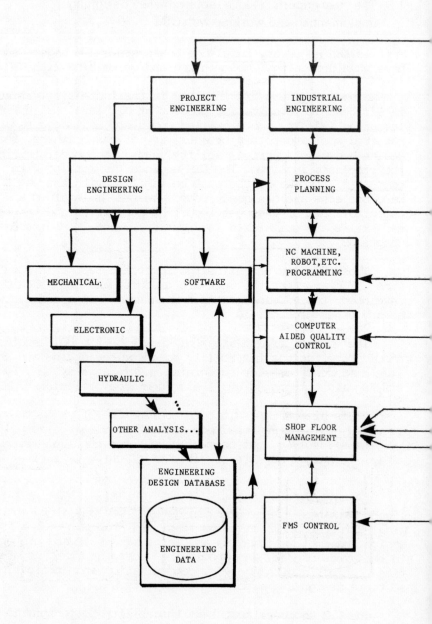

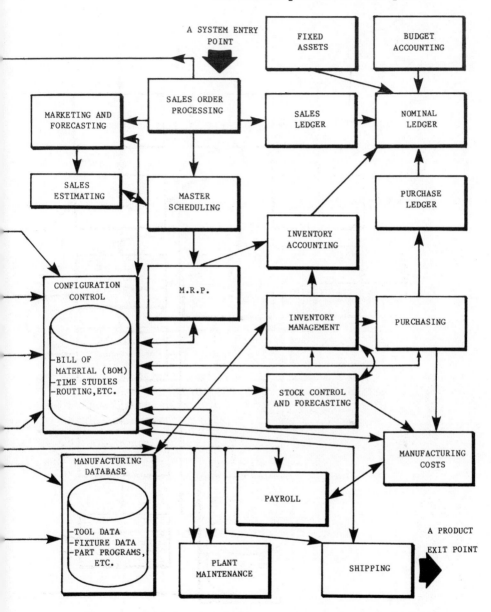

Figure 1.2 An overall CIM model integrating business data processing, Computer Aided Design (CAD), Computer Aided Manufacture (CAM) and Flexible Manufacturing Systems (FMS), incorporating machining, assembly, test and other processes.

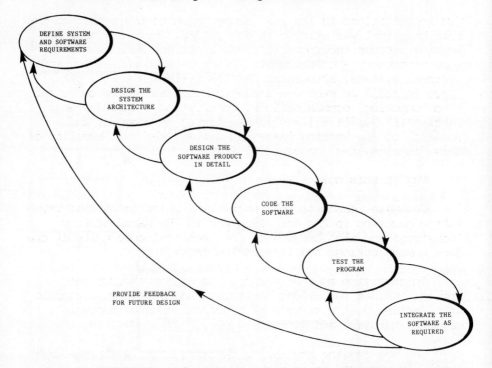

Figure 1.3 Summary of the most important software engineering activities when designing large software systems. (The arrows indicate the continuous communication requirements between different teams working on the software product.)

In general he or she must be well above average in accuracy and thoroughness in his job and must be enthusiastic about it and must co-operate with other people, otherwise he will not be able to overcome occasional setbacks, panic situations and the vast amount of work regarding information processing.

1.3   Unsolved problems in CIM

Let us emphasize that CIM is very new, there are many "unsolved problems", and that there are no "truly integrated"

CIM installations in its real sense known to the author at the time this text was written in early 1985. There are many working systems incorporating some important modules of CIM, such as factory - or corporate - wide data communication and business systems, integrated CAD/CAM systems and FMS, but there is still an enormous amount of work and money to be put into computing, communications, interfacing, knowledge engineering ([1.7], [1.8]) and manufacturing development and research to see working installations and the real benefits of this "total system" concept ([1.6]).

Why is this so?

The main reason is that there is a major industrial revolution occuring in the computing and in the manufacturing industries and also because of the technical complexity of the data communications and interfacing tasks.

To underline some of the most important factors one should consider the amount of storage and interfacing trouble when dealing with a variety of different control systems, computer hardware and software.

Can one imagine how many giga-bytes of bulk storage and memory is required to hold and process all the graphics files of drawings and manufacturing data of a large company? How many communication lines are required to interface every important area of control and how many compatible programs are required to drive the system as a whole?

The other aspect of this question is reliability. What is the reliability of current CNC machines, robots, assembly and inspection devices, computers and complex systems built from such subsystems?

In many cases we do not know, although in the case of unmanned manufacturing this is a crucial problem to be solved. For example robots haven't really worked long enough to be able to assess the full consequences of their implementation. Also there is a considerable barrier to gaining such information from manufacturers of robots, FMS machinery, etc., because of the fierce competition in this industry.

To be optimistic one should say that most computers and computer controlled equipment are more reliable than they were

a few years ago. Some robots known to the author have achieved
92-95% uptime within a period of a year. These are promising
data, but one could raise terrifying examples too.

To summarize, the major tasks to be solved in the area
of CIM are all related to integration and to providing the
possibility of further development at user level. This
involves a vast amount of hardware and software communication
and interfacing problems, database management, self-learning,
diagnosing and other expert system development needs. Powerful
16 to 32 bit microcomputers and the availability of massive
bulk stores and fast operating systems on such machines provide an excellent networking facility even for small companies,
who are not able to invest large sums of money in mainframes
and their sophisticated software systems.

The above mentioned integration problems will be more
easily solved when powerful microcomputers can be linked with
all the machinery utilized on the shop-floor and with the
factory organization as a whole and when those people who are
going to write the software for these machines not only
understand the data processing aspects, but the processes
involved too ([1.9], [1.10], [1.11], [1.12], [1.13], [1.14],
[1.15]).

References and further reading

[1.1] Paul G. Ranky: The Design and Operation of Flexible
      Manufacturing Systems, IFS (Publications) Ltd.,and North
      Holland Publishing Co. 1983. 348 pp.

[1.2] Daniel S. Appleton: The State of CIM, Datamation, Decem-
      ber 15, 1984. p. 66-72.

[1.3] "Plan for Computer Integrated Manufacturing" seminar lec-
      tures organized by the Institution of Production Engi-
      neers, London, November 1984.

[1.4] Naoaki Usui: MITI's "super" manufacturing system, Ameri-
      can Machinist, August 1984. p. 86-114.

[1.5] Duncan Holland: Strategic benefits gained from CIM, The
      Production Engineer, June 1984. p.14-16.

[1.6] Technology, a comprehensive survey of the different components of CIM, Tooling and Production, March 1984, p. 24-204.

[1.7] N. Shahla Yaghmai: Expert Systems: A tutorial, American Soc. Information Sciences, 5 September 1984. p. 297-305.

[1.8] Nick Cercone: Artificial Intelligence: underlyning assumptions and basic objectives, American Soc. Information Sciences, 5 September 1984. p. 297-305.

[1.9] P. Bunce: Planning for CIM, The Production Engineer, Vol. 64, No.2 February 1985. p. 21.

[1.10] J. Harrington: Computer Integrated Manufacturing, Krieger, 1979.

[1.11] M.Annborn: The Factory of the Future, 3rd International Conference on Flexible Manufacturing Systems, 1984, Boeblingen, W-Germany, IFS (Publications) Ltd., 1984. p. 59-73.

[1.12] M. Mollo: A distributed control architecture for a Flexible Manufacturing System, 3rd International Conference on Flexible Manufacturing Systems, 1984, Boeblingen, W-Germany, IFS (Publications) Ltd., 1984. p. 227-241.

[1.13] C. Accomazzo: The integration of an inspection cell in FMS, 3rd International Conference on Flexible Manufacturing Systems, 1984, Boeblingen, W-Germany, IFS (Publications) Ltd., 1984. p. 285-297.

[1.14] D. Van Zeeland: System software, 3rd International Conference on Flexible Manufacturing Systems, 1984, Boeblingen, W-Germany, IFS (Publications) Ltd., 1984. p. 307-311.

[1.15] Paul G. Ránky: Integrated software for designing and analysisng FMS, 3rd International Conference on Flexible Manufacturing Systems, 1984, Boeblingen, W-Germany, IFS (Publications) Ltd., 1984. p. 347-361.

CHAPTER TWO

Overview of some general purpose software sub-systems and software tools used in CIM

Because of the complexity of the required data flow in a possible CIM environment a large number of computers must communicate with each other, with users and with other machinery utilizing many, often different computer networks. The task is further complicated by the fact that in most cases different computers, with different operating systems and communications needs and interfaces must be linked together. Because speed is also an important requirement, data communications efficiency should not be overlooked, making database access and data communication tasks even harder to solve. For efficient data exchange and real-time error recovery, distributed processing with high level of local intelligence is probably the only way to provide a flexible and reliable data processing system.

Distributed processing exists if logically integrated data is physically distributed between a number of nodes, or information processors (e.g. computers, machine tool and robot controllers, etc.) in the data processing network.

This Chapter intends to give a summary of the principles of distributed processing, computer networks, local area networks and database management, the key factors of computer communication in the factory. Although not fully explored yet, it has been realized that expert systems and knowledge engineering in general can offer some hopeful signs for solving the problems of greater intelligence and reliability which are absolutely crucial in complex distributed systems. (The distributed system is reliable if the computer network and the communications software, which provide the data processing link between the nodes, are capable of processing transactions without violating consistency, even if one or more nodes break down in the network).

2.1 Distributed control and computer networks

2.1.1 Principles of distributed processing systems

In a distributed processing system different computers and controllers must communicate with each other and process each others data, and/or execute each other's programs.

Because communication takes time and costs money, low cost and efficient processing and storage methods should be used. This requires a number of conditions to be able to fulfil at each node of the computer network, of which the most important are as follows:

* Compatibility with the "outside world".

* Possibility of real-time communication and data update.

* Logical integrity and physical modularity.

* Hardware and software architecture capable of performing distributed processes.

* Safe and friendly operator interface, if the node is operated by a human operator.

Compatibility inside a subsystem is essential for the proper operation of the subsystem. Compatibility to the "outside world" is important to be able to "talk" to other nodes and to receive data from other nodes. Additional terminals, computers, CNC controllers, etc. should be capable of integrated into a distributed system. If compatibility is achieved, additional equipment and new software modules shall be possible to be located and integrated with only minimal or no modification to the original system, assuming that the designers have allowed for all possible future extensions.

If human operators can have access to the node via a user-friendly operator interface, offering checks both in terms of syntax and semantics on all data input, then it is an important component of the distributed system.

To summarize, a data processing system having distributed intelligence and an associated man-machine, machine-to-machine communication system is essential in order to achieve flexibility, system efficiency, safety and integrated processing capability.

2.1.2 Design considerations of distributed systems

When designing distributed systems it is advised to de-modularize the task into modules of manageable size such as terminals, graphic workstations, CNC controllers, or even to Programmable Logic Controllers (PLCs).

When testing such concepts and designs it is a good idea to construct a system of the selected components and try to simulate all necessary requests, both from the system as well as from the user end, at all levels of the architecture. This is the type of work which probably requires the highest level of skills when designing such systems.

If we wish to design increasingly "intelligent" systems, we require large amount of structured information and a powerful processing facility at the node level. This is the point where distributed database systems and expert systems (see later) are essential. The distributed DBMS concept is simple if we imagine for example an FMS with its machine tools, material handling system, inspection cells, robots, etc., where each cell of the system is linked to the computer network and each cell has enough local intelligence to handle its own data management tasks ([2.2]).

In most cases each processor of the distributed system requires different data types and has different data occurrences, so that the local data base is different in each case. To be able to communicate with these local databases by means of the computer network and handle each of them as a physically distributed part of a logically integrated "large" database, each cell (or in other words workstation, machine, or module) must have the system software and the necessary communications equipment.

"Well designed" distributed systems are generally more reliable than other solutions, because they limit the frequency of failures by distributing the workload among different processors and because some redundancy is always built into such systems.

High level of reliability is most often achieved by identical "warm backup" computers. These machines work only if the regularly used processor is in "trouble". In such cases they can either run parallel and handle two "halves" of the network and communicate with each other when necessary, or one can be kept as a spare and used for other less important processing tasks and to run background jobs. The major benefit of this architecture is that if one of the two machines fails, then only half the system would be affected.

If each node has local processing facility and is able to run without the distributed network for some time then there should be enough time either to repair the fault or to switch the control to the other computer. Obviously this might cause a "panic" situation and such "switches" are not simple tasks to undertake; however by preparing the system and training the staff to solve such (hopefully) occasional problems, serious breakdowns may be eliminated.

If there is a fatal hardware error and self-correcting recovery is not possible, then the node processor should disconnect itself from the distributed system and the overall control architecture must be reconfigured until the fault is repaired. After the fault has been repaired, either by self-recovery or by human intervention, the node must be rejoined to the active system. While the node was being repaired, a number of transactions would have been missed, and so its database must be updated by other nodes of the group (or network), which have remembered the missed actions.

To be able to implement this strategy, at least one of the node databases has to save actions missed by the faulty member of the group. If the communications subsystem is powerful enough to take this extra load it can also be responsible for saving these missed actions and to guarantee that files are transferred to the nodes when they have recovered.

To summarize the above outlined principles, when designing distributed processing systems one should:

* Carefully specify the overall system and user requirements.

* Demodularize the control problem into processes performed by nodes.

* Determine the communication between the nodes and then design the overall integration of the hardware and the system software.

The division of the control problems into processes should result in a set of semi-independent tasks, with a low level of interactions between them. Nodes in general should deal as much as possible with their own "domestic problems" only, since solving tasks locally increases reliability and unloads the network.

2.1.3 Distributed system software design considerations

It must be emphasized that in distributed systems hardware and software should work together in a balanced way. Software tasks should be structured into standalone procedures and functions. In most cases the tasks to be performed are clear, and lend themselves to "natural demodularization". Software modules (e.g. procedures, subroutines, Pascal units, etc.) should be self contained, thus global variables should be avoided as much as possible. The software should be designed, coded, tested and implemented as any other "well engineered" product.

An advised procedure to follow when developing large software systems and the task distribution between the developers and/or teams is outlined in Figure 2.1.

The figure indicates that any serious software engineering activity should start with Step 1: requirement analysis. The proposals should be discussed by a team, headed by the system analyst, and including people with the necessary skills.

Step 2: system design, should also consider the comments of engineers, but this, as well as Step 3: program design, are typical data processing activities and need data processing professionals.

System implementation (Step 4) is a step-by-step activity, requiring a vast amount of software engineering experience. The difficult part here is that one must combine the knowledge both relating to the processes involved as well as to data processing.

Step 5: system test, is an iterative process involving

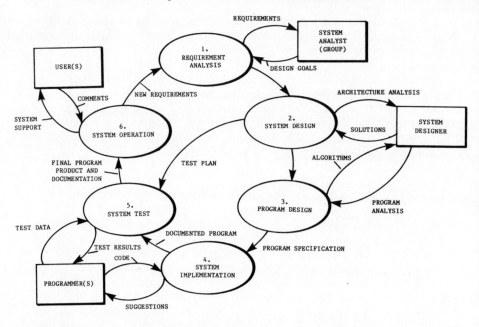

Figure 2.1 Task distribution in Computer Integrated Manu-
          facturing software design, implementation and
          support.

not only the system designers and programmers, but some future
users too. Because of the complexity of the task there is
an increasing number of test-programs generated by computer
programs. Since software reliability is extremely important,
this phase is very crucial. This is an area where expert systems have a prospering future in generating test-programs to
simulate the behaviour of large software systems and hardware.

When arriving at Step 6: system operation, one should not
overlook the importance of software maintenance and support.
This is an activity which often takes twice the amount of work
of developing the system and is often overlooked, in particular
when the resources are not adequate to cover the cost of this
service.

As the selection and/or design of the data processing network can limit or support the overall performance of the system

let us summarize some important principles relating to software development in distributed systems:

* All communication must take place through the network.

* The network driver processor should not rely on the node processors for driving network communications.

* The network should be able to diagnose itself and to prevent failure.

The choice in the node processors (i.e. usually micro-, and/or mini computers) depend on the local processing power requirements, the operating system, speed, interfacing capabilities, real-time control and data handling facilities, available programming languages, etc. It is increasingly important to select computing equipment with regard to software availability, rather than purely on the basis of hardware.

2.1.4 Computer network architectures and Local Area
      Networks (LANs)

When selecting computer networks for CIM, questions to be answered include:

* What network architecture, or configuration?

* What protocol?

* Which transmission standards?

* Which media to use?

The computer network is a collection of nodes that can communicate with each other via digital transmission lines.

The network architecture is the set of functions that should be performed by the network with its nodes, or computers. (Typical digital network architectures are shown in Figure 2.2.)

The network protocol (discussed in section 2.1.5, below) is the set of rules stating how two or more nodes should interact during a communications session.

CIM-B

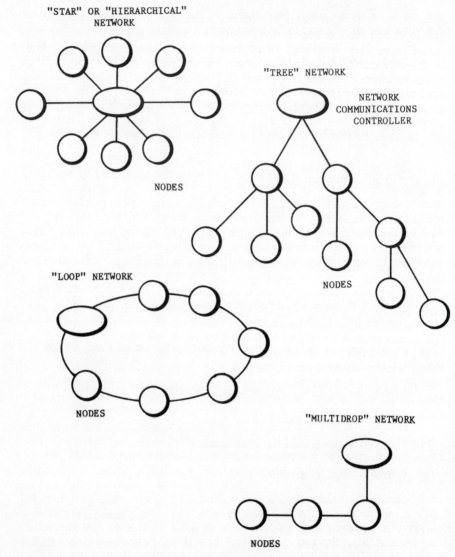

Figure 2.2 Typical distributed system architectures.

To be able to communicate with a computer or any device the first step is to ensure that the physical transmission facilities (i.e. physical lines) exist between the desired nodes.

Once the physical link is established the next step is to make sure that the network is not used as a point-to-point line. There are several different techniques available to interleave the "random" traffic of such lines, including the multi-dropping, multiple access techniques, packet switching, or fast circuit switching and different other forms of time division multiplexing (line sharing) techniques. (For further details refer to [2.1].) The last step before a succesful communication can start on the line is to make sure that the bit stream sent has correctly arrived.

Typical computer network and Local Area Network (LAN) architectures include:

* The "bus", or "open ring" structure, (Figure 2.3).

* The "star", or "hierarchical" structure, (Figure 2.4).

* The "loop" or "ring" structure (Figure 2.5).

The above listed architectures are widely used in different Local Area Networks (LANs). A LAN is a private data communications system operating in hostile environment (e.g. factory shop-floor) making use of the distributed processing concept in a limited geographical area. LANs are capable of accomplishing shopfloor communication and control between a number of different machine controllers, micro-, and minicomputers, FMS cells and workstations.

In the "bus", or "open ring" structure a master scheduler controls the data traffic. If data is to be transferred the requesting computer sends a message to the scheduler, which puts the request into a queue. The message contains an identification code which is broadcast to all nodes of the network. The scheduler works out priorities and notifies the receiver as soon as the bus is available. The identified node takes the message and performs the data transfer between the two computers. Having completed the data transfer the bus becomes free for the next request in the scheduler's queue.

The benefit of this architecture is that any computer can be accessed directly and messages can be sent in a relatively simple and fast way. The major disadvantage is that it needs a scheduler to assign frequencies and priorities to organize the traffic (Figure 2.3).

In the "star" configuration each computer at each level
has its specific assignment corresponding to the tasks to be
solved at that level. If all computers are the same only one
type of interface and communications package is required. Unfortunately in practice most "star" structures grow to an
irregular shape utilizing a variety of computers and controllers.

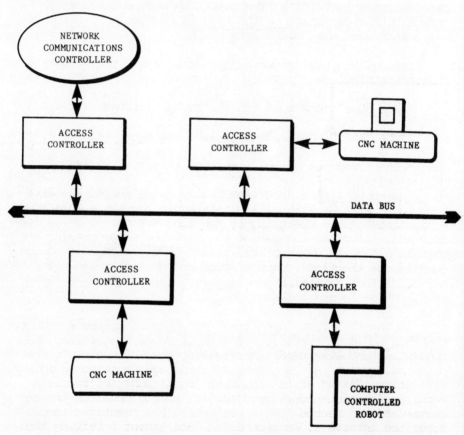

Figure 2.3 "Bus" type local area network architecture.

The "star" architecture is vulnerable in the case of a
computer switch. A further difficulty is that twisted pair
wires limit communications distance and bandwith and are
sensitive to electrical interferences (Figure 2.4).

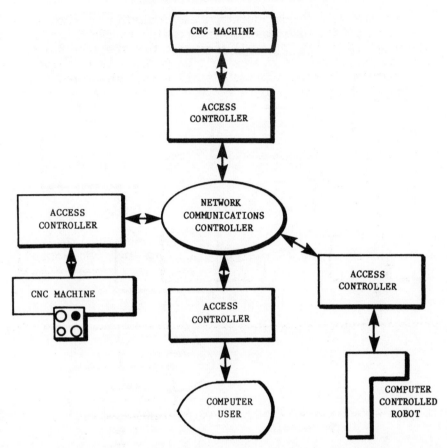

Figure 2.4 "Star" type local area network architecture.

In the "ring" architecture an intelligent interface is required for each node. The data flow within the ring can be controlled by a scheduler, or by sending the messages at pre-described intervals. As soon as an intelligent interface receives a message via the ring, it investigates it to determine whether the address in the packet matches its address. If it does match it is accepted for processing, otherwise it is sent to the next node in the network.

The major drawback of this architecture is that the network communication breaks down if any of the nodes break down. Also it is relatively slower than for example the "bus" structure, because messages usually go through many access controllers (Figure 2.5).

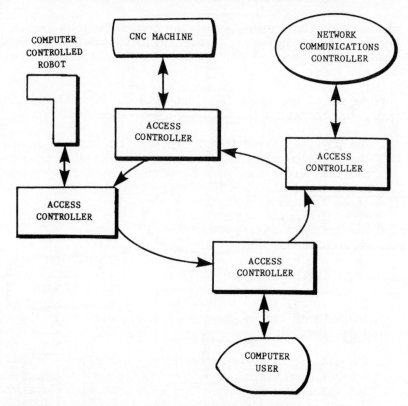

Figure 2.5 "Loop" or "ring" type local area network
        architecture.

2.1.5 Computer network protocols

The set of rules under which nodes communicate in the network
is the protocol. This consists of three elements, these being:

  * The structure of commands and responses in either field
    formatted or character string (usually ASCII) format,
    known as "syntax",

  * the "semantics", which are the requests and responses
    issued by either party, and

  * the timing of events within the network.

Because of the complexity of the communications system and because of the different communication needs of the involved nodes, CIM network architectures may utilize different protocols.

The basis of any up-to-date protocol is the ISO standard protocol, incorporating the following seven layers:

1. The physical layer (e.g. an RS 232 interface with cables for slow speed communication or the RS 422 twisted pair wires and connectors for high bandwidth communication) allows a node to send a bit stream (i.e. a message) into the network.

    The IEEE 802 standard for example specifies that a baseband LAN use a bus configuration with 50 Ohm coaxial cable and user devices are interfaced with network controllers via RS 232 interfaces.

    The physical layer is not concerned either with the contents of the message or with the way in which it is organized into larger groups of data.

2. The data link layer handles the task of correcting transmission errors, since none of the physical layers get involved in such work. It also recovers if the receiver is unable to accept data at the time required by the sender.

3. The network layer decides which outgoing line will be used to send the message to a node. Because routes are often busy, this can be a complex task for the program. The network layer gets a message sent from one node to the other, without the data link header and trailer, because they are taken off by the data link layer when it passes on the message to the network layer.

4. The transport layer looks after the packets and makes sure that they arrive without disruption to the receivers. If a packet is lost then this program provides the network independent transport service and makes sure that the packet is sent again via a perfectly working route into the network.

6. The presentation layer is responsible for performing certain decoding and conversion operations on data to match the device and network requirements.

7. Finally, the application layer is written in most cases by the user and it provides the necessary user interface to the networking system. In several cases this layer can also be purchased as a standard module of the protocol.

There are several LANs available implementing the above concept and/or protocol, including:

* DEC's Data-way, Hewlett - Packard's LAN, using coaxial cable and an Ethernet compatible protocol with a transmission rate of 10 Mbits/sec.

* Allen-Bradley's Data Highway, Gould's Modbus and Modway supporting programmable controller networks.

* General Electric's GEnet, using coaxial cable transmitting at 1M to 5M bits/sec.

There are many and there will be more than one LAN protocol standard. Because of its widespread use the Ethernet standard, originally developed by Xerox's Palo Alto Research Center, is discussed in more detail within this book (Figure 2.6/a).

Ethernet requires a 50 Ohm coaxial cable, with impedance varying no more than + or - 2 Ohms. (The maximum signal loss from one end of the cable to the other is 8.5 dB at 10 MHz). The cable consists of a 0.855 inch center conductor, made of solid tinned copper, a foam polyethylene or foam Teflon core insulator and a PVC or Teflon fluorinated ethylene propylene jacket.

On a cable segment 100 transceivers can be located, but they must maintain a gap of 2.5 meters. The transceiver drives the network and buffers signals which are received and/or transmitted in the network cable (Figure 2.6/b). In other words, it interfaces the network cable, provides drive voltages and currents for sending messages, filters noise and matches different impedance levels.

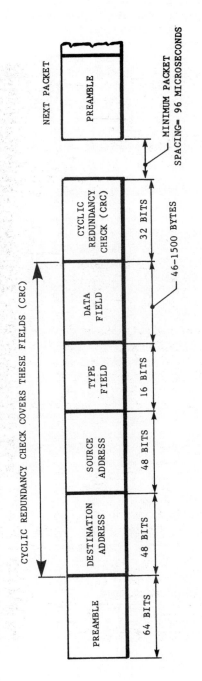

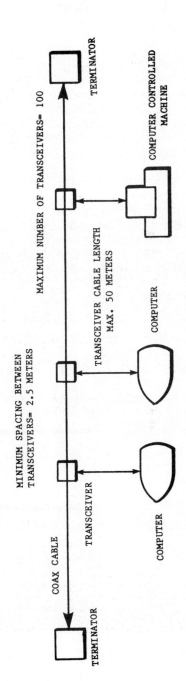

Figure 2.6/a The Ethernet specification and a typical configuration.

The serial interface circuit separates incoming messages into data and clock signals and reverses this procedure in the case of outgoing messages.

The protocol controller circuit checks for errors, decodes messages, generates cyclic-redundancy-checks (CRC) to detect bad transmissions and determines whether or not a received message is addressed to its station. The protocol controller circuits are interfaced with the workstation (e.g. a microcomputer, a CNC machine, a computer controlled robot, etc.) which generates the data to be transmitted and receives messages from other nodes.

The transmit power level in the case of Ethernet is -1.025 V DC and the voltage must always be negative on the coaxial cable. The maximum packet size is 1526 bytes, the minimum is 72 bytes. The data rate is 10 megabits/second. No device may transmit if a carrier is detected on the cable and a minimum of 9.6 microsecond waiting period must be maintained between transmissions. The network can handle up to 1024 workstations.

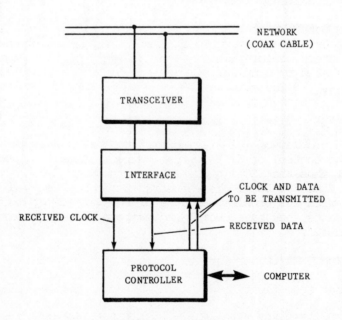

Figure 2.6/b Local Area Network (LAN) interface circuit.

The Apollo Domain System shown in Figure 2.7 indicates a possible way different LANs can work together. This configuration links together desktop minis (DN300 and DN320), midrange color workstations (DN550), high performance color graphics workstations (DN450 and DN660) used in Computer Aided Design (CAD), a domain server processor (DSP80) with its 300 Mbyte storage module disk and 500 Mbyte fixed storage drive providing network-wide peripheral support and the DSP160, 32 bit domain server processor, used as a computational server. The software system supports Ethernet (Xerox Corporation), UNIX (Bell Laboratories), MULTIBUS (INTEL Corporation) and many languages, including FORTRAN 77, Pascal, C, LISP, etc. The system can be used in a variety of different ways, including an integrated CAD/CAM system, factory management information system, etc.

Figure 2.8 illustrates CADLINC's (CADLINC Inc., Elk Grove Village, Illinois) CIM NET, which is based on their engineering graphics workstations and the Ethernet network. It allows up to 1027 engineering workstations to be linked together and share a network of a maximum length of 1600 meters. Each workstation is connected to the Ethernet coaxial cable by the transceiver cable.

The benefits of LANs can be easily realized for example in engineering design, where expensive graphics peripherals, e.g. plotters, color printers etc. and mass storage devices can be shared by a number of workstations, thus maximizing resources. (To indicate the size to which powerful networks can grow, one of Apollo's Domain installations include over 700 workstations.)

The problem of interfacing different machines working under the control of different software systems is a huge problem, thus the ISO (International Standards Organisation) and the IEEE-802 Local Area Networks Standards Committee are dealing with the problem. Because of the variety of different requirements they have decided to propose three different standards:

* The CSMA/CD, or Carrier-Sense Multiple-Access with Collision Detection, IEEE 802.3, patterned after the Ethernet specification developed by Xerox Corporation.

* The IEEE 802.4 token bus selected by General Motors to be used in the MAP (Manufacturing Automation Protocol).

Figure 2.7/a The photograph shows the DN460 graphics terminal
(in the foreground), and the DN320 and DN550
graphics terminals (left to right in the background) of Apollo Computer (UK) Ltd. The DN460
is a 32-bit supermini based high resolution
monochrome graphics workstation designed for
computer aided engineering work, including finite
element analysis. The DN320 workstation incorporates a 32-bit supermini with hardware floating
point, 1.5 to 3 Mbyte main memory, 16 Mbyte virtual address space and a high resolution bit-
mapped graphics display. The DN550 is a mid-range
workstation with high resolution graphics. All
illustrated workstations are linked to the Domain
Local Area Network (see also Figure 2.7/a.)

(Courtesy of Apollo Computer (UK) Ltd., Milton
Keynes, England.)

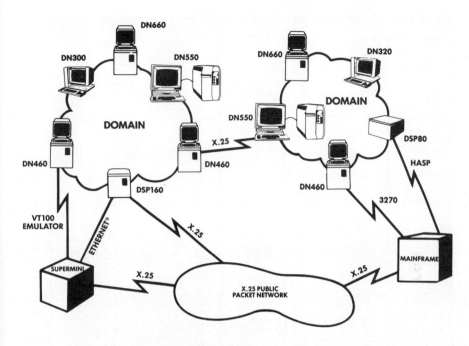

Figure 2.7/b Apollo's DOMAIN network as utilized in an integrated CAD/CAM system.

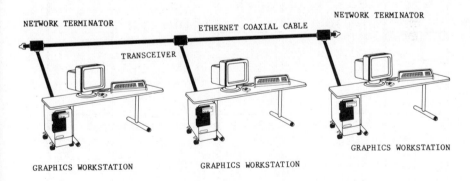

Figure 2.8 CADLINC's Ethernet compatible local area network can
link together graphics workstations for integrated
CAD/CAM work and shop-floor communication purposes.

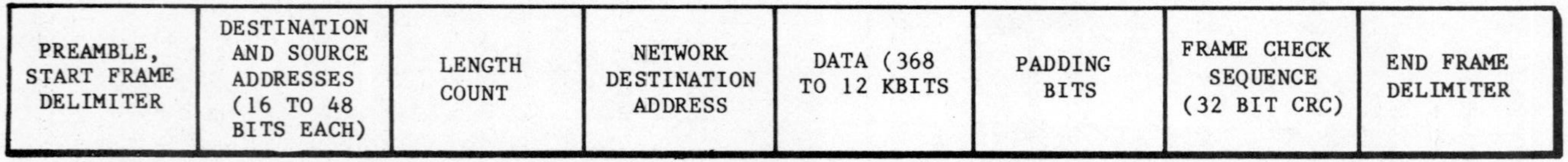

PREAMBLE, START FRAME DELIMITER	DESTINATION AND SOURCE ADDRESSES (16 TO 48 BITS EACH)	LENGTH COUNT	NETWORK DESTINATION ADDRESS	DATA (368 TO 12 KBITS)	PADDING BITS	FRAME CHECK SEQUENCE (32 BIT CRC)	END FRAME DELIMITER

Figure 2.9 The IEEE 802.3 (Carrier-sense multiple-access with collision detection, CSMA/CD) network protocol.

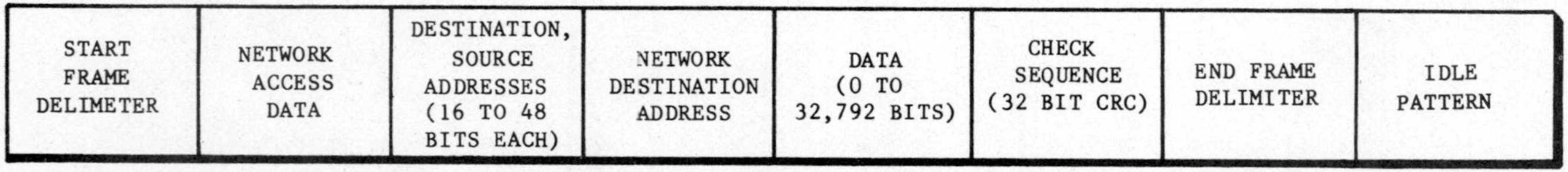

START FRAME DELIMETER	NETWORK ACCESS DATA	DESTINATION, SOURCE ADDRESSES (16 TO 48 BITS EACH)	NETWORK DESTINATION ADDRESS	DATA (0 TO 32,792 BITS)	CHECK SEQUENCE (32 BIT CRC)	END FRAME DELIMITER	IDLE PATTERN

Figure 2.10 The IEEE 802.4 token bus (General Motors Corp. Manufacturing Automation Protocol, MAP).

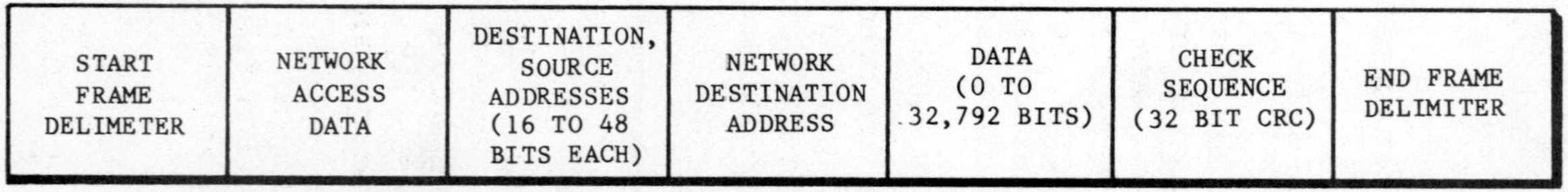

START FRAME DELIMETER	NETWORK ACCESS DATA	DESTINATION, SOURCE ADDRESSES (16 TO 48 BITS EACH)	NETWORK DESTINATION ADDRESS	DATA (0 TO 32,792 BITS)	CHECK SEQUENCE (32 BIT CRC)	END FRAME DELIMITER

Figure 2.11 The IEEE 802.5 token ring (IBM Corp. standard) network protocol.

* The IEEE 802.5 ring LAN, IBM is very likely to use
  in their LANs.

Although these proposals were not accepted as standards
at the time this text was sent to press, they are already
being used and that without them factory-wide computer-
communication would not be a reality. They are shown in detail
in Figures 2.9, 2.10 and 2.11. (For further details refer to
[2.3] and [2.4].)

2.2 Database Management Systems (DBMS)

2.2.1 Principles of database processing

The first major benefit of using database management systems
is that every time a user alters data in the data base all
other users receive the updated, i.e. the latest version, or
correct data. (Note that in this interpretation a "user" is
not only, or not necessarily a human being sitting in front of
a terminal and retrieving sales figures, but can be a machine
controller asking for a part program, or another program running on another computer and accessing the database.)

The other major benefit is that database management systems provide a standard software interface to users. This
means that different programs can access the same database,
thus consistency can be maintained at data level, even if
programs change because new, updated versions replace old
programs.

File handling and database management are often mixed
terms. This is wrong. The major difference in the two method
is that file handling does not offer "physical and logical
data independence" to each program accessing the file. In
other words, files created by independent programs do not
necessarily create compatible files, thus neither data consistency, nor data independence can be maintained. The other
major problem is that even if a file management system allows
searching in more than one file, the file structure of all
searched files must be the same . (Note that a more comprehensive discussion on this topic can be found in reference
[2.1]).)

To illustrate a file structure, Figure 2.12 shows a file
declaration in a Pascal program. It is used in a program for

34 Software sub-systems and software tools in CIM

+---+

```
CONST Max_No_Records=50;

        (* Depends on the capacity of the computer *)

        Device='PRINTER:';

        (* Output device for work sheet & results *)
TYPE CMDSET=SET OF CHAR;

        DATA_ARRAY=ARRAY[1..12] OF REAL;

        Day_Range=1..31;

        Year_Range=1984..2050;

        Month_Type=(JAN,FEB,MAR,APR,MAY,JUN,JUL,AUG,SEP,
                    OCT,NOV,DEC);

        (* Note that false year/month/day input can be

        avoided by using this TYPE declaration *)

    · Date_Rec =

        RECORD

            Day:Day_Range;

            Month:Month_Type;

            Mth:STRING[3];

            YEAR:Year_Range;

        END;

    STUDY_INFO =

    RECORD

      No_Studies:INTEGER; (* Counter for the No. of studies *)

      Name:STRING[50];     (* Name of study *)

      No_Days:INTEGER;     (* No. of days since start of study *)

      Req_Accuracy:REAL;   (* Accuracy required for results *)

      Conf:90..99;         (* Confidence level of result *)

      Z:REAL;              (* Used for statistical calculation *)
```

```
Sample file declaration of a TIME_STUDY program
                                    Page No: 2
+------------------------------------------------------------+


    No_Times:INTEGER;     (* No. of samples per day *)

    Start_Day,            (* Times of the working day: start,

                             lunch, finish lunch, finish work *)

    Start_Lunch,

    End_Lunch,

    End_Day:INTEGER;

    No_Head:INTEGER;      (* No. of headings *)

    Field_Width:INTEGER;  (* Max. field width of heading *)

    Heading:ARRAY[1..12] OF STRING[34];       (* Headings *)

    Title:STRING;         (* Title for worksheet *)

    File_Name:STRING;     (* Name of integer file for study *)

    Date:Date_Rec;        (* Date of study *)

  END;

  STUDIES=ARRAY[1..Max_No_Records] OF STUDY_INFO;

  (* Information for each study is stored in an array of

  STUDY_INFO on disk *)
```

Figure 2.12 Sample (UCSD-Pascal) file declaration indicating records and fields in a time study program.

storing data related to work measurement and time study. Because this is a file, containing records, and not a database offering a standard software interface, each program wishing to access this file (i.e. wishing to read data from, or write data into its records) must incorporate a program segment, or procedure which does the job. This creates redundancy in the programs and what is even worse, if the file structure is changed, each program accessing the file must be changed as well.

Because large data processing systems are developed more or less in an iterative way, nobody could afford the luxury of rewriting all programs as many times as new ideas and needs occur, thus the advisable solution is: use database management software, rather than file handling.

Before discussing the relational database management in detail, let us give a broad overview of some of the most important features of:

* the relational,

* the hierarchical,

* the network,

* the free format and

* the multi-user, distributed structures.

The relational model is popular among engineers and CAD/CAM system developers because it is relatively easy to design and modify its data structure. It is made up of files consisting of records, and the records of fields. Records can be combined, searched, etc. using a common field from a theoretically unlimited number of files. In other words in this model files are linked by the fields of records, thus new links between files can be established any time during the lifetime of the database.

In the hierarchical system connections between files are fixed for the lifetime of the particular database data structure. This is a major drawback if many alterations are required during its development, or use. The hierarchical model contains records as well, but the relationships between records are not established through common fields, but are defined in the data structure fixed by the Data Base Administrator when the system is implemented.

The network structure is similar to the hierarchical structure with the extension that any record type or file can be related to any other record type or file. This relationship is also known as "the many-to-many relationship" as opposed to the "one-to-many relationship".

The "free format database" is a text file, like the text part of this book, for example, since it has been written on a microcomputer. Any character, or word or a combination of these can be found by using different commands, activating different procedures to find, change, delete, insert, replace, etc. data in this text file using the text editor.

Multi-user and distributed databases can have any of the above structures. The key point is that they allow the database to be read by many users at the same time, however only one user is allowed to write into a record at the same time, to avoid data inconsistency. In the case of very large multi-user databases often sub-schemas are defined for a group of users and the database itself is only logically integrated, but physically distributed not only on many disk drives, but sometimes even geographically.

Database machines, i.e. computers designed as "hardware database processors" represent a different category. They use parallel processing techniques and are in most cases faster and more efficient than their "software database brothers".

2.2.2 The relational database architecture

The reason for discussing the relational database concept is because it is relatively easy to understand, they are more flexible than for example the hierarchical, or network models and because they are widespread mainly in engineering and in some business applications where the data structure cannot be frozen for the lifetime of the database. The major benefit of the relational concept is that it enables many changes to data links (or relations) as the database increases in size and complexity. In other words, they enable data structure changes without changing subschemas and/or application programs.

The relation is a flat file, or a two dimensional array of data elements. It is often called "a table", consisting of rows and columns, where the rows are representing the records (or segments) and the columns are the fields within the records. (Note that an example is shown in section 2.2.3).

Regarding to the internal data storage format, it is important to know that the relational view exists at a level at which the user accesses the data, and not necessarily at the physical level. Thus user interfaces to such databases could be written in many high level languages. The physical implementation of the database can considerably affect the speed and the overall performance of the system, thus this is an important parameter to test.

38   Software sub-systems and software tools in CIM

Because of the large amount of data to be handled in the business side of CIM , because data access speed is a very important factor and because such data structures are usually more stable compared to the engineering databases, often the faster hierarchical (CODASYL standard), or network type databases are used. When comparing relational database systems with the CODASYL structure, the following points are recommended for careful consideration:

* In both cases the Database Administrator defines the data structures and data types, organizes them into records and files, finds the necessary access routes for the users and generally makes sure that both the group of users as well as the system are satisfied.

* CODASYL structures usually cannot be easily restructured without some effect on existing programs, whereas the access path in relational databases can always pass through matching fields, thus they can be extended.

* Simplicity is a very strong argument in any large database setup. To understand relational systems is usually easier than the CODASYL structure.

* Query techniques in relational database systems aim to be less procedural, which means that the query is more or less a "proper" English question, or sentence rather than a large selection of nested and linked procedures used in a purpose written program.

* From the programming point of view, relational techniques often simplify the host program, because access paths are simpler, but on the other hand they can often confuse the user when trying to dig out the complex interrelationships between tables.

* Access speed can often be a problem with relational databases, also in real-time applications, thus it is crucial to clarify the necessary time constraints be fore a large amount of work is invested in designing data structures. Some of these problems can also be solved at the data structure level by normalizing record structures and by eliminating dependencies among field types.

(For further details on database architectures and related problems refer to [2.1] and [2.5] to [2.10]).

2.2.3 Case study: data structure implementation using the ORACLE relational Database Management System

The purpose of this case study is to explain more about relational database management systems and to give an example of the way a simple data structure is set up and could be used. It is not our aim to give a full discussion of the ORACLE (TM) relational database management system, nor to teach its use or the related software packages.

The problem we shall solve in this case study is relatively simple, but very important in the practice for example when controlling the part flow and manufacturing sequence in FMS.

Let us assume that we have an FMS consisting of a number of machines, or cells, each of them capable of handling one tool magazine at a time. The contents of the tool magazines can be changed of course and they are interchangeable between machines (Figures 2.13/a and b).

The question we would like to ask from our database is very simple and practical: Which machine(s), or cell(s) have the appropriate tool mix in their tool magazines to process a given part program?

This question could be asked by an FMS production manager, or by the FMS scheduling program itself, before finalizing the sequence in which the certain part, or parts will be processed in the system (Note that this aspect of the case study is discussed in detail in Chapters 6 and 7). The problem to be solved is practical, because tool loading into tool magazines, as well as tool magazine changing takes time and costs money when operating the FMS, thus it should be everybody's interest to minimize changes of tool drums, and/or their contents (Figure 2.13/c).

To solve our problem we shall use the ORACLE relational database management system, running on a powerful minicomputer. (Further details on this database management software can be found in [2.11].) Before creating our little data structure and "navi-gating" in it, let us introduce some of the main features and available software tools of this system:

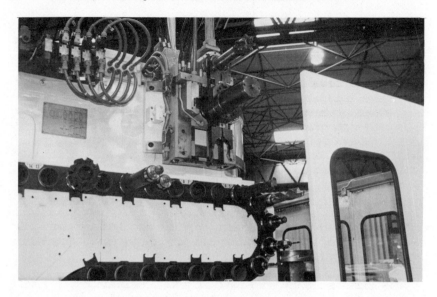

Figure 2.13/a Chain type tool magazine (Courtesy of Csepel
Machine Tool Company, Budapest, Hungary).

Figure 2.13/b Drum type tool magazines (Courtesy of Yamazaki
Machinery Co., Japan).

* Integrated application development management, allowing users to develop their own "end-user" packages without the need for using traditional programming languages.

* Interactive Application Generator, which is an interactive facility, tayloring screens for quick on-line data input, query and update, without the need for any high level programming.

* Interactive Report Writer, creating reports automatically using a few commands only.

* SQL free format, on-line database query language.

* Integrated Data Dictionary, providing access to all information regarding to tables, fields, users, programs, access priviliges and others.

The first step in solving our problem is to set up the tables (or files), (see Figure 2.13/d) containing the necessary (and other) information about our cells, the tool magazines and the tools in the part programs (note that FMS part prog-ramming is discussed in Chapters 5 and 6).

The first table we have created was the "MACHINE" table. This is a general purpose cell description containing all important information relating to a machining center capable of changing tools, tool magazines and palletized workparts, and providing three axis continuous path control and a rotary indexing table. The name of the table is "MACHINE" consisting of the following type of fields:

1. MACHINE_ID, a 16 character long alphanumeric name, containing the machine identifier.

2. MAX_POWER, maximum machine power in [KWatts] is stored as a real number, (ORACLE can define character, number, date, integer and money attributes).

3. X_MIN to Z_MAX describe the motion range limits of the machine in milimeters.

4. ROT_INDEX stores the size of increment the table is capable of rotating on the X, Y table of the machine.

5. POSERR_X_Y_Z contains the positioning error regarding to the relevant axis.

Figure 2.13/c Drum type (FTS Flexible Tooling System) tool magazine and integrated tool loading/unloading manipulator of Karl Hertel Ltd., Warwickshire, UK.

```
UFI> CREATE TABLE MACHINE (
  2  MACHINE_ID CHAR(16),
  3  MAX_POWER NUMBER(5,2),
  4  X_MIN NUMBER(8,4),
  5  X_MAX NUMBER(8,4),
  6  Y_MIN NUMBER(8,4),
  7  Y_MAX NUMBER(8,4),
  8  Z_MIN NUMBER(8,4),
  9  Z_MAX NUMBER(8,4),
 10  ROT_INDEX NUMBER(8,4),
 11  POSERR_X NUMBER(6,4),
 12  POSERR_Y NUMBER(6,4),
 13  POSERR_Z NUMBER(6,4),
 14  CONTROLLER CHAR(16),
 15  MAGAZINE CHAR(16));

Table created

UFI> CREATE TABLE MAGAZINE (
  2  MAGAZINE_ID CHAR(16),
  3  T1 CHAR(8),
  4  T2 CHAR(8),
  5  T3 CHAR(8),
  6  T4 CHAR(8),
  7  T5 CHAR(8),
  8  T6 CHAR(8),
  9  T7 CHAR(8),
 10  T8 CHAR(8),
 11  T9 CHAR(8),
 12  T10 CHAR(8),
 13  T11 CHAR(8),
 14  T12 CHAR(8));

Table created

UFI> CREATE TABLE PROGTOOL (
  2  PROG_NAME CHAR(16),
  3  T1 CHAR(8),
  4  T2 CHAR(8),
  5  T3 CHAR(8),
  6  T4 CHAR(8),
  7  T5 CHAR(8),
  8  T6 CHAR(8));

Table created
```

Figure 2.13/d Relational data structure created using ORACLE, to describe an FMS cell, or machine (MACHINE), to identify its tool magazine (MAGAZINE) and to store the number of tools required in a part program (PROGTOOL). The purpose of this structure is to be able to automatically find matching tool magazine contents on FMS cells using a tool list extracted from the FMS part program.

6. CONTROLLER identifies the type of CNC control used. (Note that this is normally a field which points to a record identifier, in which more details are given of the controller).

7. Finally, MAGAZINE stores the magazine identifier currently being utilized on the cell. (In this data structure for simplicity we assume that each cell can have only one tool magazine for automated access at a time. To extend the data structure to describe more than one magazine on any cell at the same time represents no problem.)

Note that because of the very nature of relational database management systems, this table does not define, or rigidly fix any further relationship between its data, as a hierarchical system would do, and that any field can be related to any field of any other table by means of a query, as we shall see later...

The second file we need must contain information about the tools located in their magazines, thus we have set up the "MAGAZINE" table, identifying twelve tool locations. In our example there are only 12 locations in the tool magazines, simply because this saved some typing. In a real situation machines often have over fifty tools or even more (usually up to 240 "block tools") if the Sandvik Coromant "block tooling" concept, or if the Hertel FTS (Flexible Tooling System) is used.

It must be noted that in the case of CNC machining centres and FMS "physical" as well as "logical" tool codes are often utilized and because of this, the controller needs to decode tool codes, as well as assign "physical" tool code values to those "logical" codes used in the part programs. As a matter of interest, tool codes could also contain some coded information about their set length and diameter values, and other important data such as the equvivalent of 100% tool life. To simplify this decoding problem we shall use only one simple code, identifying a single tool in the tool magazine.

The third and last table we set up is called "PROGTOOL". It contains the tool codes used in different part programs. Again, for simplicity we use only six fields describing six tools, although in a real situation a part program could easily require the access of over twenty tools, or more.

Now, that we have created all required tables, let us think again of what we really wish to do: we want the database to find the cell on which there is the appropriate magazine to execute a particular part program. (And of course if there is none, we shall have to change a tool magazine on one of the appropriate cells, containing the tools listed in the "PROGTOOL" table.)   In other words:

1. First we need to find the matching magazine to an inquiery given in our "PROGTOOL" table, (see Figure 2.14; this is a typical "one-to-many" relationship).

2. Then we must find the appropriate cell containing this magazine, (see Figure 2.15; this is a "one-to-many-to-one" relationship).

This is the point where we need to know something about the database again. ORACLE offers a non-procedural, "English-like" query language, called SQL. This language interface is also used by programmers when writing query programs in COBOL, FORTRAN or "C". It can be used to set up tables interactively, to edit them, to delete them, to put them or the table contents in a required order, etc. We demonstrate the way it can be used for searching (or selecting) data.

Typically in our application a program should have been written, rather then the interactive query language used, however by following this interactive solution we can not only see some of the structure of such languages, but also the way our high level program and SQL code (i.e. ORACLE language program) would work together.

To select the appropriate magazine we need to input some data (see Figure 2.16/a) and describe our "one-to-many" type query, using the SQL language (see Figure 2.16/b). The result of this search is obviously: "MAG0", what we have expected.

Now we must find the machine, or FMS cell this magazine is attached to. The query for this is listed in Figure 2.17/a. As one can see only the first line differs from the listing in Figure 2.16/b, offering an other "FOR" cycle-like search. The result of this run (see Figure 2.17/b) is obviously that "MAG0" can be found on "CELL1".

There are many more aspects of databases we did not discuss, but by giving a simple example of a typical appli-

```
UFI> CREATE TABLE MACHINE
  2   MACHINE_ID CHAR(16),
  3   MAX_POWER NUMBER(5,2),
  4   X_MIN NUMBER(8,4),
  5   X_MAX NUMBER(8,4),
  6   Y_MIN NUMBER(8,4),
  7   Y_MAX NUMBER(8,4),
  8   Z_MIN NUMBER(8,4),
  9   Z_MAX NUMBER(8,4),
 10   ROT_INDEX NUMBER(8,4),
 11   POSERR_X NUMBER(6,4),
 12   POSERR_Y NUMBER(6,4),
 13   POSERR_Z NUMBER(6,4),
 14   CONTROLLER CHAR(16),
 15   MAGAZINE CHAR(16));

Table created

UFI> CREATE TABLE MAGAZINE
  2   MAGAZINE_ID CHAR(16),
  3   T1 CHAR(8),
  4   T2 CHAR(8),
  5   T3 CHAR(8),
  6   T4 CHAR(8),
  7   T5 CHAR(8),
  8   T6 CHAR(8),
  9   T7 CHAR(8),   :
 10   T8 CHAR(8),
 11   T9 CHAR(8),
 12   T10 CHAR(8),
 13   T11 CHAR(8),
 14   T12 CHAR(8));

Table created

UFI> CREATE TABLE PROGTOOL (
  2   PROG_NAME CHAR(16),
  3   T1 CHAR(8),
  4   T2 CHAR(8),
  5   T3 CHAR(8),
  6   T4 CHAR(8),
  7   T5 CHAR(8),
  8   T6 CHAR(8));

Table created
```

Figure 2.14 Graphical representation of the "one-to-many"
relationship in our data structure.

```
UFI> CREATE TABLE MACHINE (
  2   MACHINE_ID CHAR(16),
  3   MAX_POWER NUMBER(5,2),
  4   X_MIN NUMBER(8,4),
  5   X_MAX NUMBER(8,4),
  6   Y_MIN NUMBER(8,4),
  7   Y_MAX NUMBER(8,4),
  8   Z_MIN NUMBER(8,4),
  9   Z_MAX NUMBER(8,4),
 10   ROT_INDEX NUMBER(8,4),
 11   POSERR_X NUMBER(6,4),
 12   POSERR_Y NUMBER(6,4),
 13   POSERR_Z NUMBER(6,4),
 14   CONTROLLER CHAR(16),
 15   MAGAZINE CHAR(16));

Table created

UFI> CREATE TABLE MAGAZINE (
  2   MAGAZINE_ID CHAR(16),
  3   T1 CHAR(8),
  4   T2 CHAR(8),
  5   T3 CHAR(8),
  6   T4 CHAR(8),
  7   T5 CHAR(8),
  8   T6 CHAR(8),
  9   T7 CHAR(8),    :
 10   T8 CHAR(8),
 11   T9 CHAR(8),
 12   T10 CHAR(8),
 13   T11 CHAR(8),
 14   T12 CHAR(8));

Table created

UFI> CREATE TABLE PROGTOOL (
  2   PROG_NAME CHAR(16),
  3   T1 CHAR(8),
  4   T2 CHAR(8),
  5   T3 CHAR(8),
  6   T4 CHAR(8),
  7   T5 CHAR(8),
  8   T6 CHAR(8));

Table created
```

Figure 2.15 Graphical representation of the "one-to-many-to-one" relationship in our data structure.

```
TABLE STRUCTURE                               DATA
---------------------------------------------------------------

TABLE MACHINE
*************

 1   MACHINE_ID    CELL1           CELL2           CELL3
 2   MAX_POWER     15.00           22.00           15.00
 3   X_MIN         0.0000          0.0000          0.0000
 4   X_MAX         650.0000        800.0000        650.0000
 5   Y_MIN         0.0000          0.0000          0.0000
 6   Y_MAX         850.0000        1250.0000       850.0000
 7   Z_MIN         -150.0000       0.0000          -150.0000
 8   Z_MAX         950.0000        1100.0000       950.0000
 9   ROT_INDEX     0.1             0.0             0.05
10   POSERR_X      0.01            0.018           0.005
11   POSERR_Y      0.01            0.018           0.05
12   POSERR_Z      0.015           0.02            0.01
13   CONTROLLER    CINCINNATI      CINCINNATI      CINCINNATI
14   MAGAZINE      MAG0            MAG1            MAG2


TABLE MAGAZINE
**************

 1   MAGAZINE_ID      MAG0         MAG1            MAG2
 2   T1               T9001        T8085           T2109
 3   T2               T9004        T6800           T3456
 4   T3               T9008        T8087           T6652
 5   T4               T9002        T8086           T1254
 6   T5               T8081        T2626           T8972
 7   T6               T9102        T1584           T3212
 8   T7               T7532        T5472           T2341
 9   T8               T6739        T5670           T2000
10   T9               T3480        T7800           T2001
11   T10              T6811        T4496           T2002
12   T11              T2329        T4097           T2000
13   T12              T1002        T4797           T2002


TABLE PROGTOOL
**************

 1   PROG_NAME        MILL1        BORE1           DRILL1
 2   T1               T3480        T6800           T8081
 3   T2               T9004        T2626           T9102
 4   T3               T1002        T2329           T7800
 5   T4               T7352        T1002
 6   T5                            T5555
 7   T6
```

Figure 2.16/a Sample data loaded into the data structure
             containing machine, tool magazine contents data
             and a tool list extracted from the FMS part
             program. The Figure illustrates graphically the
             "one-to-many" relationship established by means
             of a query.

```
SELECT MAGAZINE_ID FROM MAGAZINE WHERE

T1=(SELECT T1 FROM PROGTOOL WHERE PROGNAME='MILL1') OR

T2=(SELECT T1 FROM PROGTOOL WHERE PROGNAME='MILL1') OR

T3=(SELECT T1 FROM PROGTOOL WHERE PROGNAME='MILL1') OR

T4=(SELECT T1 FROM PROGTOOL WHERE PROGNAME='MILL1') OR

T5=(SELECT T1 FROM PROGTOOL WHERE PROGNAME='MILL1') OR

T6=(SELECT T1 FROM PROGTOOL WHERE PROGNAME='MILL1') OR

T7=(SELECT T1 FROM PROGTOOL WHERE PROGNAME='MILL1') OR

T8=(SELECT T1 FROM PROGTOOL WHERE PROGNAME='MILL1') OR

T9=(SELECT T1 FROM PROGTOOL WHERE PROGNAME='MILL1') OR

T10=(SELECT T1 FROM PROGTOOL WHERE PROGNAME='MILL1') OR

T11=(SELECT T1 FROM PROGTOOL WHERE PROGNAME='MILL1') OR

T12=(SELECT T1 FROM PROGTOOL WHERE PROGNAME='MILL1'));
```

Figure 2.16/b The query using ORACLE compares the required list of tools of program MILL1 with the matching tool magazine contents of all machines in the Flexible Manufacturing System.

```
SELECT MACHINE_ID FROM MACHINE WHERE MAGAZINE=(

SELECT MAGAZINE_ID FROM MAGAZINE WHERE

T1=(SELECT T1 FROM PROGTOOL WHERE PROGNAME='MILL1') OR

T2=(SELECT T1 FROM PROGTOOL WHERE PROGNAME='MILL1') OR

T3=(SELECT T1 FROM PROGTOOL WHERE PROGNAME='MILL1') OR

T4=(SELECT T1 FROM PROGTOOL WHERE PROGNAME='MILL1') OR

T5=(SELECT T1 FROM PROGTOOL WHERE PROGNAME='MILL1') OR

T6=(SELECT T1 FROM PROGTOOL WHERE PROGNAME='MILL1') OR

T7=(SELECT T1 FROM PROGTOOL WHERE PROGNAME='MILL1') OR

T8=(SELECT T1 FROM PROGTOOL WHERE PROGNAME='MILL1') OR

T9=(SELECT T1 FROM PROGTOOL WHERE PROGNAME='MILL1') OR

T10=(SELECT T1 FROM PROGTOOL WHERE PROGNAME='MILL1') OR

T11=(SELECT T1 FROM PROGTOOL WHERE PROGNAME='MILL1') OR

T12=(SELECT T1 FROM PROGTOOL WHERE PROGNAME='MILL1'));
```

Figure 2.17/a Nested query using ORACLE compares the required list of tools of program MILL1 with the matching tool magazine contents of all machines in the FMS, as well as searches and finds the machine identifier of the cell onto which the found tool magazine is currently mounted.

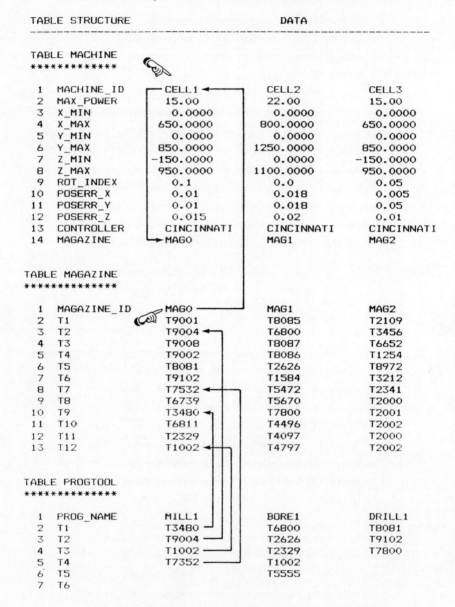

```
TABLE STRUCTURE                              DATA
---------------------------------------------------------------

TABLE MACHINE
**************

  1   MACHINE_ID     CELL1            CELL2         CELL3
  2   MAX_POWER      15.00            22.00         15.00
  3   X_MIN           0.0000           0.0000        0.0000
  4   X_MAX         650.0000         800.0000      650.0000
  5   Y_MIN           0.0000           0.0000        0.0000
  6   Y_MAX         850.0000        1250.0000      850.0000
  7   Z_MIN        -150.0000           0.0000     -150.0000
  8   Z_MAX         950.0000        1100.0000      950.0000
  9   ROT_INDEX       0.1              0.0           0.05
 10   POSERR_X        0.01             0.018         0.005
 11   POSERR_Y        0.01             0.018         0.05
 12   POSERR_Z        0.015            0.02          0.01
 13   CONTROLLER     CINCINNATI       CINCINNATI    CINCINNATI
 14   MAGAZINE       MAG0             MAG1          MAG2


TABLE MAGAZINE
**************

  1   MAGAZINE_ID    MAG0             MAG1          MAG2
  2   T1             T9001            T8085         T2109
  3   T2             T9004            T6800         T3456
  4   T3             T9008            T8087         T6652
  5   T4             T9002            T8086         T1254
  6   T5             T8081            T2626         T8972
  7   T6             T9102            T1584         T3212
  8   T7             T7532            T5472         T2341
  9   T8             T6739            T5670         T2000
 10   T9             T3480            T7800         T2001
 11   T10            T6811            T4496         T2002
 12   T11            T2329            T4097         T2000
 13   T12            T1002            T4797         T2002


TABLE PROGTOOL
**************

  1   PROG_NAME      MILL1            BORE1         DRILL1
  2   T1             T3480            T6800         T8081
  3   T2             T9004            T2626         T9102
  4   T3             T1002            T2329         T7800
  5   T4             T7352            T1002
  6   T5                              T5555
  7   T6
```

Figure 2.17/b Sample data loaded into the data structure
              containing machine, tool magazine contents data
              and a tool list extracted from the FMS part
              program. The Figure illustrates graphically the
              "one-to-many-to-one" relationship established by
              means of a nested query.

cation one can understand their impact on man-machine and
real-time data communication and data search. In the case of
large FMS systems, (see some examples in Chapter 6) where
there are hundreds of tool magazines containing a total of two
to three thousand block tools, or tools, the above described
relatively simple problem would require several hours, or
longer to be solved manually and only a few seconds or a
minute using the data base management system.

2.3   The potential of expert systems in CIM

The phrases:"expert systems" and "AI-Artificial Intelligence"
ligence" have already been used a few times in this text and
will again be used because of their importance in the development of CIM technology. What is new in expert systems and
why is so much money spent on research in this area? The reason is that above a certain level of system complexity "normal", or "ordinary" programs, i.e. computer programs using algorithmic principles are not capable of solving the task and
because humans are not available, or are slow in making the
correct decisions at the required speed, only "expert computer
programs" can help.

What are these complex problems? What is knowledge and
what kind of expert systems could be utilized in CIM ? A typical area which can be relatively easily imagined and where
expert systems have a solid ground in CIM is for example fault
analysis. But there are many other fields as well where expert
systems under development can be essential in the design and
operation phases of complex systems, such as FMS, CAD/CAM and
CIM.

2.3.1 What is an expert system?

Instead of approaching the impossible task of defining expert
systems, or computer programs which are capable of demonstrating specialized problem solving expertise, or skills, let us
explain what expert systems are by listing a few impor tant
features and attributes of such programs.

Expert systems are applied in fields where decision making is difficult because of the complexity of the task and
because the reasoning cannot be expressed in an algorithmic
way. Typical areas include: diagnosing FMS faults and computer

breakdowns, designing truly user-friendly man-machine interfaces, creating test programs for integrated circuits and complex electronic logic boards, designing computer configurations for estimated future use, etc. Human experts can achieve
outstanding performance in solving problems because they are
knowledgeable and experienced. But how can a computer program
learn similar expertise even only in a narrow domain? Very
briefly: expert performance depends on the knowledge, thus
collecting, storing, retrieving and duplicating knowledge in
computer (or machine) readable format are the key points.

Knowledge itself consists of symbolic descriptions that
characterize certain relationships in a domain and of procedures which can manipulate the symbolic descriptions. Knowledge engineering aims to build computer programs which are
capable of problem solving using expert knowledge.

The method expert systems follow in achieving outstanding
performance in decision making is known as "symbolic reasoning". According to Newell [2.12] "the most fundamental contribution so far of artificial intelligence and computer
science to the joint enterprise of cognitive science has been
the notion of a physical symbol system, i.e., the concept of
a broad class of systems capable of having and manipulating
symbols, yet realizable in the physical universe".

One important conclusion of these words is that the representation of knowledge in a "rule base" or by other means is
one of the most crucial tasks to be solved.

Many expert systems use a database of symbols and a long
term memory comprising a set of rules consisting of an ordered
pair of symbol strings. In this case the rules are selected
by matching the predicat(s) on the left hand side with those
in the database. The most important feature of this database
is that unlike in conventional, or procedural computer
languages data and control information are stored together.
This is often referred to as "unity of data and control
store".

If the rule set is predetermined the interpreter program
after a rule has been executed continues with the search for
matching with the next rules, or alternatively starts again
with the first. The interpreter is a program acting on the
rule base following a "select-execute" cycle.

An expert system can typically be data driven, (antecendent), where the matching ocurs with the left hand side of the rule, (see Figure 2.18/a and b), or goal driven (consequent) where the matching is performed on the right hand side of the rule (Figure 2.19). The data driven system attempts to find and "fire" the appropriate rule(s) from left-to-right, (called "forward chaining"), whereas the goal driven system follows a method, often referred to as "backward chaining", aiming to find appropriate rules for a certain goal (i.e. from right-to-left).

These two different rule execution methods are important, since they will influence the application areas and the usage of the expert system.

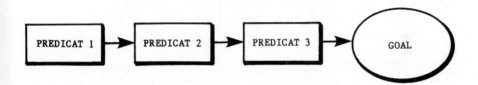

Figure 2.18/a Forward chaining model.

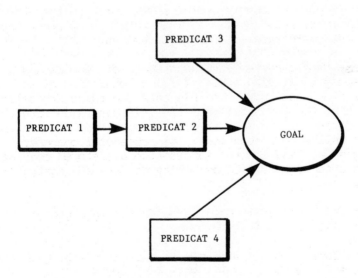

Figure 2.18/b Forward chaining model.

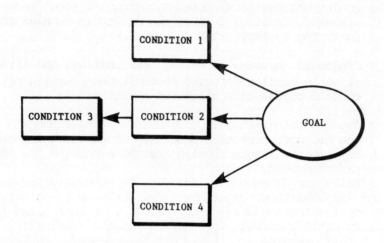

Figure 2.19 Backward chaining model.

2.3.2 Types of expert systems and their possible applications

Although expert system history is less then 25 years old,
there are already programs which have invented new chemicals
(PROSPECTOR), which help to design properly working large computer configurations (the R1 program) and other programs
mainly in the medical/chemical analysis and diagnostics field
(DENDRAL,PUFF, CADUCEUS, etc.) and in expert scheduling at
factory level ([2.12] and [2.13]).

Types of expert systems include:

* Design systems capable of creating new combinations
  and configurations of electronic, mechanical, structu-
  ral, etc. systems meanwhile satisfying certain predefined constraints and relationships between components.

* Planning systems involved in decision making, forecasting, project planning.

* Expert control systems, utilizing Adaptive Control (AC)
  techniques and capable of adapting their behaviour to
  the environmental conditions in scheduling components
  in FMS, in selecting cutting parameters and tools for
  five axes milling of super hard materials, etc.

* Manufacturing expert systems selecting optimum cutting conditions, assembly motions of robots, robot task planning, routing, dynamic scheduling in complex manufacturing systems, etc.

* Diagnosis systems observing malfunctions and irregularities in complex software systems, when testing printed circuit boards and integrated circuits, etc.

* Repair systems capable of advising their users on how the diagnosed problem should be tackled, considering all important aspects of the problem.

Finally let us emphasize that skills necessary for designing and implementing expert systems are not identical to those applicable in general programming. It requires excellent computing skills, as well as expertise in the domain(s) involved. Partly because of this, partly because of the lack of many techniques in other areas of computer sciences and data processing in general, expert systems will not emerge on a wide basis in industry for some time.

The fascinating thing about future expert systems is that they will be capable of gathering, structuring and storing knowledge at least in a very narrow domain in their knowledge base, a kind of database, and that this knowledge can be duplicated and distributed in a very short time. The possible impact of this as one can imagine is fascinating as well as dramatic both in terms of their application possibilities in CIM as well as regarding their effect on our education system and social life ([2.12] to [2.15]).

References and further reading

[2.1] Paul G. Ránky: The Design and Operation of Flexible Manufacturing Systems, IFS(Publications) Ltd., and North-Holland Publishing Co. 1983. 348 pp.

[2.2] N.A. Duffie: An approach to the design of distributed machinery control systems, IEEE Transactions on Industrial Applications, 1982. Vol.IA-18, No.4, p. 435-441.

[2.3] Rodney J. Heisterberg: Fundamentals of factory communications, Tooling and Production, November 1984. p. 79-85.

[2.4] Bruce A. Loyer: LANs: The Coming Revolution in Factory Control, Machine Design, November 22, 1984. p. 127-130.

[2.5] James Martin: Managing the database environment, Prentice-Hall, Engle-wood Cliffs, NJ, 1983.

[2.6] C.J. Date: Database: A primer, Reading MA, Addison-Wesley, 1983.

[2.7] Charles M. Eastman: Database facilities for engineering design, Proceedings of the IEEE, Vol. 69, No. 10, October 1981. p. 1249-1263.

[2.8] M. Zloof: Query by example, Proceedings 1975 AFIPS NCC, 1975. p. 431-438.

[2.9] R. W. Taylor and R. Frank: CODASYL database management systems, ACM Comp. Surveys, Vol.8, No.1, 1976. p. 67-104.

[2.10] A. L. Scherr: Distributed data processing, IBM Systems Journal, Vol.17, No.4, 1978. p. 324-343.

[2.11] ORACLE relational database management system, Terminal User Guide, SQL/UFI Reference Manual, Interactive Application Facility Guides, Published by ORACLE Corporation, USA, Marketed in the UK under licence by CACI, Oriel House, 26 The Quadrant, Richmond, Surrey.

[2.12] Frederic Hayes-Roth and others, ed.: Building expert systems, Addison-Wesley Publishing Co., 1983.

[2.13] Melvin K. Simmons: Artificial intelligence for engineering design, Computer-Aided Engineering Journal, April 1984. p. 75-84.

[2.14] J. McDermott: R1: a rule based configurer of computer systems, Report No.:CMU-CS-80-119, Carnegie-Mellon University, Pittsburgh, PA, 1980.

[2.15] Max Schindler: Artificial intelligence begins to pay off with expert systems for engineering, Electronic Design, August 9, 1984. p. 107-167.

[2.16] Shimon Ando and Tatsuo Goto: Current status and future of intelligent robots, IEEE Trans. on Industrial Electronics, Vol.IE-30, No.3, August 1983. p. 291-298.

[2.17] P. T. Rayson: A review of expert systems principles and their role in manufacturing systems, Special FMS issue of ROBOTICA (ed. Paul G. Ranky), by Cambridge University Press, in print 1985.

CHAPTER THREE

The business data processing system of CIM

As already emphasized in the introduction, (see again Figure 1.1 in Chapter 1) Computer Integrated Manufacturing in this book is considered to be a general concept for integrating the business system, Computer Aided Design (CAD), Computer Aided Manufacture (CAM) and Flexible Manufacturing Systems (FMS, FAS, etc.) by means of distributed processing systems, .computer networks, databases and expert system programs.

In this Chapter we attempt to give an overview of some of the components of the business data processing system by discussing:

* customer order processing tasks,

* Bill Of Material (BOM),

* Material Requirements Planning (MRP) and control,

* Master Production Scheduling (MPS),

* factory level capacity planning and control,

* inventory control and purchase order control,

* product costing tasks.

As indicated in Figure 3.1 the business system in CIM consists or could consist of even more modules, but since they are partly well known, or well covered in other books and publications (see [3.1] to [3.6]) we shall not deal with them in this text.

Figure 3.1 A Computer Integrated Manufacturing system model indicating the business data processing system and its links.

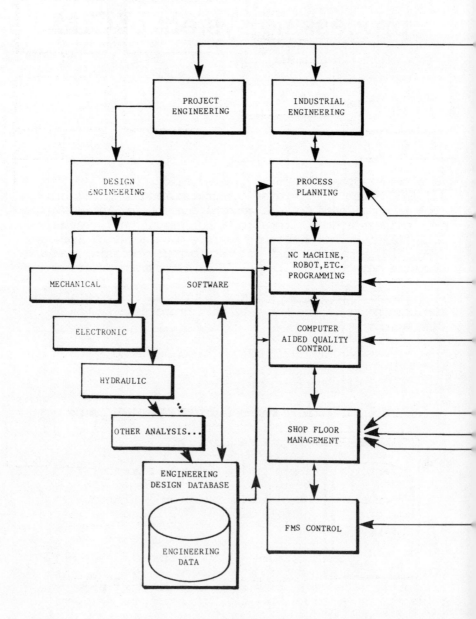

THE BUSINESS DATA PROCESSING SYSTEM

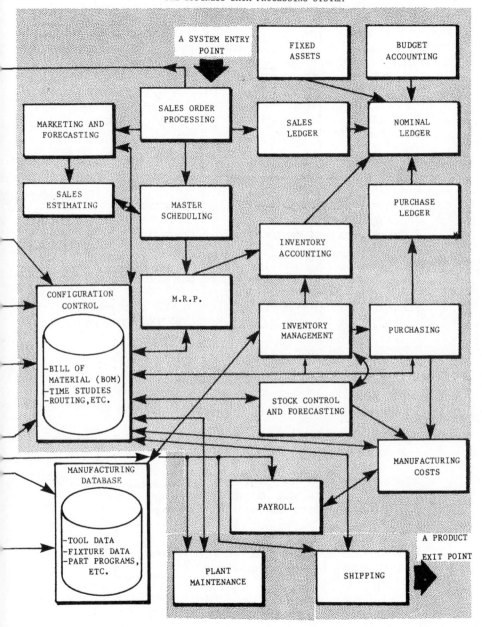

Since almost "everything relates to everything" in CIM, the emphasis in our discussion is put naturally on integration and on the fact that static production management systems (thanks to the factory wide data logging facilities and the computer networks) are becoming more and more real time, thus the available data to be processed by the different modules of the business system is more accurate. Because there are many links between the subsystems this is a complex topic. To simplify the understanding of the material many figures are shown in each section of this chapter illustrating important data flow and the interrelationships between different files and subsystems.

To illustrate the benefits of the above mentioned links let us list some macro level queries, transactions and data groups an up-to-date business data processing system should offer to managers in a manufacturing company:

* Master data of parts, bills of material, process layouts.

* Suppliers, buyers, currencies information.

* Manufacturing system, cell, tool and fixture data, product codes and their current status, parts lists.

* Markets, budgets.

* Customer orders, customer details.

* Part inventories, purchase orders and items.

* Current updates on the above listed vast amount of information.

It must not be overlooked that to be able to enjoy the benefits of computers and generally the huge amount of accurate and up-to-date business information provided by them one needs

1. Well educated and trained people who understand the benefits of such systems, and

2. equipment (including software) capable of performing the required operations and transactions.

Management used to the "crisis solving type" of operations needs time and adequate training to understand and learn about decision making processes using computers. It is also very important that the top management of the company supports computer installations and provides the necessary time for the middle management to learn how to use the benefits of the fast decision making systems.

It is also crucial to understand that although computers can process data at a very fast speed, current software products do not have the ability to learn from the environment and "tune their mathematical models" according to the requirements of the factory. Because of this there will always be some exceptions that must be handled to some extent separately. To achieve a successful integration it is important that the number of these exceptions do not exceed the manager's ability to review them and react accordingly.

From the data processing system point of view this, as well as the other subsystems of CIM, must make full use of the data logging and computer-to-computer communications facilities established by the distributed processing system as described in the previous chapter.

3.1 Customer order processing data flow

Since the manufacturing company exists to make products and to deliver them to customers on time, let us begin our discussion with order processing. When the customer places an order he or she opens an "entry point" in the business system (see Figure 3.1). The order must be entered into the data base and it must be processed, during which customer shipping, billing, part information, available discount, delivery terms, etc. data must be dealt with.

As a response, the business system would execute pricing and invoicing programs and would provide the stock picking list and the shipping information to the order processing department.

The benefit of using the business information system in this case is that at any time the customer order status can be determined and that the complete sales knowledge is available to all subsystems of the company as soon as the customer order data is in the database. Typically the macro level queries, transactions and data groups of the customer order processing module should be able to:

* Execute searches on customer data, on parts and purchase orders.

* List customers and parts in different orders and combinations, update customer lists.

* List customer enquiries, their status, billing and shipping addresses.

* List the commercial condition of customers, shipping bill information, invoicing conditions and customer status.

* List totals of orders, charges, comments, instructions on orders.

* Provide order addresses, commissions on order, status of order, invoicable data on order, invoices against orders, details of an invoice and item status totals of an order.

* Display prices, charges, commission, manufacturing instructions, status availability of the order per item.

* List suppliers, update list of suppliers, delete obsolete suppliers.

* Amend the address, the invoicing and/or commercial conditions, commissions, charges or the status of an existing order.

* Amend the price, commission charges, status, manufacturing dates, etc. of an existing item.

* Add, amend, delete items of the planned manufacturing order, etc.

     To illustrate the effect a new order makes on the manufacturing company refer to Figure 3.2. This figure shows a simplified model, in this form applicable only to small companies, but explains clearly the most important interrelationships between the basic modules affected.

     As can be seen order data is utilized for long range planning, and for short-range requirements planning. Without

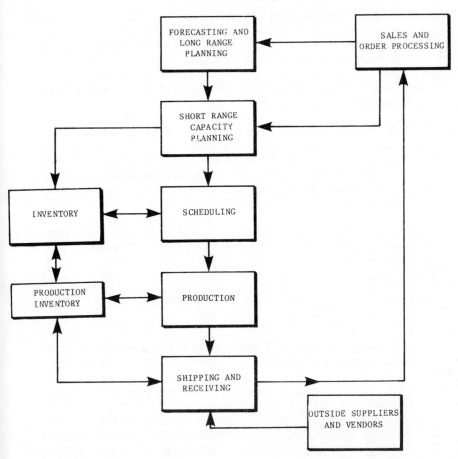

Figure 3.2 Simplified production control system model.

forecasts it is impossible to accomplish long-range capacity plans, but order data is also required for short range requirements planning providing information for capacity planning and for scheduling production and equipment.

Capacity planning is necessary to establish data on the required machines, personnel, equipment and inventory to accomplish the order, or in general to manufacture goods. (Note that long-range capacity planning is discussed in this Chapter, since it affects the business as a whole, whereas production oriented capacity planning is introduced in Chapter 8, because it affects directly the FMS).

An important concern of capacity planning is the amount of inventory to be held. The control of inventory means comparing what is actually on hand with the desired quantity. In up-to-date manufacturing companies inventory and work-in-progress (WIP) is held at a minimum level to keep inventory holding costs, related capital investment and taxes as low as possible.

Even in the case of small companies scheduling is considered to be a long-range and a short-range activity. The master schedule is a long-range activity, affecting the business as a whole, thus discussed in this Chapter. It can identify production goals for a year, for several months, or weeks ahead, depending on the nature of the business the company is in. Production scheduling on the other hand is a short term activity, sensitive to dynamic changes in the actual production system, for example FMS, or FAS and is discussed in Chapter 7.

3.2 Material Requirement Planning (MRP)

The objectives of Material Requirement Planning are to plan and release orders, to focus on orders actually requiring attention, and to ensure that all parts, manufactured and purchased, are available to meet finished production schedules, based on orders. MRP provides the flow of information throughout the whole company so that both planning and detailed execution activities can be controlled to achieve the same objectives. It integrates long-range business strategies, short-range tactical plans, master and detailed schedules and measurements of whether activities meet objectives ([3.7], [3.8], [3.9], [3.10]).

In short, MRP answers the question: "When should a manufacturing and/or purchase order be placed and for what quantity?".

The input for MRP is the Master Production Schedule (MPS) (discussed in section 3.4) containing the finished product requirements, i.e. quantities of each deliverable end item and the dates they are needed. MRP requires also accurate Bills Of Material (BOM) data, inventory status information, production and order status and lead time values for all parts (Figure 3.3). (The BOM file contains the product structure and the

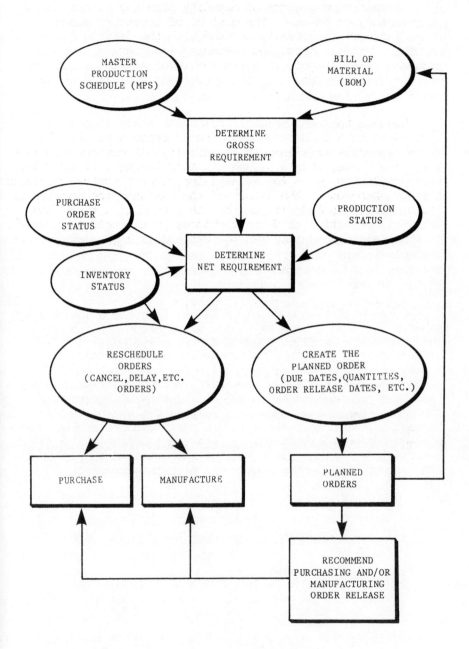

Figure 3.3 The basic functions and operation of Material
Requirements Planning (MRP).

number of subassemblies. Note that BOM is discussed in detail
in section 3.3.)

The long-range MRP strategies and plans can extend several years into the future with the aim of providing guidance
for the particular business in the future. They involve detailed activities and data analysis, including:

* aggregate forecasts,

* potential market analysis,

* evaluation of possible products,

* resources, etc.

The long-range plans are important since they affect the
company's future competitiveness. Because of this in most
companies long-range plans are made annually and adjusted
during the year only as necessary.

Other activities covered by MRP include the adjustment of
master production schedules to actual requirements and short-
range tactical planning. These activities include:

* Gross requirements determination, resulting from
  independent demands created by the master schedule,
  and from dependent demand generated by the MRP program itself (see again Figure 3.3).

* Net requirements determination, which is a calculation
  done for each time period determining the available
  (i.e. current inventory) the following way:

  Net req.= Gross requirements - Available inventory

  Where: Available inv.= On hand stock + Released orders
                        - Allocated stock to be withdrawn

* Rescheduling orders. Since requirement dates and quantities are recalculated, open order quantities and due
  dates must be analyzed to reschedule orders.

* Creating the planned order by calculating the order
  quantity then the date on which it will be available in
  stock, to cover calculated net requirements.

* product forecasting, and new product introductions, etc.

Material Requirement Planning systems work in one of two ways:

1. Either they completely recalculate the master schedule item inventory requirements and replan all released orders, based on the current database contents, i.e. using new BOM files, new order policies, etc. known as the "schedule regeneration method", or

2. They process only the changes which occur from one regeneration of the schedule to the next, known as the "net change method". This method minimizes the time spent on processing the new, up-to-date MRP and provides a continuous replanning facility.

With the "schedule regeneration method" all previous planning is re-evaluated and the entire Master Schedule is reprocessed against the up-to-date database. Because of the large amount of data processing work involved this method is costly to run frequently. But there is also less need for this in CIM compared to the "crisis driven companies" because of the availability of accurate information, because of the better and more consistent organization and the reasonably lower scrap percentage by utilizing FMS, FAS, etc., and their real-time quality control systems with on-line feedback data.

The "net change generation method" is performed frequently. In order to utilize it effectively the business management system must be designed to react rapidly to any type of changes resulting from

* changes in the Master Production Schedule item quantity, and/or due dates,

* engineering design changes, including: new or deleted items on a bill of material, changes in quantities, design modifications, etc.

* inventory transactions that result in a change to the available inventory quantity,

* changes in inventory policy, which effect the safety stock, the order size and lot size calculation methods, changes in lead time,

* revised release order due dates,

* a revised order quantities, etc.

The plan regeneration process continues downward from the point of change, modul by modul until the planned order quantity and the due date is no longer affected, or until the lowest level of disturbance is reached.

To summarize the application areas of the above discussed two different MRP calculation methods one must realize that requirements planning is normally done following the "net change" method, but "schedule regeneration" is unavoidable:

* when making an initial explosion,

* if the master schedule changes significantly,

* if there are major changes in the inventory, or in the inventory policy, and finally

* if the cummulative errors in the system become highly disruptive.

Measurements, done as much as possible in real-time, are important in the MRP strategy because they ensure that plans have been met. Major measurement points include comparisons of actual sales with the forecast, actual production with the schedules, actual inventory with the planned, etc. To see the major areas where measurement is crucial refer to the list given below:

* sales compared to forecasts,

* production compared to the plan,

* inventory compared to the budget,

* materials purchases compared to the planned.

Since all necessary Master Production Schedule (MPS) information as well as the BOM (Bill Of Material) files are available in the MRP database detailed schedules are usually generated automatically. The usual timescale for the master production schedule is one year, but in most cases the MPS is reviewed in shorter horizons . However it must be emphasized

that the length of these reviews largely depends on the
product and the company itself.

Finally the typical macro level queries, transactions and
data groups of the MRP software module should:

* give basic and full description data of part, management, lead time, order data on part,

* provide price summary information on part, cost makeup
  data of part, stock location data of part, routings
  of part, options for part routing, full data on part
  and manufacturing routing options,

* list items on a manufacturing order, external priorities of the manufacturing order, status of the manufacturing orders and/or items,

* perform material availability checks on an order,

* check scheduling dates, status of all operations on a
  manufacturing order,

* provide progress reports on a manufacturing order.

3.3 Bill Of Material (BOM)

A bill of material is a description of a designed product in
which the relationship of component parts, assemblies and
subassemblies to the final product is expressed either graphically (Figure 3.4/a) or by means of a structured list (Figure
3.4/b). To be able to represent a finished, or designed
product structure each assembly, sub-assembly, etc. must be
uniquely identified by a number.

The parts-usage lists contain the same information as the
BOM file, but they present the products in which each assembly
and sub-assembly appears.

In a manufacturing company utilizing up-to-date technology, the BOM file and the parts-usage lists are created by

MANUFACTURING LEVEL

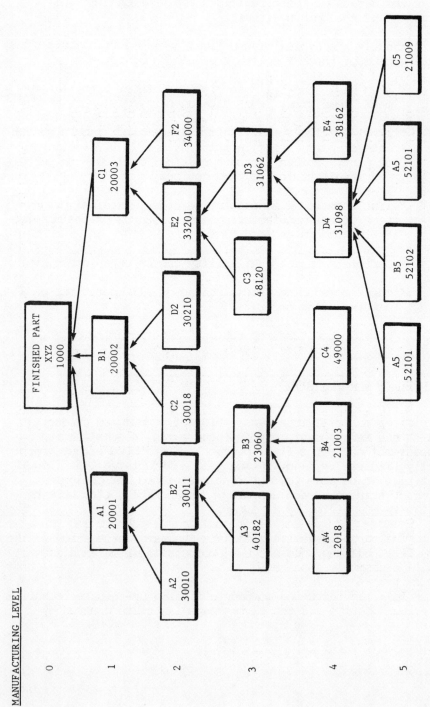

Figure 3.4/a Graphic representation of a Bill Of Material (BOM) file.

Level	Part number	Description	Quantity/unit	Total quantity
0	10000	Finished part		1000
1	20001	Component A1	1	1000
*2	30010	Component A2	1	1000
*2	30011	Component B2	1	1000
**3	40182	Component A3	1	1000
**3	23060	Component B3	1	1000
****4	12018	Component A4	1	1000
****4	21003	Component B4	1	1000
****4	49000	Component C4	1	1000
1	20002	Component B1	1	1000
*2	30018	Component C2	1	1000
*2	30210	Component D2	1	1000
1	20003	Component C1	1	1000
*2	33201	Component E2	1	1000
*2	34000	Component F2	1	1000
**3	48120	Component C3	1	1000
**3	31062	Component D3	1	1000
****4	31098	Component D4	1	1000
****4	38162	Component E4	1	1000
*****5	52101	Component A5	2	2000
*****5	52102	Component B5	1	1000
*****5	21009	Component C5	1	1000

Figure 3.4/b Standard list of Bill Of Material shown in Figure 3.4/a.

the CAD (Computer Aided Design) system and are accessed and utilized by the following CIM modules:

* CAD, when creating and/or redesigning the product, or assembly,

* manufacturing, or industrial engineering when expanding the bill of material to include routings, facilities planning data, etc.,

* production control and planning subsystems when selecting appropriate manufacturing facilities (e.g. FMS) from the equipment database,

* process engineering subsystems, when generating CNC part programs, robotized assembly and test programs, etc.,

* purchasing modules to order parts and material from suppliers,

* MRP, as discussed above,

* accounting programs when performing cost analysis for products based on the bill of material,

* customer services sub-system, when parts are replaced in the product,

* maintenance department, when the product is repaired, and by some other, less important subsystems.

Macro level queries, transactions and data groups of this software module should be able to:

* add, modify, delete a part from a bill of material, or copy an existing bill of material,

* provide a single level bill of material of a part,

* list all levels and/or lowest level of bill of material of a part,

* list single level, highest level and/or all levels "where used?" bills of material for a part,

* list basic data, text data and lead time of parent
  component per component entry, etc.

3.4   Master Production Scheduling (MPS)

Scheduling is a process that relates specific events to
specific times or to a specific span of time. Scheduling in
general involves the order and timing of assigning resources
(i.e. facilities, machines, employes, services, etc.) to
specific orders. Scheduling is of extreme importance in any
complex activity, such as project planning, factory management, production planning and control, etc., since inefficient
scheduling can create:

* delays of some orders through the system,

* low utilization levels of exsisting resources,

* disorder and even "panic" if scheduling faults generate
  major queues the system is unable to cope with; and
  other type of breakdowns.

Scheduling can deal with the various activities relating
to orders, the project plan, the shop, the allocation of the
workforce, the warehouse and in general with all activities of
the factory ([3.11], [3.12], [3.13], [3.14]) (Figure 3.5).

Scheduling in the manufacturing industry means the allocation of jobs to be processed on the specified machines in a
given time span.

A job is a product (i.e. a workpart, or component) to be
manufactured (i.e. milled, turned, ground, forged, cleaned,
inspected, painted, assembled, etc.) through single or multiple stages. At each stage (i.e. on each machine, or FMS
station, or FMS cell) an operation or more operations are
executed. The sum of operations, or processes performed at
each stage in the proper order produces the component, or job.

Depending on the level scheduling is utilized, it can be
long range and short range. It can be applied to the company
as a whole, or to a shop, to one of the FMS systems, to a
single processing cell (i.e. machine) or to a mixture of some
of the listed different levels.

Scheduling is usually more effective if applied at different levels and if certain real-time data feedback is provided between the different organizational levels to be able to consider certain changes and variations in the production plan. During the fine tuning phase, scheduling deals with individual sequencing of jobs on machines in order to achieve certain management objectives.

Typical management objectives in this sense could be the achievement of the:

* minimum throughput times for certain jobs,

* optimum utilization of the existing capacity,

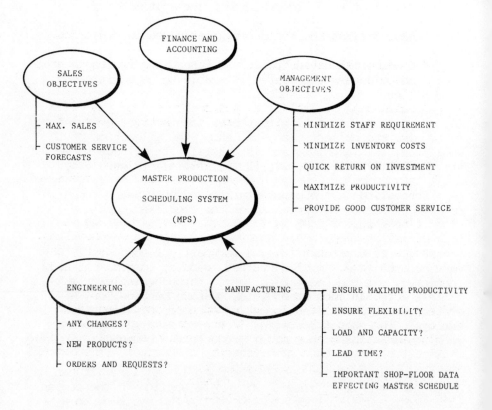

Figure 3.5 Master Production Scheduling (MPS) information sources and different objectives such systems need to address.

* maximum production rate,

* optimum production rate (defined here as the maximum production rate, considering the highest level of reliability)

Input data for the Master Production Scheduling system usually includes:

* allocating capacity requirements to orders,

* allocating and checking capacity requirements via different process routes,

* assigning job priorities.

Output data of the scheduling system usually includes:

* loading allocation (i.e. matching capacity requirements and orders),

* sequencing of jobs (i.e. the assignment of job priorities to the specified shops, FMS systems, machines, or processors),

* dispatching and control data (i.e. executing established sequences of jobs on assigned FMS cells, or systems and reviewing the status of different orders).

If the flow of jobs is identical for all jobs, we may talk of a flow-shop, and accordingly of flow-shop scheduling. (This type of manufacturing and scheduling method is typical for the mass production industry).

If for each job the sequence of the required machines differ, we may talk of a job-shop and the scheduling method we employ is job-shop scheduling. As an example, one could imagine a machining shop milling and turning several different machine tool components of different batch sizes. (The actual scheduling process must be proceded by a planning phase, considering the selection of the necessary operations to be performed on the part and their routing).

Job-shop scheduling combines loading and sequencing of jobs. Loading means the the assignment of machines to one or more operations that must be performed on the component. Sequencing determines the order of processing the job, or various jobs assigned to a particular machine or machines.

The major disadvantage of the job-shop scheduling method is that it is off-line, since it applies for a fixed period, throughout which it is valid in its unchanged form and that it is executed utilizing manually or partly computer controlled machines and the material handling systems.

CIM should enable FMS technology, enabling to cope with dynamic changes without disrupting production plans. (Note that FMS operation control is discussed in Chapter 6 and FMS scheduling in Chapter 7).

Having introduced the most important functions of MPS, herebelow we discuss some of the relating data processing and organizational aspects in more detail.

Figure 3.6 indicates the basic relationship between MPS, the manufacturing and the sales activity indicating the way a manufacturing company achieves its sales objectives.

In this model the customer service includes the processing and acknowledgement of orders, realistic shipment promises and competitive lead times.

The MPS is essential to be able to provide realistic dates of product shipment, by integrating marketing and manufacturing efforts and data and by establishing the optimum size inventory. If the incoming demand rate and the products to be manufactured vary a lot only flexible, computer based systems can cope with the decision making process. ([3.12], [3.13]).

If for example in the case of a fixed lead time order system the demand is abnormally high the MPS will be overloaded. On the other hand if the demand is relatively low, capacity will be under-utilized, thus profit will be lost. In both cases if the MPS can be adjusted to the actual market needs, the company will produce goods at a lower cost and eventually will be more successful.

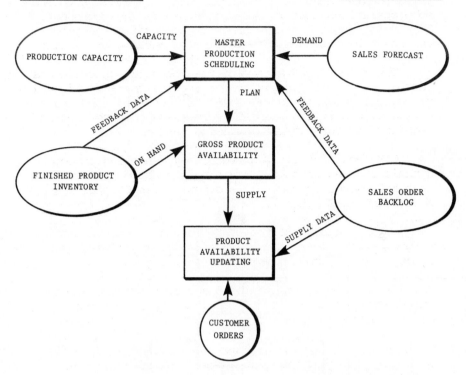

Figure 3.6 The relationship between the Master Production
          Scheduling system and the manufacturing and sales
          activities.

The properly designed MPS architecture communicates with
the subsystems as indicated in Figure 3.7 and relies on some
real-time information provided by shop floor control computers. This model allows monitoring of the production as well
as taking account of inventory sizes and real-time disturbances, in order to generate a realistic schedule as far as
possible.

To be able to support MPS development, product availability reporting, FMS cell efficiency calculations, capacity
requirements, lotsize analysis, maintenance policy planning,
sales order processing and other manufacturing activities and

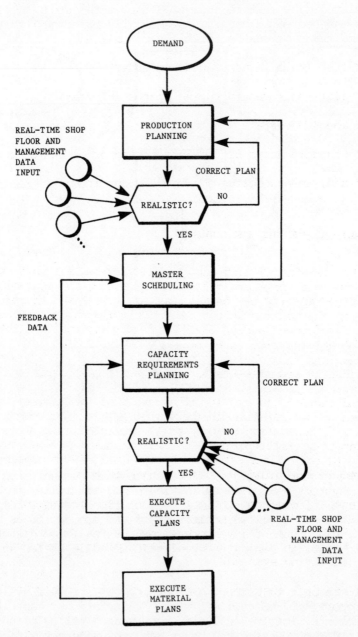

Figure 3.7 Closed loop master scheduling and manufacturing
control model making use of real-time data collec-
ted on-line from the shop-floor and from management
data bases.

the related data must be organized into structured logical
groups and stored in data bases generally offering the enquiries and transactions as follows:

* list items in manufacturing order,

* list their manufacturing status information,

* check materials availability,

* schedule dates of a manufacturing order and/or item,

* add, amend, delete a new MPS order item.

3.5 Capacity planning and control

Capacity planning is a crucial decision because it affects
both short and long-term managerial planning and control.
Productive capacity can be measured in units, and it refers
either to the maximum output rate for products or services or
to the development and/or manufacturing resources available in
each operating period (i.e. shift) (Figure 3.8).

Capacity requirements planning means the comparison of
the available capacity with the required capacity. The
required capacity is the sum of work hours, or minutes
required to perform the planned operations.

Capacity control provides a load summary of the manufacturing system, or any other equipment used in the company
involved in designing and fabricating the products.

Capacity requirements planning in general can be
utilized:

* to design plants (i.e. evaluate plant sizes) and manufacturing systems,

* to check whether or not the existing capacity is large
  enough to take on a job or not,

* to check the utilization levels of existing and planned
  manufacturing systems, machines or any other equipment

(e.g. CAD work-stations, etc) that are required in the fabrication process, together with

* operational scheduling to evaluate different routings and precedence rules generated by managers or the scheduling program, together with

* production balancing programs to cut Work In Progress (WIP) that can rise sharply if for example two machines supply components for a third but do not work in balance.

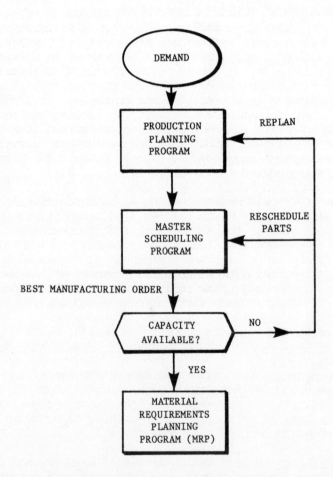

Figure 3.8 The basic capacity planning algorithm in relationship to master scheduling and Material Requirements Planning (MRP).

Managers and shop supervisors are always interested in the production rate required to finish their overdues. The load summary in conventional production systems is usually never level, nor will it match exact targets in any given week other than by coincidence. Up-to-date capacity control is a matter of moving from a reactive attitude to an active mode, in which upcoming workload is carefully analyzed, planned and executed utilizing flexible production facilities with random scheduling methods (see Chapters 6 and 8).

3.6 Inventory control and purchase order control

Inventory items are supplies of raw material, components and assemblies manufactured and/or purchased from outside vendors. Inventory is often divided into the following groups of items:

* raw material inventory identifies all items which are partially finished or raw components waiting to be processed,

* maintenance repair and operating supplies are important to run the machines or system by which the products are manufactured,

* Work In Progress (WIP) identifies parts which are partially processed only and are waiting to be completed, and

* finished goods inventory, representing the final product, ready for sale or shipment.

Flexible production systems working together with other components of CIM can dramatically reduce inventory. By producing single parts on order, in other words "just-in-time", at lower costs then making them in large batches, besides other benefits, inventory and in particular WIP is reduced to a fraction compared to conventional methods. It is important to underline that one of the methods by which FMS investments and CIM development can be justified is inventory analysis.

Obviously this is not a simple task, because of the very low level, or no inventory, and because other sub-systems such

CIM-D

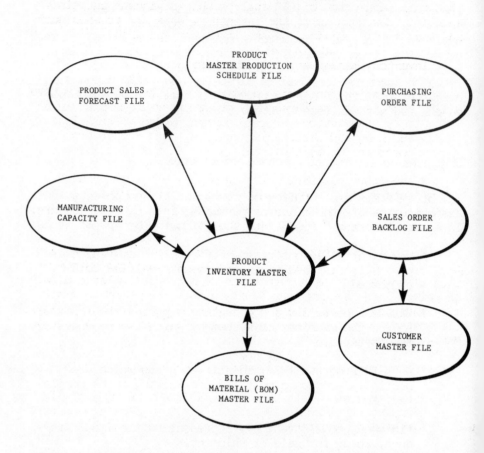

Figure 3.9 Some of the most important data connections between the product inventory master file and the other modules of the business system.

as order processing, MRP, scheduling, capacity planning, manufacturing, shipping, etc. in the factory as well as outside the factory (represented by vendors and suppliers) rely on each other and must work accurately together as a quartz clock (Figure 3.9).

Inventory management is concerned with co-ordinating the relevant departments of the conventional factory, or the above

listed software systems of CIM. "Co-ordination" here means
that this sub-system of CIM must be able to answer questions
such as: "How much is in the inventory, where is it, how much
and what should be added to the inventory and when?".

Inventory management incorporates:

* accounting functions, concerned with the administrative
  aspects of the warehouse, order entry processing inventory transactions, record current stock information,
  auditing and history information, purchase orders, customer shipment, etc., and

* inventory planning activity which is concerned with
  future requirements on the basis of forecasting include
  the current inventory status, material requirements,
  marketing and other less important modules.

The basic principle is that stock-on-hand should be
exactly as much as required. If stocks are very high production planning has less constraints because of the large size
of buffer large stocks represent. On the other hand to minimize inventory costs stocks should be reduced, or practically
eliminated. To keep these two opposing demands in balance the
MRP system must:

* continuously post all stock transactions, and

* be integrated with the other indicated modules of CIM
  and react to real-time changes in the operation control
  system of the factory (see again Figures 3.1 and 3.3).

In the case of expensive components the sales plan ought
to be the departure point for MRP, whereas parts or raw material of lower value may be planned statistically, utilising
usage history data. (Statistical forecasting technics can be
utilized most successfully to parts which have developed a
prredictable demand pattern.)

Macro level queries and transactions offered by the
inventory module usually include:

* list summary of inventory,

* list manufacturing, purchasing and/or MPS order cover
  for item, or product,

* material availability check for a part,

* customer, purchase, manufacturing order requirements list for a part,

* stock count and difference list of the inventory,

* inventory movement operations, including: unplanned issue or receipt of a part, transfer of stock to an other manufacturing division, or FMS, etc., physical transfer of stock inside the warehouse, and the

* inventory analysis module, including list inventory movements, inventory usage analysis, evaluation of different inventory policies, status, balance, etc.

3.7 Product costing

The inability to accurately determine the components of the product cost often result in false profit reports and generally misguides the management about the profitability of different products, divisions, or departments of the firm.

The important point is that any activity related to product design, manufacture, sales, test, marketing, etc., any component built into the product, or any process carried out on the product, costs money. In other words each department, or sub-system of CIM must have its own costing function and the total cost must be built up proceeding from the lowest level of the bill of material item of the product.

(Let us illustrate the most important components of the product cost in Figure 3.10.) Here again CIM can revolutionize design, production, test, etc., since the integrated CAD/CAM system data base linked to other modules of CIM can advise the designer and the production engineer on economical alternatives.

The computer controlled factory must react rapidly to changes in the product cost. Product cost alterations can occur because of a variety of different reasons, including:

* changes in the energy prices,

* vendor price changes,

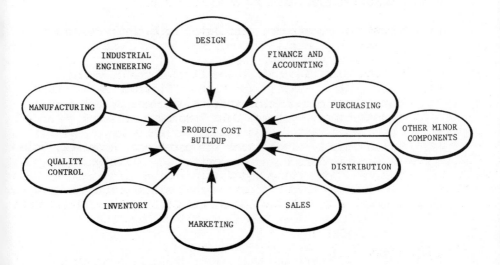

Figure 3.10 Typical product cost components and information
        sources in manufacturing companies.

* exchange rate changes,

* changes in the manufacturing process and available
  facilities,

* design changes,

* changes in Bill Of Material,

* overhead rate changes,

* labour rate changes, etc.

The ignorance of maintaining product cost information
results in loosing the availability to assess the impact on
profit of these changes, thus corporate level and strategic
decisions become unfounded. The availability of up-to-date and
accurate information on cost matters help to guide the company
at all levels to make profitable products, which should be the
result of CIM.

References and further reading

[3.1] David. D. Butterworth and James E. Bailey: Integrated Production Control Systems. Management, Analysis, Design. John Wiley and Sons, 1982.

[3.2] H.J.J. Kals, F.J.A.M Van Houten: Flexible Manufacture Based Production Information Management System (PIMS), Manufacturing Systems, Vol.12 No.3. p. 187-196.

[3.3] Brian Conolly: Techniques in Operational Research, Volume 2, Models, Search and Randomization. Ellis Horwood Limited, John Wiley and Sons.

[3.4] Frederick S. Hillier and Gerald J. Lieberman: Introduction to Operations Research, Holden Day, Inc. San Francisco, 1981.

[3.5] Elwood S. Buffa and William H. Taubert: Production-inventory systems, Planning and Control. Richard D. Irwin Inc., Georgetown, Ontario, 1972.

[3.6] Harold J. Steudel: Computer Aided Process Planning: past, present and future. International Journal of Production Research, 1984. Vol. 22, No.2, p. 253-266.

[3.7] William Maxwell and others: A Modelling Framework for Planning and Control of Production in Discrete Parts Manufacturing and Assembly Systems, Interfaces, 13; 6, December 1983. p. 92-104.

[3.8] John Gallimore: How to make MRP really work, The Production Engineer, March 1984. p. 22.

[3.9] J.K. Lenstra and A.H.G. Rinnooy Kan: New directions in Scheduling theory, Operations Research Letters, March, 1984. Vol.2, Number 6, p. 255-259.

[3.10] Edward T. Grasso and Bernard W. Taylor: A Simulation-based Experimental investigation of supply/timing uncertainty in MRP systems, International Journal of Production Research, 1984. Vol. 22, No.3, p. 485-497.

[3.11] C. Walter: Manufacturing Process Control Specification with Functional Programming, IFAC Information Control Problems in Manufacturing Technology, Maryland, USA, 1982., Session 2: Manufacturing Process Control, p. 21 - 26.

[3.12] Juan J. Gonzalez and Gary R. Reeves: Master Production Scheduling: A Multiple-objective Linear Programming Approach, International Journal of Production Research, 1983. Vol. 21, No.4, p. 553-562.

[3.13] R. Malko: Master Scheduling: a key to results, Proceedings of the Nineteenth Annual Conference of the American Production and Inventory Control Society, 1976.

[3.14] Alfred Buchel: Stochastic Material Requirements Planning for Optional Parts, International Journal of Production Research, 1983. Vol. 21, No.4, p. 511-527.

[3.15] Willem Dijkus: Fifth generation terminals, Data Processing, Vol. 26, No.2, March 1984., p. 43-45.

[3.16] William I. Bullers at. all: Artificial Intelligence in Manufacturing Planning and Control, AIIE transactions, December 1980., p. 351-363.

CHAPTER FOUR

Computer Aided Design (CAD)

In general terms Computer Aided Design (CAD) means computer
assistance whilst a human operator converts his or her ideas
and knowledge into a mathematical and graphical model represented in a computer. The key point is that there is an
increasingly accurate conversion from a model created and
represented in the human's brain to the "machine's brain" in
the form of digital data.

The conversion process results in a computer model
represented and stored in digital format, so that further
analysis, graphics manipulation, etc. can be made on it at
different CAD workstations without the need for describing
the model every time a new analysis, or graphics manipulation
is required. This makes the method attractive to designers
because they can create, modify, test, simulate, copy and save
different solutions, and to their managers because productivity and design reliability can be increased drammatically.

CAD is one of the core modules of the "total CIM concept". Figure 4.1 indicates that CAD should be integrated into
the manufacturing company's data processing system, providing
the possibility of integration with subsystems such as project
engineering and planning, NC/CNC part programming, robot programming, FMS programming, tool design, the MRP system and
other indicated modules. (Note that CAM and integrated CAD/CAM
are discussed in the next Chapter.) If CAD is analyzed and
evaluated bearing this "company wide" approach in mind and
utilized by able people then it can be of benefit to most
industries.

During the conversion of the human's image to a computer
representation the machine should provide as much help as

COMPUTER AIDED DESIGN

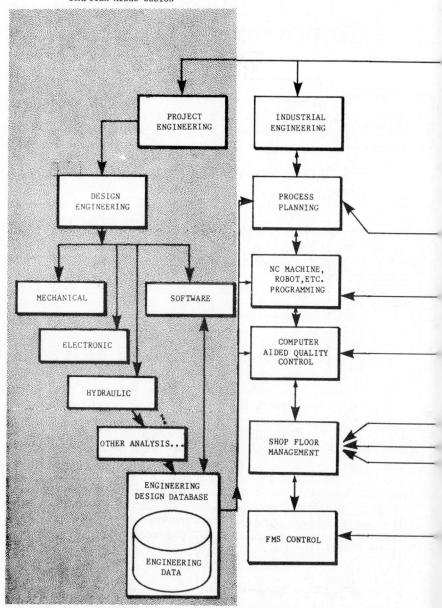

Figure 4.1 A Computer Integrated Manufacturing system model indicating Computer Aided Design and its links to the rest of the system.

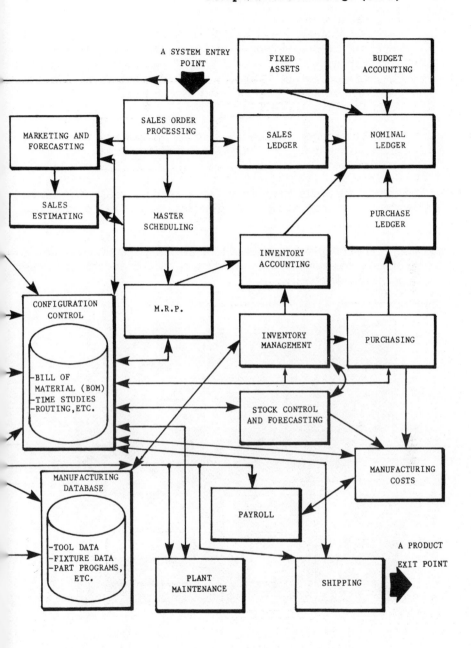

possible by means of:

* graphics input/output facilities (such as a "mouse", a light pen, a digitizer, a set of function keys, software menus, graph plotter, graphic printer, a keyboard and one or more graphics and alphanumeric displays, etc.),   (Figures 4.2 and 4.3),

* graphics manipulation facilities in two and three dimensions (e.g. translation, scaling, rotation, animation, etc.),

* model analysis facilities (e.g. stress analysis, heat transfer analysis, vibration analysis, finite element modelling and analysis, etc.),

and many others, depending on the available resources, the application, the model, the CAD system, the operator and other less important factors.

Figure 4.2 The DEC-VAX station 100 supports high-level graphics functions and text manipulation for a variety of general purpose and technical applications. (Courtesy of Digital Equipment Corporation.)

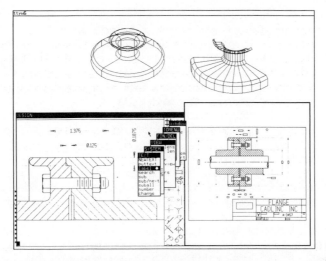

Figure 4.3 The CADLINC multi-window capacity. The screen can
be broken up into a maximum of eight windows. This
allows the programmer or user to access different
commands and/or files at any time during the
session. A new window is opened using the "pop-up"
menus and the cursor (see the little arrow in the
front of the screen) driven by a "mouse".

(Courtesy of CADLINC Incorporated, Elk Grove
Village, Illinois, USA.)

CAD is a very large and rapidly growing field, thus
within this text we shall give a brief overview of the benefits and the cost at which the benefits can be achieved by
utilising CAD, and discuss some aspects of computer graphics
and design with case studies, including:

* two dimensional spline interpolation and plot using the
UCSD-p. system with the "Turtlegraphics" unit and
Pascal,

* three dimensional graphics manipulation of a wire frame
robot model using the UCSD-p. system with "Turtlegraphics" and Pascal,

* the GKS and the IGES graphic standards, and

* graphic programming using the McAuto Unigraphics system
developed by McDonnel Douglas Corporation, St. Louis,
Missouri, USA.

4.1   The benefits and cost of Computer Aided Design

CAD is one of the core modules of the CIM concept and it should be developed and utilized with a "company wide" approach. The trend is that CAD is no longer a design and analysis tool for electronic and mechanical designers only, but is available to many other professions, including architecture, building, chemistry, medicine, fashion and textile design, etc.

Because CAD is expanding and penetrating into many new areas it is quite obvious that Computer Aided Manufacture, Flexible Manufacturing Systems and CIM as such are not limited in any sense to the machine tool, or the electronic assembly industries, but are general purpose tools, or "know-how", available to fashion designers, architects, design engineers and in general to any industry wishing to adapt quickly to rapidly changing conditions of its type of business. To illustrate this point and to show some interesting CAD models and a large variety of different applications, refer to Figures 4.4 to 4.11.

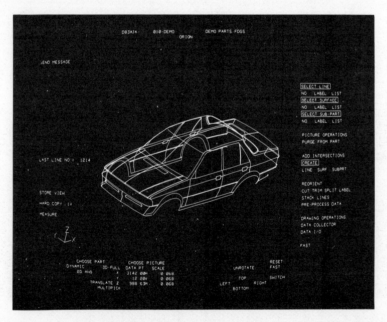

Figure 4.4/a Interactive Computer Aided Design at Ford Motor Company Ltd., (Courtesy of Ford Motor Company Ltd., Brentwood, Essex, UK.)

Before the early 1970s there was a complete reliance on designers and draughtsmen for producing detailed manufacturing drawings. Today CAD and its related subsystems represent a multi-million dollar business. The reason why this industry has developed so fast is because the technical and price gap is gradually narrowing between the "ideal or perfect solution" and the "realistic proposal" ([4.1] to [4.10]).

Figure 4.4/b Interactive Computer Aided Design at Ford Motor Company Ltd., Light pen input. (Courtesy of Ford Motor Company Ltd., Brentwood, Essex, UK.)

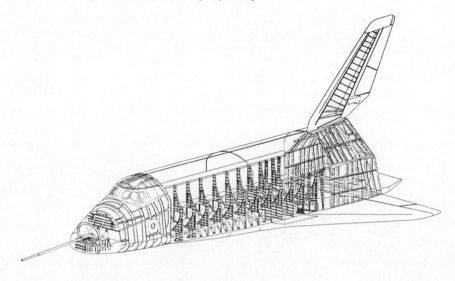

Figure 4.5 The USS Enterprise Space Shuttle produced by M and S
         Computing, Huntsville, Alabama on a Versatec
         printer/plotter.

Before purchasing a CAD workstation, or a larger system
consisting of several possibly different workstations linked
to a host computer or to a local area network, one must decide
its basic configuration.

Basically these can be one of the following:

* the so called "elaborate workstation", linked to a minicomputer or a mainframe (see again Figure 4.2) offering
  limited local computing power and basic graphics input/
  output facilities, such as a high resolution display, a
  keyboard and a digitizer tablet, or a "mouse", and maybe
  an optional hardcopy device, or plotter, or

* a "common workstation", offering similar facilities as
  the above, but with more screens, (perhaps an alphanumeric and a graphics screen combined), and generally
  with more input/output facilities (see Figure 4.4
  again) and finally

* the "standalone workstation" which is a powerful 16
  to 32 bit CPU micro-, or minicomputer (see again Figu-

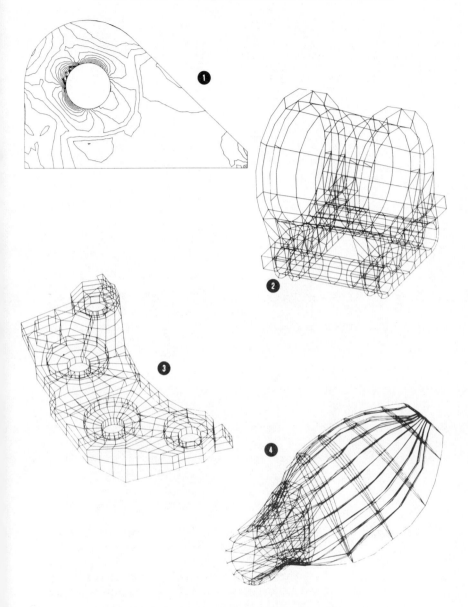

Figure 4.6/a Sample applications using the LUSAS Finite Element
System. (Courtesy of Dr. Paul Lyons, FEAL, 25
Holborn Viaduct, London.) 1. Principle stress
contours for lifting hook, 2. Robot support component, 3. Gearbox housing, 4. Composite propeller
blade.

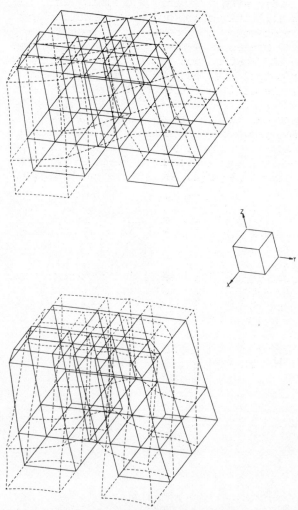

Figure 4.6/b Simulation of acoustic mode shapes inside a lorry
cab, using the PAFEC finite element modelling
package at Trent Polytechnic, Nottingham. The
finite element method enables the design engineer
to use a computer to carry out accurate stress and
vibration analysis on almost any shape, size and
complexity of structure under loads. The feedback
achieved enables the designer to modify his or her
design as many times as required and find the best
solution. (Special thanks are expressed to Mr.
S.T.W Keiller, Department of Mechanical Eng. for
offering the computer generated plots.)

res 2.7 and 2.8), with a Graphic Processing Unit (GPU)
for handling all interactive graphics functions and
to speed-up processing in general, a reasonably large
memory, e.g. 1 Mbytes to 4 Mbytes, hard disk storage
10 Mbytes to 100 Mbytes, and interactive graphics
input and/or output devices such as the light pen,
the "mouse", the digitizer tablet, the plotter, etc.

These machines can be used stand-alone as well as linked into a LAN, providing the possibility of sharing
processing capacity as well as hardware, and/or expensive peripherals (e.g. an accurate flat bed plotter,
large Winchester disk unit, etc.) (See again Figures
4.3, 2.7 and 2.8).

To justify some important CAD benefits in the case of
manufacturing applications one should consider the following
points:

* New designs can be created faster and at higher quality
  if not only draughting, but model analysis tools, such
  as finite element, vibration, heat transfer, functional
  analysis, etc. are utilized.

* New designs created on the CAD system will be more
  economic if the CAD system is linked into the distributed processing network of the company, offering
  direct feedback data from CAM, FMS and various business
  system modules ensuring that the design takes into
  account manufacturing realities at an economic cost.

* Existing designs if stored in computer readable format
  can be modified faster and at a lower error rate than
  without CAD.

* Detail draughting, assembly drawings, technical illustration, and Bill Of Material file can be automatically,
  rather than manually generated by a program.

* Product variations can be minimized by employing database management tools and applying existing components
  in new designs, called-up from the database. Parametric design programming techniques can also contribute
  to minimize product variations and to create components
  belonging to a part family, thus reducing tooling,
  fixturing and testing costs and supporting macro programming in FMS.

* From the marketing point of view lead times can be reduced through faster throughput of drawings and quotations can be produced faster and at better quality.

To achive partially or fully the above listed benefits a major capital investment is often required. Costs to be analyzed regarding the CAD installation should incorporate:

* equipment related one-off expenses, such as the hardware, the building costs of the CAD terminal and computer rooms if the system must be air conditioned,

* the initial operating system and other system and applications software,

* training costs of engineers and operators,

* system hardware/software maintenance and running costs,

* additional software and hardware costs to satisfy the "growing appetite" for new options, analaysis software, color graphics, additional workstations and inter- faces to newly linked equipment in different depart- ments, additional disk drives because of the increased database and data storage (e.g. disk capacity) needs, etc.,

* system analyst and programmer staff expenses, repre- senting a substantial sum in the case of relatively large installations in particular, or if the current staff is inexperienced in computing, and other less important cost factors.

In the early days, and in some cases at present as well, the programmer had to develop both the graphics and the appli- cations software, which considerably slowed down the actual design activitiy and required computer graphics expertise from machine tool design, electronics and other engineers. Thus there was a need, which still exists, to develop widely used, portable graphics packages and languages.

Several different approaches are used, and these can be classified into the following categories:

1. Special purpose graphics packages, designed to be

used in a particular application area only where the
programmer does not need to know much about computer
graphics.

2. General-purpose graphics packages, isolating the
   application programmer of device dependencies by
   means of a graphics software interface. To use such
   systems properly, the designer must understand computer graphics. (Examples include the Tektronix Plot 10
   or the Calcomp subroutine packages.)

3. Extending existing high level languages with graphics,
   providing an attractive way of producing powerful
   application software. Languages such as FORTRAN,
   Pascal, BASIC, APL, etc. have graphics extensions,
   which unfortunately are not compatible with each other
   and are not portable. However, there are some very
   good approaches, such as for example the Turtlegraphics unit in the UCSD-p version IV.1 system, which
   provides a portable graphics interface for over a
   dozen of different processors. (To demonstrate some
   features of such language extensions refer to the
   first two case studies shown in the next section of
   this Chapter.)

4. Graphics languages, offering advantages similar to
   existing extensions of high level languages.
   Furthermore, the user here is not constrained to work
   within the philosophy and structure of an exsisting
   language. (Refer to the McAuto Unigra-phics case study
   at the end of this section.)

Because of the variety of different applications, because
of speed and because of the vast amount of graphics data storage economy, efficient ways of creating graphics models are
important. In the latest approach, solid modellers not only
contain sufficient information to produce hidden surface drawings, but are also "clever" enough to prevent accidental creation of unrealistic combinations of such primitives. This
feature can be of use not only when designing, but also when
for example simulating tool path, or the assembly of components stored in a geometrical database, or when designing the
way the product is going to be packaged and in many other
cases.

These and other tasks need CAD (and CAM) systems which
are capable of learning from their users and from the data

104   Computer Aided Design (CAD)

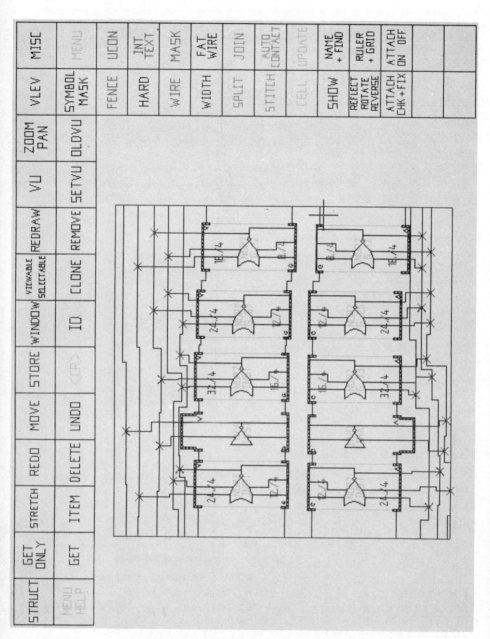

Figure 4.7/a Integrated circuit logic design on a Calma CAD system. (Please note that since this Figure is a reproduction of a color master some lines and symbols are faded.) (Courtesy of Calma (UK) Ltd.)

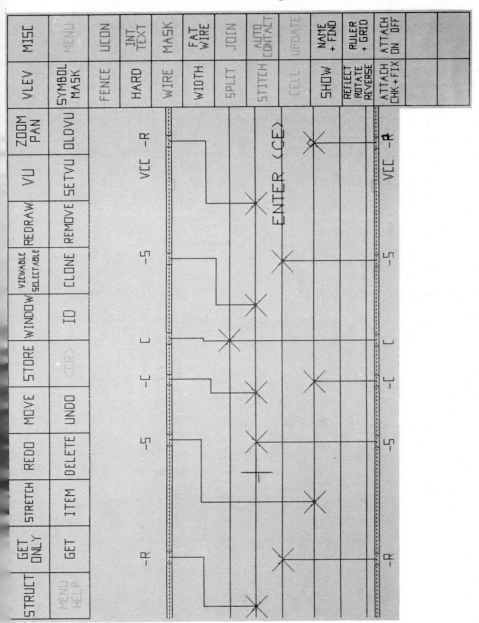

Figure 4.7/b Integrated circuit symbolic design phase on a Calma CAD system. (Please note that since this Figure is a reproduction of a color master some lines and symbols are faded.) (Courtesy of Calma (UK) Ltd.)

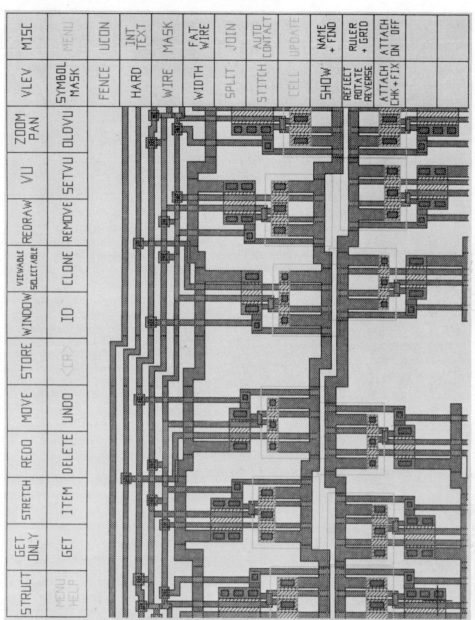

Figure 4.7/c Integrated circuit mask design phase on a Calma
CAD system. (Please note that since this Figure is
a reproduction of a color master some lines and
symbols are faded.) (Courtesy of Calma (UK) Ltd.)

they have processed. This additional knowledge is not fed back
into the design process at present, although application
programs contain a certain amount of knowledge in the form of
data, algorithms and formulas. Decisions made in most CAD/CAM
systems are currently deterministic, because the offered
choices do not represent intelligence.

In our view a CAD/CAM system is "intelligent" if it can
self-determine which rule to fire, i.e which decision to take
and if its operation is based upon the experience gained in
the past both from failures and successful solutions, which
are stored in the form of rules in the system's knowledge
base. At present CAD/CAM users have to take too many subjec-
tive decisions on the basis of deterministic programming
methods, and so the expert system should act as an adviser in
cases where the user needs help, or assistance. As soon as
expert systems are sophisticated enough, we shall have to
agree that they will often make better decisions than the
average designer, although there will still always be a
portion of the most creative work left for us as well.

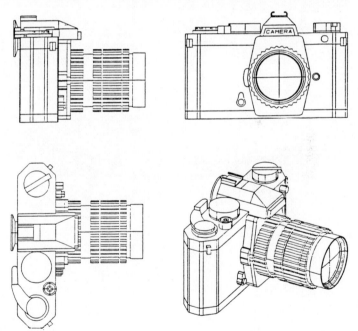

Figure 4.8 "Four visions of a camera". Generated by the MEDUSA
package, plotted on a Nicolet Zeta plotter. (MEDUSA
is a trademark of Cambridge Interactive Systems
Ltd., UK.)

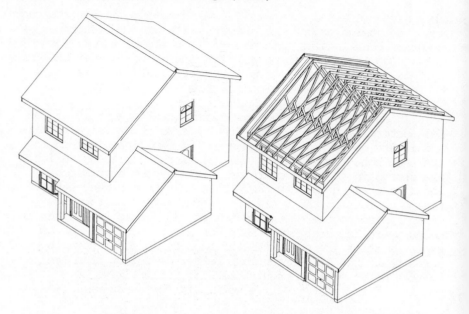

Figure 4.9 Computer aided architectural design of a building. (Courtesy of ACROPOLIS Graphic Modeling and Draughting Services, BOP Computing Services Ltd., UK.)

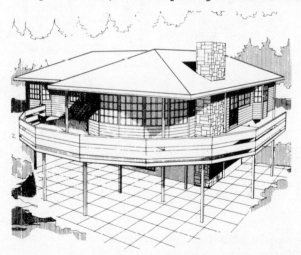

Figure 4.10 Automatic generation of perspective view and shadows from plan input and from any selected position. (Zeta 2000 Architectural Design and Draughting, Redland Construction Software Ltd., Farnham, Surrey, England.)

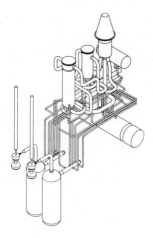

Figure 4.11 Views of a boilerhouse with hidden lines removed.
Drawn by the ACROPOLIS graphic modeling package.
(Courtesy of ACROPOLIS Graphic Modeling and
Draughting Services, BOP Computing Services Ltd.)

4.2   Case study: spline interpolation and plot

The goal of this case study is to demonstrate the graphics
extensions of a high level language (in this case UCSD-Pascal)
and to give a useful set of routines for those who wish to
incorporate them into their own programs. Smooth curve fitting
in general is in the interest of many areas of CAD and CAM,
including mechanical and arcitectural design, statistical data
analysis, NC programming, production engineering, test
enginering, etc.

To fully understand the case study one should be familiar
with at least some Pascal and the Turtlegraphics procedures of
the UCSD-p. system. The program segments shown are self
documented, but for further reading please refer to [4.35] and
[4.36].

The example shown is an application where measured data
is interpolated and displayed. Although the data relates to a
time study program (see again the file structure in Figure
2.12) where job times can be plotted against time, the
procedures are general purpose and can be used in many other
applications where two dimensional spline interpolation is
required.

4.2.1 The mathematical model of the program

A spline curve fit generates a smooth curve passing through all of the defined data points. The curve is constructed so that it takes a shape that minimizes the curvature of the line. The principle is analogous to the mechanics theory of bending a flexible beam, which subject to forces at discrete points, takes the shape which minimizes energy. The curve is therefore continuous at each data point and a similar type of curve is obtained to that drawn using a flexicurve.

Spline program routines (see Figure 4.12) provide a good curve fitting tool and its advantage over other methods of curve fitting is that it highlights small peaks in the data which may be significant but can be disguised with other methods.

The spline interpolating program is defined by the following criteria:

1. The curve is piecewise cubic, that is, the coefficients of the polynomial are different between each interval of the curve, i.e. between each pair of data points.

2. The curve passes through all the given data.

3. The first and second derivates are continuous at each node point. That is the slope and rate of change of slope for the curve is the same each side of a data point. This is impossible for two end points and hence these have to be treated as a special case.

The spline can be represented by the equation:

$$\frac{d_{i+1}}{6} f''(x_{i-1}) + \frac{(d_{i-1}+d_i)}{3} f''(x) + \frac{d_i}{6} f''(x_{i+1})$$

$$= \frac{y_{i+1} - y_i}{d_i} - \frac{y_i - y_{i-1}}{d_{i-1}}$$

Where: $i = 1, 2, 3, \ldots n$
         n = the number of data points

x_i, y_i = the co-ordinates of the data points, and
d_i = $x_{i+1} - x_i$

This equation forms a set of linear algebraic equations for $f''(x_i)$ which can be evaluated.

In the case of i=1 and i=n this is not possible because there are no values existing for i-1 and i+1 respectively. Thus only n-2 equations can be formed and n unknowns are present. This is solved by applying end conditions to the spline and some approximation of what is happening at the end of the curve required. In the case of the natural spline the end points are pinned and no bending moment exists at this point. This allows for the minimum curvature of the spline (see Figure 4.13).

As no bending moment is exerted then:

$f''(x_i) = f''(x_n) = 0$

This condition is chosen by letting LAMDA = 0 when CALC_SPLINE is called. A value of LAMDA = 1 assumes that the second derivate is constant for the curve between the end intervals of the spline and that the function is quadratic there. LAMDA may be chosen between these extremes and thus the end conditions are the combination of the above.

4.2.2 The main procedures of the program

The spline curve fit is generated by the CALC_SPLINE procedure (see also the self-documented listings provided in Figure 4.12).

CALC_SPLINE calculates the second derivates of the curve at every data point. These are the values of $f''(x_i)$. The second derivates are calculated for data points 2 to n - 1. The second derivates for points 1 to n depend on the value of LAMDA (see also the source code listing in Figure 4.12).

The curve is displayed by the PLOT procedure, which splits the x axis into 100 equal parts. For each part the value of the function is calculated thus giving the y value. These x and y values are then plotted which gives a smooth

Figure 4.12 General purpose spline curve fitting and graphics routines written in UCSD-Pascal.

```
General purpose spline curve fitting and graphics routines
                                                  Page No: 1
+----------------------------------------------------------------+


(*********************** CALC_SPLINE ***********************)


PROCEDURE Calc_Spline(Lamda:REAL);

(* Calculates second derivatives of curve at each data point *)

(* Lamda = 0 implies that the ends of the spline are free i.e.

            a natural spline.

   Lamda = 1 implies that the second derivative is constant in

            the end intervals of the spline, and that the func-

            tion is a quadratic there.

   Lamda may be one or the other or a combination of the above *)


VAR A,B,C,R:ARRAY[1..20] OF REAL;

                   (* Temporary arrays for calculation *)

     TEMP:REAL;      (* Temporary variable *)
BEGIN

   DER1:=NPOINTS-1;

   DER2:=NPOINTS-2;

         (* DER1,DER2,DERI are integers, used to derive

            second derivate. NPOINTS is an integer,containing

            the number of points in the graph *)
   C[1]:=X[2]-X[1];

   FOR I:=2 TO DER1 DO

      BEGIN

      C[I]:=X[I+1]-X[I];

         (* X and Y are real arrays, containing the x

            and y input co-ordinates of data points *)

      A[I]:=C[I-1];

      B[I]:=2*(A[I]+C[I]);

      R[I]:=6*((Y[I+1]-Y[I])/C[I]-(Y[I]-Y[I-1])/C[I-1])
```

General purpose spline curve fitting and graphics routines
                                                    Page No: 2
+--+

```
      END;

   B[2]:=B[2]+Lamda*C[1];

   B[DER1]:=B[DER1]+Lamda*C[DER1];

   FOR I:=3 TO DER1 DO

      BEGIN

         TEMP:=A[I]/B[I-1];

         B[I]:=B[I]-TEMP*C[I-1];

         R[I]:=R[I]-TEMP*R[I-1]

      END;

   FUNCDER2[DER1]:=R[DER1]/B[DER1];

         (* FUNCDER2 is a real array, containing the second

            derivates of the function *)

   FOR I:=2 TO DER2 DO

    BEGIN

     DERI:=NPOINTS-I;

     FUNCDER2[DERI]:=(R[DERI]-C[DERI]*FUNCDER2[DERI+1])/B[DERI]

    END;

   FUNCDER2[1]:=Lamda*FUNCDER2[2];

     FUNCDER2[NPOINTS]:=Lamda*FUNCDER2[DER1];

   END; (* Calc_Spline *)


   (***************************** SPINTERPOL ********************)

   PROCEDURE Spinterpol;

   (* This procedure interpolates value on curve for time given *)


   VAR DXM,DXP,DEL:REAL;    (* Temporary variables *)

       J:INTEGER;           (* Loop variable *)

   BEGIN

      DER1:=NPOINTS-1;

      I:=1;
```

114   Computer Aided Design (CAD)

+--+

```
   FOR J:=1 TO DER1 DO IF RESINT>=X[J+1] THEN I:=I+1;

                 (* RESINT is a real, containing the result

                    of the interpolation *)

   DXM:=RESINT-X[I];

   DXP:=X[I+1]-RESINT;

   DEL:=X[I+1]-X[I];

   Result:=FUNCDER2[I]*DXP*(DXP*DXP/DEL-DEL)/6

           +FUNCDER2[I+1]*DXM*(DXM*DXM/DEL-DEL)/6

           +Y[I]*DXP/DEL+Y[I+1]*DXM/DEL

END; (* Spinterpol *)


(***************************** SCALES ***********************)

PROCEDURE Scales(Min_X,Max_X,Min_Y,Max_Y:REAL; Title:STRING);


(* Sets up the screen to plot the spline fitted curve on.
   Note that this routine uses a file containing Start_Morn,
   End_Morn, Start_Aft, End_Aft time study data - see also
   Figure 2.12. The routine can be used as a general scale
   procedure *)
VAR Space:REAL;
    I,No_Divs,Start,z:INTEGER;
    Num:STRING;
    Ind:BOOLEAN; (* Indicates if file name is already known *)


BEGIN
    Xlgth:=Max_X-Min_X; (* Xlgth and Ylgth are reals, containing
                           the length of axes *)
    Min_Y:=0;
    Ylgth:=Max_Y;
    DISPLAY_SCALE(Min_X-0.11*Xlgth,Min_Y-0.21*Ylgth,
             Max_X+0.11*Xlgth,MAX_Y+0.16*Ylgth);
```

```
PENCOLOR(1);

PENMODE(O);

MOVETO(Min_X-O.1*Xlgth,Min_Y-O.20*Ylgth);

PENMODE(1);

MOVETO(Min_X-O.1*Xlgth,Max_Y+O.15*Ylgth);  (* Draw frame *)

MOVETO(Max_X+O.1*Xlgth,Max_Y+O.15*Ylgth);

MOVETO(Max_X+O.1*Xlgth,Min_Y-O.20*Ylgth);

MOVETO(Min_X-O.1*Xlgth,Min_Y-O.20*Ylgth);

PENMODE(O);

MOVETO(Min_X,MAX_Y+Ylgth/40);              (* Draw axes *)

PENMODE(1);

MOVETO(Min_X,Min_Y);

MOVETO(Max_X+Xlgth/40,Min_Y);

Ind:=TRUE;


FOR I:=TRUNC(Start_Morn/100) TO TRUNC(End_Morn/100) DO

(* Label X axis *)

BEGIN

        IF Ind AND (I*100<>Start_Morn) THEN I:=I+1;

        (* If do not start at whole hour, jump to

           next hour *)

        Ind:=FALSE;

        PENMODE(O);

        MOVETO(I*100,Min_Y);

        PENMODE(1);

        MOVETO(I*100,Min_Y-Ylgth/50);

        PENMODE(O);

        MOVETO(I*100-Xlgth/50,Min_Y-Ylgth/15);

        PENMODE(1);

        STR(I*100,Num);

        IF I<>TRUNC(End_Morn/100) THEN WSTRING(Num,1,1);
```

116   Computer Aided Design (CAD)

```
        (* Write file time only if it will not overlap

            End_Morn *)

    END;

PENMODE(0);

MOVETO(Time_Hrs(End_Morn),Min_Y);

PENMODE(1);

MOVETO(Time_Hrs(End_Morn),Min_Y-Ylgth/50);

PENMODE(0);

MOVETO(Time_Hrs(End_Morn)-Xlgth/50,Min_Y-Ylgth/15);

STR(ROUND(End_Morn),Num);

WSTRING(Num,1,1);

MOVETO(Time_Hrs(End_Morn)-Xlgth/50,Min_Y-Ylgth/8);

STR(ROUND(Start_Aft),Num);

WSTRING(Num,1,1);


FOR I:=TRUNC(Start_Aft/100+1) TO TRUNC(End_Aft/100) DO

(* Label Y axis *)


    BEGIN (* This section of the program allows to display

             numbers on the graphics screen *)

        PENMODE(0);

        MOVETO((I-1)*100,Min_Y);

        PENMODE(1);

        MOVETO((I-1)*100,Min_Y-Ylgth/50);

        PENMODE(0);

        MOVETO((I-1)*100-Xlgth/50,Min_Y-Ylgth/15);

        PENMODE(1);

        STR(I*100,Num);

        IF I*100>Time_Hrs(Start_Aft)+50 THEN WSTRING(Num,1,1);

    END;


IF Ylgth>=5
```

General purpose spline curve fitting and graphics routines
                                                    Page No: 6
+---+

```
    THEN BEGIN

           Ind:=TRUE;

           Space:=0.5;

           REPEAT

               IF Ind THEN BEGIN

                           Space:=Space*2;

                           Ind:=FALSE;

                       END

                   ELSE BEGIN

                           Space:=Space*5;

                           Ind:=TRUE;

                       END;

               UNTIL Ylgth/Space<=10

           END

       ELSE Space:=1;

    No_Divs:=TRUNC(Ylgth/Space);

    IF (Min_Y/Space<>TRUNC(Min_Y/Space)) AND (Min_Y/Space>0)

       THEN BEGIN

               Start:=TRUNC(Min_Y/Space)+1;

               No_Divs:=No_Divs-1;

           END

       ELSE IF Min_Y/Space>=0 THEN Start:=TRUNC(Min_Y/Space)

                           ELSE BEGIN

                                   Start:=TRUNC(Min_Y/Space);

                                   No_Divs:=No_Divs-1;

                               END;

    FOR I:=Start TO No_Divs+Start DO

       BEGIN

           PENMODE(0);

           MOVETO(Min_X,I*Space);

           PENMODE(1);
```

```
General purpose spline curve fitting and graphics routines
                                              Page No: 7
+------------------------------------------------------------------+

          MOVETO(Min_X-Xlgth/50,I*Space);

          PENMODE(0);

          MOVETO(Min_X-Xlgth/12,I*Space-Ylgth/50);

          PENMODE(1);

          z:=TRUNC(Space);

          STR(I*z,Num);

          WSTRING(Num,1,1);

      END;


    PENMODE(0);

    MOVETO(Min_X-Xlgth/12,Max_Y+0.04*Ylgth);  (* Label graph *)

    WSTRING('Job time',1,1);

    MOVETO(Min_X+0.4*Xlgth,Min_Y-0.19*Ylgth);

    WSTRING('Time data recorded',1,1);

    MOVETO(Min_X+0.2*Xlgth,Max_Y+0.09*Ylgth);

    WSTRING(Title,1,1);

    MOVETO(Min_X+0.75*Xlgth,Min_Y-0.19*Ylgth);

    WSTRING('Press <CR> to cont.',1,1);


END;



(*********************** PLOT  ***************************)
PROCEDURE Plot(Title:String);
(* Plots curve on screen *)


BEGIN

    (*$I-*)    (* Input/Output error handled from program *)

    REPEAT

      PAGE(OUTPUT);

      WRITE('Input the plot file name:');

      Stringin(69,10,File_Name,8); (* This is a string input

                              routine *)

      RESET(F,File_Name);
```

Case study: spline interpolation and plot   **119**

```
General purpose spline curve fitting and graphics routines
                                            Page No: 8
+-----------------------------------------------------------------+

        IOerror:=IORESULT;

        IF IOerror<>0 THEN IOERR;

    UNTIL IOerror=0;

    (*$I+*)    (* Input/Output error handling back to default *)

    Read_Data; (* This is an input routine not shown here to

                      update plot file *)

    IF NPOINTS<2 THEN BEGIN       (* Plot graph only if there is

                                      sufficient amount of data *)

            PAGE(OUTPUT);

            GOTOXY(1,10);

            WRITE('Warning ! Less than two data points are ');

            WRITE('present, therefore no graph can be drawn');

            FOR I:=1 TO 3000 DO;   (* Wait to read message *)

            EXIT(Plot);                (* Exit routine *)

                    END;

    Calc_Spline(1);

    PAGE(OUTPUT);

    GOTOXY(10,10);

    WRITE('The points of the curve is now being calculated');

    FOR K:=1 TO 100 DO    (* Calculates 100 points on curve *)

      BEGIN

        RESINT:=X[1]+K*(X[NPOINTS]-X[1])/100;

        SPINTERPOL;

        XCALC[K]:=RESINT;   (* XCALC and YCALC are real variables,

                              containing the calculated points of

                              the curve *)

        YCALC[K]:=Result;

      END;

    MAX_Y:=Y[1];

    FOR I:=1 TO 100 DO                  (* Find max. value of Y *)

        BEGIN
```

General purpose spline curve fitting and graphics routines
                                                      Page No: 9
+---+

```
          IF MAX_Y<YCALC[I] THEN MAX_Y:=YCALC[I];

      END;

   Graphics_Screen;

   Fill_Screen(0,0);                (* Clear screen *)


   Scales(Start_Morn,End_Aft,0,Max_Y,Title);

   (* Sets scales for plotting area *)


   Ratio:=Ylgth/Xlgth;

   PENMODE(0);

   MOVETO(X[1],Y[1]);

   PENMODE(1);

   FOR I:=1 TO 100 DO MOVETO(XCALC[I],YCALC[I]);  (* Plot curve *)

   FOR I:=1 TO NPOINTS DO                    (* Draw crosses *)

      BEGIN

         PENMODE(0);

         MOVETO(X[I]-Xlgth/100,Y[I]);

         PENMODE(1);

         MOVETO(X[I]+Xlgth/100,Y[I]);

         PENMODE(0);

         MOVETO(X[I],Y[I]+Ratio*Aspect_Ratio*Xlgth/100);

         PENMODE(1);

         MOVETO(X[I],Y[I]-Ratio*Aspect_Ratio*Xlgth/100);

      END;


   READLN;

   Text_Screen;  (* Switch back to textmode *)

END; (* Plot *)

(***************************************************************)
```

curve through the data points when joined up with small
straight line-increments (Figure 4.13).

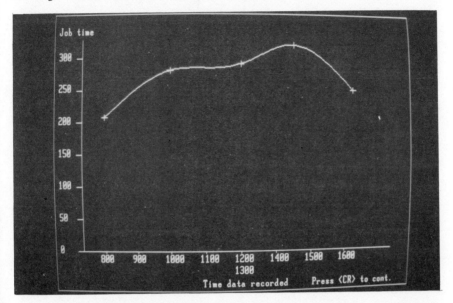

Figure 4.13 Sample output of the spline interpolation program.
The photograph taken from the screen shows job
time versus time data recorded values displayed by
a timestudy program using the spline interpolation
routines as a unit.

4.3 Case study: three dimensional graphics manupulation of a robot arm

This case study and the computer program is concerned with the
demonstration of the translation and rotation of simple three
dimensional graphics objects using homogeneous co-ordinate
systems and matrix manipulation technics. Such transformations
can be found in CAD systems both within a graphics package
and/or in the application program, and so their understanding
is important.

This program has been developed to simulate a robot (Figure 4.14) designed by the author and currently under development by himself and his students at Trent Polytechnic Nottingham.

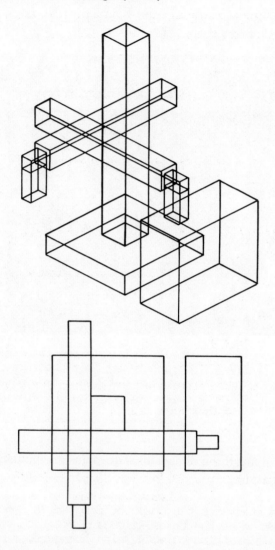

Figure 4.14 Perspective view and top view "dual-vision plot"
        showing two different locations of the cartesian
        robot model developed for simulation purposes.

    The program can also be used to generate motion co-
ordinates for this small and accurate device (Figure 4.15).
The robot design is modular, allowing many axes to be built
using the same linear module. The bearings mounted on each
side provide a smooth and accurate run of each arm module.

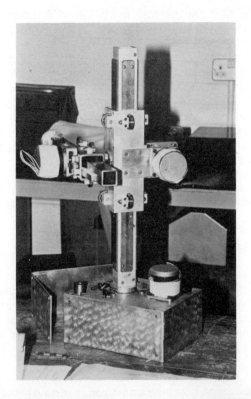

Figure 4.15 Microcomputer controlled robot designed by the author, under development at Trent Polytechnic, Nottingham, for teaching purposes.

Currently the device has only three axis of freedom, these being arm up/down, in/out and rotate base, but using further modules the number of controlled axes can be extended. (For example the whole robot can be put on a further linear axis etc.) The gripper flange interface of this device is compatible with the Automated Robot Hand Changer's interface (discussed in detail in [4.11]) used on the Polytechnic's Puma 560 robot reducing the number of robot tools required. The main purpose of designing and building this low cost model was to allow students to carry out practical development work on every hardware and software aspect of a robot.

The implementation language in this case study is again the well structured Pascal and the graphics extension (the Turtlegraphics unit) of the UCSD-p. system. The self documented procedures can be modified and used in many other applications and languages.

To be able to follow the source code listings let us give a brief summary of the mathematics of two dimensional (2D) and three dimensional (3D) graphics translation, scaling and rotation of a point and an object, or shape. (Further details on 2D and 3D graphics manipulation can be found in references [4.12] to [4.14]).

4.3.1 Two dimensional transformations: translation, scaling and rotation

For the P(x, y) point which is to be shifted, or "translated" in the xy plane by Dx and Dy values we can write:

$$x' = x + dx \quad \text{and} \quad y' = y + dy,$$

where dx and dy are parallel to the x and y axis respectively (Figure 4.16).

Expressing the points and the transformation in vectorial form we can write:

$$P = [x \quad y], \quad P' = [x' \quad y'], \text{ and transformation}$$

$$T = [dx \quad dy],$$

thus

$$[x'\ y'] = [x\ y] + [dx\ dy]$$

or in a more compact form,

$$P' = P + T$$

Using this principle not only a point but also a shape, or as we shall see a three dimensional object can be translated simply by applying the required dx, dy or in the case of 3D dx, dy and dz translation to each of its end points.

Points can be scaled. For this operation let us introduce Sx, the scaling factor parallel to the x axis, and Sy for the y axis. Similarly to the translation of a point in the two dimensional xy plane, (Figure 4.17) for scaling we can write

$$x' = x * Sx \quad \text{and} \quad y' = y * Sy \quad \text{(note that the * sign means multiplication)}$$

Defining S as a two dimensional matrix $S = \begin{bmatrix} Sx & 0 \\ 0 & Sy \end{bmatrix}$

in matrix form we can write

$$[x'\ y'] = [x\ y] \begin{bmatrix} Sx & 0 \\ 0 & Sy \end{bmatrix}$$

or in a more compact form scaling means

$$P' = P * S$$

It is needless to say that scaling can be applied to a two dimensional shape or to a three dimensional object, by scaling each end point of the shape or object. (In the portion of the routines shown in this case study scaling is not used but is equally important to translation or rotation).

Two dimensional rotation through a   angle about the origin (O) as shown in Figure 4.18 can be described as follows:

From the OAP triangle $x = r * \cos\alpha$ and $y = r * \sin\alpha$

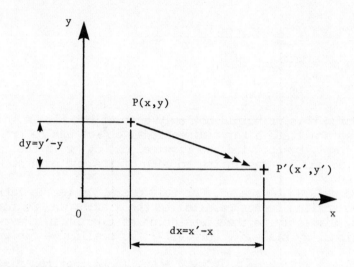

Figure 4.16 Two dimensional translation of a point.

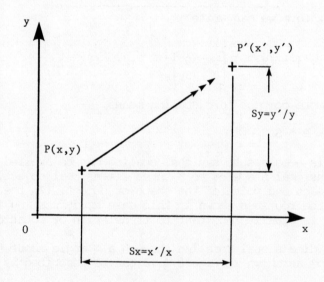

Figure 4.17 Two dimensional scaling of a point (Note that Sx must not be equal to Sy, and that in the above example both Sx and Sy are > 1.)

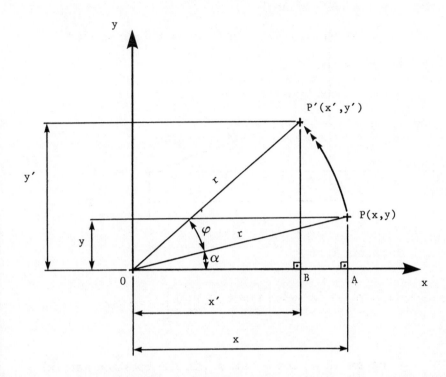

Figure 4.18 Two dimensional rotation of a point. The important equations derived from this Figure are as follows:

$$x = r\cos\alpha \qquad\qquad y = r\sin\alpha$$

$$x' = r\cos(\alpha + \phi), \qquad y' = r\sin(\alpha + \phi);$$

$$x' = r\cos\alpha\,\cos\phi - r\sin\alpha\sin\phi$$

$$y' = r\cos\alpha\sin\phi + r\sin\alpha\cos\phi$$

and from the OBP' triangle

$$x' = r * \cos(\alpha + \phi) \quad \text{and} \quad y' = r * \sin(\alpha + \phi)$$

then we can write that

$$x' = r * \cos\alpha \cos\phi - r * \sin\alpha \sin\phi$$

and

$$y' = r * \cos\alpha \sin\phi + r * \sin\alpha \cos\phi$$

thus

$$x' = x * \cos\phi - y * \sin\phi$$

and

$$y' = x * \sin\phi + y * \cos\phi$$

and in matrix form

$$[x' \ y'] = [x \ y] \begin{bmatrix} \cos\phi & \sin\phi \\ -\sin\phi & \cos\phi \end{bmatrix}$$

$$\left(\text{where} \quad R = \begin{bmatrix} \cos\phi & \sin\phi \\ -\sin\phi & \cos\phi \end{bmatrix} \text{ is the rotation matrix}\right)$$

or in a compact form rotation about the origin can be described as

$$P' = P * R$$

To summarize two dimensional translation (T), scaling (S) and rotation (R) can be described in matrix form as

$$P' = P + T$$

$$P' = P * S$$

$$P' = P * R$$

Because we would like to handle all three transformations in a homogeneous way, i.e. we would like to avoid addition in the case of translation and use only multiplication as in the other two cases, we express the P point in a slightly different way:

P(W*x, W*y, W)   where W > 0 or < 0 scale factor.

Selecting W = 1 the translation equation

$$P' =   P   +   T$$

can be represented in matrix form as:

$$[x'\ y'\ 1] = [x\ \ y\ \ 1] * \begin{bmatrix} 1 & 0 & 0 \\ 0 & 1 & 0 \\ Dx & Dy & 1 \end{bmatrix}$$

and similarly scaling

$$[x'\ y'\ 1] = [x\ y\ \ 1] * \begin{bmatrix} Sx & 0 & 0 \\ 0 & Sy & 0 \\ 0 & 0 & 1 \end{bmatrix}$$

and rotation as

$$[x'\ y'\ 1] = [x\ y\ \ 1] * \begin{bmatrix} \cos\phi & \sin\phi & 0 \\ -\sin\phi & \cos\phi & 0 \\ 0 & 0 & 1 \end{bmatrix}$$

4.3.2 Three dimensional transformations

Having introduced the matrix representation in the previous section now we can see the benefits. Selecting a right handed co-ordinate system (Figure 4.19) and extending the results of the two dimensional transformations we can write for 3D translation:

$$T(dx,\ dy,\ dz) = \begin{bmatrix} 1 & 0 & 0 & 0 \\ 0 & 1 & 0 & 0 \\ 0 & 0 & 1 & 0 \\ dx & dy & dz & 1 \end{bmatrix}$$

for 3D scaling:

$$S(Sx, Sy, Sz) = \begin{bmatrix} Sx & 0 & 0 & 0 \\ 0 & Sy & 0 & 0 \\ 0 & 0 & Sz & 0 \\ 0 & 0 & 0 & 1 \end{bmatrix}$$

and finally for 3D rotation:

$$R(Rx, Ry, Rz) = \begin{bmatrix} \cos\phi & \sin\phi & 0 & 0 \\ -\sin\phi & \cos\phi & 0 & 0 \\ 0 & 0 & 1 & 0 \\ 0 & 0 & 0 & 1 \end{bmatrix}$$

4.3.3 The 3D robot modeling program

The 3D robot modeling program uses 3D translation and rotation of simple wire frame models as shown in Figure 4.19.

The robot can be rotated in both directions, the arm can be moved "up" and "down" or "in" and "out" and a component can be picked up and manipulated by using simple character based commands.

The program would remember taught "pickup" and "place" locations, thus a simple pick-and-place operation can be demonstrated graphically. The graphic routines also include perspective representation of the wire frame model. Note that all important routines are given in the form of self documented UCSD-Pascal procedures listed in Figure 4.20 and that photographs taken from the screen of the running program are shown in Figures 4.19 and 4.21 to 4.24, demonstrating a pick-and-place simulation.

4.4 The GKS graphics standard

The heart of any CAD and integrated CAD/CAM system system is the component database, holding the geometrical description of the product which is input to all other programs and subsystems of CIM, such as the finite element analysis package, the kinematic and kinetic analysis package, the layout

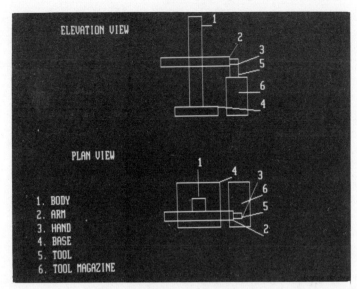

Figure 4.19/a The components of the three-dimensional robot model used in this case study. (Photographs taken from the screen.)

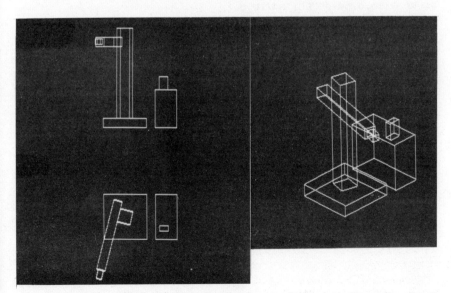

Figure 4.19/b Demonstration of the arm "up" and "down" and the arm "in" and "out" features and the isometric view option. (Photographs taken from the screen.)

132   Computer Aided Design (CAD)

Figure 4.20 Three dimensional manipulation routines of a wire
          frame robot.

```
THREE DIMENSIONAL MANIPULATION ROUTINES OF A WIRE FRAME ROBOT
                                                    Page No: 1
+------------------------------------------------------------------+


   PROCEDURE SCETCH;

                       (* DRAWS AN ISOMETRIC VIEW OF THE ROBOT *)


   VAR RESULTS : ARRAY [1..8,1..3] OF REAL;

                       (* ARRAY FOR DRAWING ROBOT ON THE SCREEN *)
   BEGIN
     FOR N:= 1 TO 8 DO            (* FOR EACH ROW AND COLUMN...*)
       FOR M:= 1 TO 3 DO
       RESULTS[N,M]:=0;           (* CLEAR THE RESULTS MATRIX *)
   DISPLAY_SCALE(0,0,50*ASPECT_RATIO,50);

                               (* SET UP AXES ON DISPLAY *)
   VIEWPORT(0,0,50*ASPECT_RATIO,50);

                               (* SET UP VIEWPORT *)
    IF PLACE THEN     (* IF THE "PLACE" POSITION IS LOCATED... *)
     BEGIN

        IF PICK THEN D:=8     (* IF THE OPERATION IS "PICK"

                             THEN INCLUDE THE PICK MATRIX.

                             NOTE THAT "D" CONTAINS A USER

                             SELECTED OPTION *)

               ELSE D:=7;    (* OTHERWISE DO NOT INCLUDE PICK

                             BUT USE PLACE *)

     END


   ELSE

     D:=6;             (* INCLUDE NEITHER PICK NOR PLACE *)

     FOR I:= 1 TO D DO  (* DO FOR ALL APPROPRIATE MATRICES *)
     BEGIN
```

+--+

```
FOR N:= 1 TO 8 DO          (* FOR ALL ROWS *)

  BEGIN

    CASE I OF              (* THIS IS A LOOP VARIABLE *)


    1:BEGIN                  (* BODY *)
      RESULTS[N,1]:=WORK1[N,1]+25;

                            (* TRANSFER FROM STORAGE MATRIX TO

                               WORKING MATRIX *)
      RESULTS[N,3]:=WORK1[N,3]+10;

                               (* NOTE: ADD 25 TO X AXIS AND 10

                                  TO Z AXIS TO DISPLAY ROBOT BODY

                                  IN THE APPROPRIATE LOCATION *)
    END;


    2: BEGIN        (* ARM *)
        RESULTS[N,1]:=WORK2[N,1]+25;

        RESULTS[N,3]:=WORK2[N,3]+10;
      END;


    3: BEGIN        (* HAND *)
        RESULTS[N,1]:=WORK3[N,1]+25;

        RESULTS[N,3]:=WORK3[N,3]+10;
      END;


    4: BEGIN        (* TOOL *)
        RESULTS[N,1]:=WORK4[N,1]+25;

        RESULTS[N,3]:=WORK4[N,3]+10;
      END;
```

+---+

```
5: BEGIN      (* TOOL TABLE *)

   RESULTS[N,1]:=WORK6[N,1]+25;

   RESULTS[N,3]:=WORK6[N,3]+10;

   END;


6: BEGIN      (* BASE *)

   RESULTS[N,1]:=WORK7[N,1]+25;

   RESULTS[N,3]:=WORK7[N,3]+10;

   END;


7: BEGIN      (* PLACE *)

   RESULTS[N,1]:=PLACE5[N,1]+25;

   RESULTS[N,3]:=PLACE5[N,3]+10;

   PEN:=1;

   END;


8: BEGIN     (* PICK *)

   RESULTS[N,1]:=PICK5[N,1]+25;

   RESULTS[N,3]:=PICK5[N,3]+10;

   END;


 END;   (* OF CASE *)

END;    (* OF: "IF 1 TO N..." *)

IF NOT DUMP THEN         (* HARDCOPY PLOT NOT REQUIRED *)

             BEGIN

                 PENMODE(3);

                 PENCOLOR(0);    (* PEN UP *)

                 MOVETO(RESULTS[1,1],RESULTS[1,M]);

                 (* MOVE TO CO-ORDINATE POSITION *)

                 PENCOLOR(1);
```

The GKS graphics standard **135**

THREE DIMENSIONAL MANIPULATION ROUTINES OF A WIRE FRAME ROBOT
                                                    Page No: 4
+---+

```
                         (* PEN DOWN... START DRAWING *)
              MOVETO(RESULTS[2,1],RESULTS[2,M]);

              MOVETO(RESULTS[4,1],RESULTS[4,M]);

              MOVETO(RESULTS[3,1],RESULTS[3,M]);

              MOVETO(RESULTS[1,1],RESULTS[1,M]);

              MOVETO(RESULTS[5,1],RESULTS[5,M]);

              MOVETO(RESULTS[6,1],RESULTS[6,M]);

              MOVETO(RESULTS[8,1],RESULTS[8,M]);

              MOVETO(RESULTS[7,1],RESULTS[7,M]);

              MOVETO(RESULTS[5,1],RESULTS[5,M]);

              MOVETO(RESULTS[6,1],RESULTS[6,M]);

              MOVETO(RESULTS[2,1],RESULTS[2,M]);

              MOVETO(RESULTS[4,1],RESULTS[4,M]);

              MOVETO(RESULTS[8,1],RESULTS[8,M]);

              MOVETO(RESULTS[7,1],RESULTS[7,M]);

              MOVETO(RESULTS[3,1],RESULTS[3,M]);

        END (* DRAWING LOOP *)


            ELSE
            BEGIN
             REWRITE(PLOTIT,'REMOUT:');
                    (* OPEN PLOT FILE, PLOTIT *)
              BEGIN
                WRITELN(PLOTIT,'=*');  ... ETC...


     ...ETC. NOTE THAT THIS PART IS PLOTTER DEPENDENT,
            THUS NOT LISTED HERE...


                 WRITELN(PLOTIT,.... LAST PLOT LINE...
```

```
THREE DIMENSIONAL MANIPULATION ROUTINES OF A WIRE FRAME ROBOT
                                              Page No: 5
+-------------------------------------------------------------+


               CLOSE(PLOTIT,NORMAL); (* CLOSE PLOT

                                         FILE *)


           END;

         END;

        END;


      IF NOT DUMP THEN   (* NO HARDCOPY REQUIRED *)


               BEGIN

         PENCOLOR(0);

         MOVETO(1,46);

         WSTRING('AN ISOMETRIC VIEW OF THE ROBOT ',1,1);

         MOVETO(1,44);

         WSTRING('IN IT''S PRESENT POSITION',1,1);

         MOVETO(0,0);

         WSTRING('PRESS <RETURN> TO CONTINUE',1,1);

         READLN;

         FILLSCREEN(0,0);   (* CLEAR GRAPHICS SCREEN *)

               END;
END;

(*************************************************************)


PROCEDURE ROTZ;

             (* ROTATES 3D IMAGE ABOUT VERTICAL Z AXIS *)

VAR RZ: ARRAY[1..4,1..4] OF REAL;

             (* ARRAY FOR ROTATING ABOUT Z AXIS *)

BEGIN

  FOR N:= 1 TO 8 DO        (* FOR EACH ROW *)
```

THREE DIMENSIONAL MANIPULATION ROUTINES OF A WIRE FRAME ROBOT

+---+

```
   FOR M:= 1 TO 4 DO         (* FOR EACH COLUMN *)

   C[N,M]:=0;                (* INITIALIZE   RESULTS MATRIX *)

   FOR M:= 1 TO 4 DO         (* FOR EACH COLUMN OF THE

                                ROTATION MATRIX *)

    FOR I:= 1 TO 4 DO        (* AND FOR EACH COLUMN OF THE *)

    RZ[M,I]:=0;              (* INITIALIZE MATRIX *)

    RZ[1,1]:=COS(ZR); RZ[1,2]:=SIN(ZR);

                             (* CALCULATE LOCATIONS *)

    RZ[2,1]:=-SIN(ZR); RZ[2,2]:=COS(ZR);

    RZ[3,3]:=1; RZ[4,4]:=1;

   FOR N:= 1 TO 8 DO          (* FOR EACH ROW AN COLUMN *)

    FOR M:= 1 TO 4 DO

    FOR I:= 1 TO 4 DO

    C[N,M]:=C[N,M]+(PP[N,I]*RZ[I,M]);

         (* MULTIPLY EACH ELEMENT OF THE WORKING MATRIX BY THE

            ROTATION MATRIX *)

    FOR N:= 1 TO 8 DO        (* FOR EACH ROW AND COLUMN *)

   FOR M:= 1 TO 4 DO

   PP[N,M]:=C[N,M];          (* PUT RESULTS BACK INTO

                                THE WORKING MATRIX *)

END;

(*************************************************************)

PROCEDURE ROTX;    (* ROTATES ABOUT THE X AXIS (HORIZONTAL

                      ON THE SCREEN) *)

VAR RX : ARRAY[1..4,1..4] OF REAL;

                      (* ARRAY FOR ROTATION ABOUT X AXIS *)

BEGIN

 FOR N:= 1 TO 8 DO           (* FOR EACH ROW AND COLUMN *)

  FOR M:= 1 TO 4 DO
```

+--+

```
    C[N,M]:=0;                  (* SET RESULTS MATRIX TO ZERO *)

   FOR M:= 1 TO 4 DO           (* FOR EACH ROW AND COLUMN OF *)

   FOR I:= 1 TO 4 DO           (* THE ROTATION MATRIX *)

   RX[M,I]:=0;                 (* SET MATRIX TO ZERO *)


   RX[2,2]:=COS(R1); RX[2,3]:=SIN(R1);   (* CALCULATE MATRIX

                                            ELEMENTS *)

   RX[3,2]:=-SIN(R1); RX[3,3]:=COS(R1);

   RX[1,1]:=1; RX[4,4]:=1;


  FOR N:= 1 TO 8 DO            (* FOR EACH ROW AND COLUMN *)

  FOR M:= 1 TO 4 DO

  FOR I:= 1 TO 4 DO


  C[N,M]:=C[N,M]+(PP[N,I]*RX[I,M]);

  (* MULTIPLY EACH ELEMENT OF THE WORKING MATRIX BY THE

     ROTATION MATRIX *)

 FOR N:= 1 TO 8 DO

  FOR M:= 1 TO 4 DO

  PP[N,M]:=C[N,M];        (* TRANSFER THE RESULTS TO THE

                             WORKING MATRIX *)

END;

(***********************************************************)



PROCEDURE TRANS;

     (* TRANSLATES ARM BODY AND HAND UP, DOWN, IN AND OUT *)



VAR T : ARRAY[1..4,1..4] OF REAL;

                (* ARRAY FOR TRANSLATION VALUES *)
```

THREE DIMENSIONAL MANIPULATION ROUTINES OF A WIRE FRAME ROBOT
                                                    Page No: 8
+---+

```
BEGIN

 FOR N:= 1 TO 8 DO      (* SET UP RESULTS MATRIX *)

  FOR M:= 1 TO 4 DO

  C[N,M]:=0;

 FOR M:= 1 TO 4 DO      (* SET UP TRANSLATION MATRIX *)

  FOR I:= 1 TO 4 DO

  T[M,I]:=0;

  T[1,1]:=1; T[2,2]:=1; T[3,3]:=1; T[4,4]:=1;

  T[4,1]:=X; T[4,3]:=Z;

                        (* INPUT MATRIX ELEMENTS *)

 FOR N:= 1 TO 8 DO

  FOR M:= 1 TO 4 DO

   FOR I:= 1 TO 4 DO

  C[N,M]:=C[N,M]+(PP[N,I]*T[I,M]);

  (* MULTIPLY EACH ELEMENT OF THE WORKING MATRIX BY THE

      TRANSLATION MATRIX *)

 FOR N:= 1 TO 8 DO

  FOR M:= 1 TO 4 DO

  PP[N,M]:=C[N,M];      (* TRANSFER RESULTS TO THE

                           WORKING MATRIX *)

END;

(*************************************************************)

PROCEDURE ISOM;   (* THIS PERFORMS THE NECESSARY PROCEDURES

                     SO AS TO TRANSFORM THE ROBOT INTO AN

                     ISOMETRIC VIEW *)

BEGIN

 FOR A:= 1 TO 9 DO

  BEGIN

    PAGE(OUTPUT);  (* CLEAR TEXT SCREEN *)
```

THREE DIMENSIONAL MANIPULATION ROUTINES OF A WIRE FRAME ROBOT
                                                    Page No: 9
+---+

```
      GOTOXY(22,16);

      WRITELN('CALCULATING MATRIX TRANSFORMATIONS');

                              (* MESSAGE TO USER *)

      R1:=3.142/6;      (* ROTATE THROUGH 30 DEGREES *)

      ZR:=3.142/4;      (* ROTATE THROUGH 45 DEGREES *)

      ROTZ;             (* ROTATE ABOUT Z AXIS *)

      ROTX;             (* ROTATE ABOUT X AXIS *)

      FINISH;        (* TRANSFER RESULTS TO STORAGE MATRIX *)

   END;

    GRAPHICS_SCREEN;    (* INITIALIZE GRAPHICS SCREEN *)

    SCETCH;             (* CALL SCETCH TO DRAW ISOMETRIC VIEW

                           SEE LISTED BEFORE THIS ROUTINE *)

  FOR A:=1 TO 9 DO

  BEGIN               (* REVERSES ALL ROTATIONS MADE TO

                         ACHIEVE THE ISOMETRIC VIEW *)

    FILLSCREEN(0,0);    (* CLEAR GRAPHICS SCREEN *)

    TEXT_SCREEN;        (* BACK TO THE TEXT SCREEN *)

    PAGE(OUTPUT);       (* CLEAR TEXT SCREEN *)

    GOTOXY(22,16);

    WRITELN('RETURNING MATRIX TRANSFORMATIONS');

                        (* MESSAGE TO USER...*)

    START;

    R1:=-3.142/6;       (* NOTE: ROTATING IN OPPOSITE SENSE *)

    ZR:=-3.142/4;

    ROTX;

    ROTZ;

    FINISH;

   END;

 END;
```

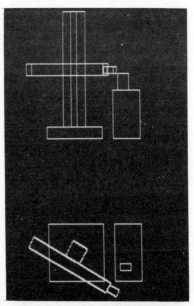 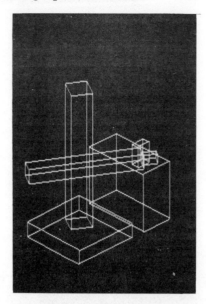

Figure 4.21 The following sequence illustrates a part of a pick-
         and-place operation simulated by the program.
         (Approach pickup location. Photographs taken from
         the screen.)

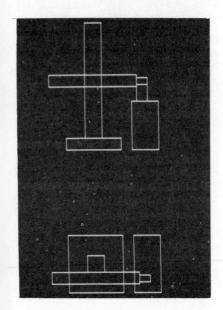 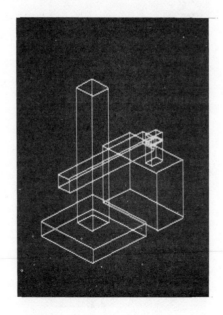

Figure 4.22 Pickup component from table.

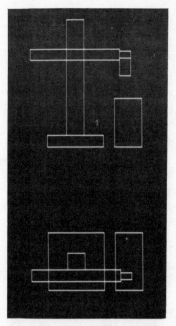
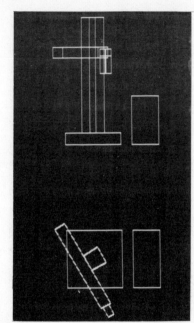

Figure 4.23 Raise and rotate the arm.

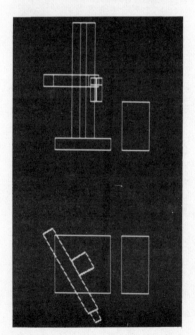
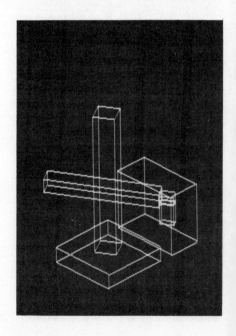

Figure 4.24 Place component at the desired location.

designer, drafting and technical illustration package, the NC part programming package, the robot programming and co-ordinate measuring machine programming package, the MRP system, simulation packages, and others depending on the design and its further use.

To be able to achieve at least some of the above listed level of integration, the component database must contain:

* the final shape of the component, (as built up from three dimensional wire-frame, or preferably solid primitives), including the final shape and size with all necessary part dimensions and tolerances,

* Bill Of Material (BOM),

* the material the component is made of,

* the function of the component,

* the required manufacturing, test and assembly procedure to be carried out on the component,

* the classification code of the component, if part classification is used.

It is obvious that the data structure of this graphics database, its user friendliness and the portability of the graphics system are very important for vendors developing graphics software for a variety of different graphics workstations as in the case of CIM. This has been realized in the past two decades, but unfortunately the first sign of an agreement has only been reached in 1982, with the publication of GKS [4.15].

The reasons why a graphics standard had to emerge include the need for exchanging graphics data and the need for a clear distinction in a graphics software system between those parts that deal with modeling and those parts dealing with viewing. The main objectives of the GKS standard are to allow easy portability of graphics systems between different computers, workstations and graphic terminals, and to allow different CAD programs to be written and transferred from one installation to the other simply by setting up an interface at different workstations in different ways.

To achieve these objectives GKS offers

* Device independence, meaning that the standard does not assume that any of the devices used for input or output have any particular features or restrictions. (Not to limit any new developments there are functions defined by the application implementor to customize display screens, to set up printer/plotter interfaces, etc. However this feature limits portability i.e. interchangeability of software using different hardware.)

* Language independence, meaning that all names used are in English, or other natural languages if translated.

* Standard display management, offering a complete suite of display management functions providing overall control of the cursor movement and other features.

* Graphics functions, for 2D (two dimensional) applications. (Note that work is currently being undertaken to extend the 2D standard into 3D.)

The GKS graphics workstation assumes a graphics display oriented device with fair amount of intelligence and some input/output devices, such as a disk store, a storage tube or refreshing tube display, plotter, etc. as illustrated in Figure 4.25. The Figure illustrates the device drivers which interpret the graphics commands of the GKS package for each different device type. Among the devices virtual devices, or "metafiles" (i.e. devices which have no graphics capability in the normal sense, for example a disk unit), are also permitted. In true sence "metafiles" describe and store pictures and/or picture generation sequences on mass storage devices, such as disk, etc.

GKS incorporates the possibility of handling different co-ordinate systems in which the graphics images (i.e. pictures) are defined. The application system in GKS always works in a rectangular window, or "world co-ordinate system". The window defines also the scaling factor to be used to map the created picture into the GKS systems common internal co-ordinate system, known as "normalized device co-ordinates", or NDC. Windows and viewports can then work in this co-ordinate system.

GKS offers two sets of routines for defining the user created pictures. The first set allows the picture to be

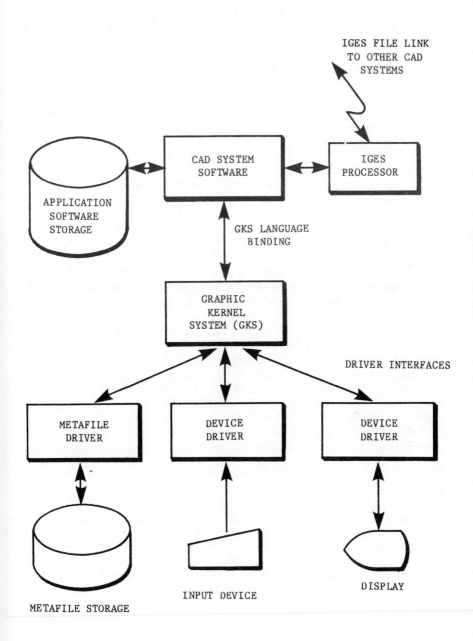

Figure 4.25 The "complete" GKS implementation in a CAD workstation, with the IGES file transfer interface.

defined by so called "primitive functions" and the second set includes the "attribute functions" defining the appearance of the image, e.g. line-type, color, etc.

The available set of "primitive functions" are as follows:

* POLYLINE, or draw a set of connected straight line vectors.

* POLYMARKER, or draw a set of markers or shapes.

* FILL AREA, or draw a closed polygon with specified interior in-fill.

* TEXT, which draws characters, or a string of characters.

* The GDP (Generalized Drawing Primitive) function to specify the implementation dependent standard drawing facilities, such as the circle, the ellipse, etc.

.GKS is installed on a general purpose computer in the form of subroutines, or procedures. Standard "language binders" define exactly the computer language functions and their parameters for the application interface. Language binders are available for ANSI FORTRAN and are in preparation for Pascal, C and ADA. (For further details on GKS refer to the following references: [4.15] to [4.20].)

4.5 The IGES graphics standard

GKS's main concern is to offer graphics portability on different devices at the cost of increased overhead, whereas IGES, or the Initial Graphics Exchange Specification is a CAD system independent data format, or protocol, used for describing product design geometry and manufacturing information for mainly graphics data exchange purposes between two or more CAD workstations.

The product definition data is stored in an IGES file which has the following sections:

* START section, which is a human-readable message in text for the person who is going to convert the file between different CAD workstations.

* GLOBAL section, incorporating details such as product name, author, company, copyright, etc. of the sender.

* DIRECTORY ENTRY section, containing of 2 records, each built up of 20, 8-character long fields for each entry.

* PARAMETER DATA section, storing the number of records per entity (being geometric e.g. co-ordinate pairs of an arc, or non-geometric, e.g. a pointer), depending on the actual entity.

* TERMINATE section, providing the sub-totals of records in each section for data transmission check purposes.

The major shortcomings of IGES include:

* It is very complex and wordy; for example a simple line segment transfer requires a minimum of three records of data.

* IGES files are an estimated five times larger than an equivalent picture file.

* there is a shortage in defining geometric entities for example for electronic designers and for applications involving finite element methods, etc, not to mention 3D solid modeling.

The conclusion is that although IGES does not allow product definitions to be transferred in its full interpretation, it does offer at least the transportation of entities consisting mainly of graphics data, in a format in which it exists in one way or an other in almost every CAD system. It must be emphasized that the problems are encountered because data to be transferred is structured and/or coded graphics data, rather then simple free format text.

It must also be noted, that this kind of data transfer between CAD workstations is almost a one-off procedure, rather than a regular activity as with text files using Local Area Networks, although some vendors use different LANs, but do not

follow the IGES format because they find it slow and too
complex. However most companies offer the IGES interface as an
option, even if their networks do not use it on a regular
basis because of the above mentioned reasons.

Although there are some encouraging demonstrations of
different CAD systems talking to each other using the IGES
format, while a universal standard format is not fully developed and tested between different CAD systems and other
machines within CIM, the slow and ineffificient but working
solution is to transfer data in terms of ASCII text files via
an RS 232c serial interface and decode them by user written
programs running on the receiver's machine. (The main task
when using LANs or other form of digital communication for
such purposes is to write the application layer, unless
supplied by the vendor as discussed in Chapter 2, under
section 2.1.5.)

For further information on IGES refer to the following
references: [4.20] to 4.25].

4.6   Three dimensional and solid modeling

Before discussing a case study using the McAuto Unigraphics
CAD/CAM system let us clearify the difference between three
dimensional graphics and the increasingly utilized solid
modeling methods.

Three dimensional (3D) models can be generated by wire
frames, made up of interconnected lines. (As an example refer
to the 3D robot modeling case study in section 4.3 and to the
Unigraphics case study in section 4.7). They are easy to
create and provide accurate definition of surface discontinuities, but they do not contain information about the surfaces
themselves. Through hidden line removal these models often
appear to be solids, but in fact they represent only a shell
of the designed image.

Solid modeling incorporates the design and analysis of
virtual parts and assemblies created from primitives of solids
stored in an image database. Solid modelers can create
alternative project plans by varying colour, perspective and
the position of objects. Some systems offer the programming of
light, over three million color variants to choose from and a

specific color and shine value which can be assigned offering
100% realistic images to materials which compose a structure
of cast iron, steel, plastic, rubber, glass, concrete, etc.

Of all different representation modes of solids let us
briefly discuss:

1. The translational sweep representation, where a solid
   is represented as the volume swept by a 2D image when
   it is translated along a line. Translation may include
   rotation as well, when a 2D image is rotated around a
   line which serves as an axis.

2. Constructive solid geometry handles primitives of
   solids, which are bounded intersections of closed half
   -spaces, defined by planes or shapes i.e. $F(x,y,z)>=0$,
   where F is an analytical function. More complex solids
   can be built by composition and decomposition using
   set operations, such as union, intersection and difference of solid bodies, or extrusion. Extrusion operators allow the creation of faces which are then swept
   to create solids or holes in solids. The Boolean and
   the extrusion operations are generally intermixed
   during the design process to provide maximum flexibility.

3. Boundary representations allow the definition of
   solids by means of their enclosing surfaces (i.e.
   boundaries). This representation handles surfaces,
   curves, lines and points, which can be then used to
   create boundaries.

Solid modelers use raster graphics to produce solid
images. Each of these images contain millions of different
picture elements to create realistic (i.e. hidden surface)
images. To demonstrate some of these effects solid images
created in color, are shown in black-and-white in Figures
4.26, 4.27 and 4.28.

Solid modeling is a fascinating technology, but is not
developed yet for many relatively simple industrial applications. Current fields of development include:

* The interface between the designer and the solid
  modelling package, analyzing the way and methods solid
  images are defined and manipulated by the user. Solu-

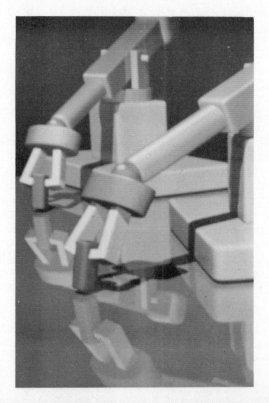

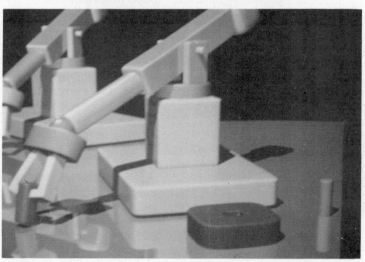

Figure 4.26 Sample solid models of industrial robots.

Figure 4.27 Images generated by a solid modeling system for
computer graphics animation purposes. (Courtesy of
Electronic Arts Limited, Primrose Hill, London
NW3 3AJ.)

tions include textual languages, menu driven packages, icon driven interfaces and others.

* Geometry representation and and computation methods, since solid modelers are still far too slow and stored images and designs occupy far too much storage area. An other problem is resolution and the accuracy of represen tation. (For example two surfaces which intersect may ha approximate representations which do not intersect.)

* Applications software, and new applications areas are practically infinite and range from simple draughting, tool-path simulation in CNC machining, off-line robot task planning in assembly and other applications, fin- ite element analysis and other applications including art design and animation (Figure 4.28).

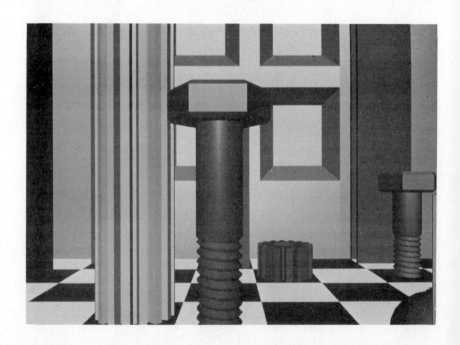

Figure 4.28 Computer art using solid modeling. (Courtesy of Electronic Arts Limited, Primrose Hill, London NW3 3AJ.)

4.7   Case study: Three dimensional design, using the McAuto
Unigraphics system

The goal of this case study is to demonstrate what powerful
minicomputer or mainframe driven CAD (and to some extent
integrated CAD/CAM; see also next Chapter) systems can offer
and how they are used in a narrow domain. The sample runs,
plots and the system are discussed as being used running on a
VAX 11/780 computer and six graphics workstations at Trent
Polytechnic Nottingham. No effort is made within this text,
or in the case studies represented in this text to cover all
facilities of this large and complex system. (For further
information refer to references [4.31] to [4.34].)

The Unigraphics CAD/CAM software is an integrated, or
partly integrated family of draughting, graphics design and
analysis packages and manufacturing engineering programs
providing design aids in areas including:

* Two and three dimensional geometric construction and
  modeling.

* Three dimensional view independent construction,
  (meaning that the user can define planes and co-
  ordinate systems to create geometry on planes not
  parallel to the screen).

* Extended geometric surface definition.

* Transformation and editing of entities.

* File management.

* Creation and editing of cutter location (CL) data for
  NC programming.

* Finite element analysis.

* Solid modeling, and other modules.

Typical CAD modules of the software system include:

* UGRAF, the basic Unigraphics module which can be used
  for three dimensional interactive geometry construction,
  analysis, modification and draughting.

154   Computer Aided Design (CAD)

* GFEM, the Graphics Finite Element Module.

* The Graphics Semantics Module (GSM) to produce
  schematic drawings and the associated Bill Of Material
  (BOM) files.

* GRIP, the Graphics Interactive Programming module,
  providing a high level graphics programming language.

* The User Function Module (UFUNC) to allow the suspen-
  sion of Unigraphics and the execution of a user-ins-
  talled system.

* Solid modeling.

After this very brief overview let us use some modules of
the system and illustrate what we are doing by making a series
of screen dumps and plots as we develop the design on the
screen.

Note that the manufacturing of the part designed in the
frame of this case study is described in the next Chapter in
section 5.8.

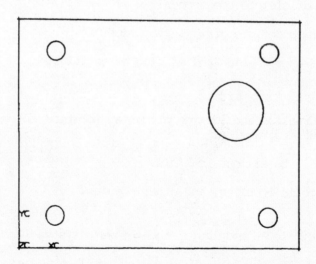

Step 1. Initial boundary design of a component. Note the x,y,z
        co-ordinate system employed in order to describe the
        part geometry.

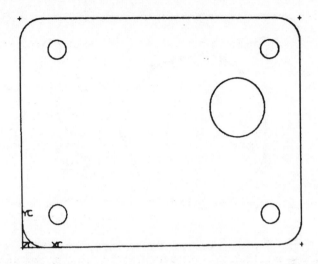

Step 2. The corners are rounded. Note the construction lines and the center points at the rounded corners.

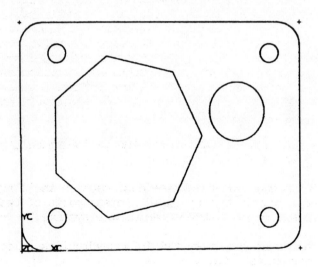

Step 3. Polygon drawn.

CIM-F*

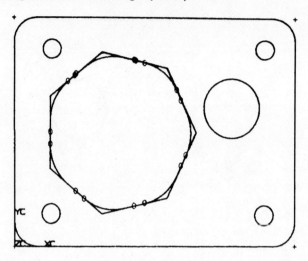

Step 4. Polygon filleted. Note the way the boundary was selected by the operator and marked by the system.

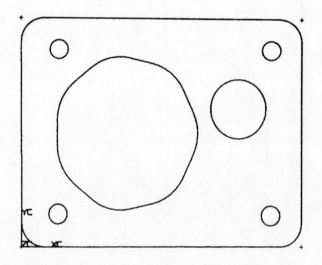

Step 5. Construction lines removed.

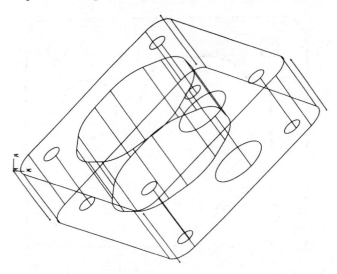

Step 6. The three dimensional wire frame model of the component.

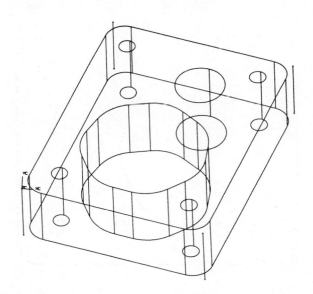

Step 7. A different three dimensional view of the same wire frame model.

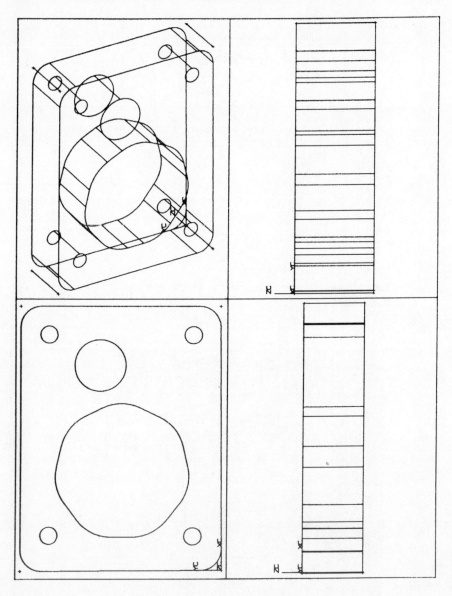

Step 8. All important views of the part. Note the co-ordinate
system indicated in each view.

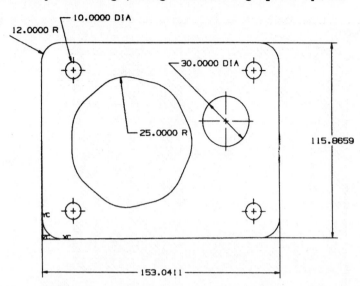

Step 9. Automatic dimensioning occurs after indication of
position by operator.

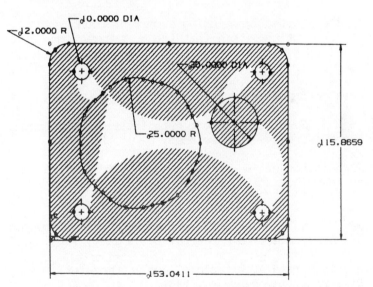

Step 10. This is what happens if the boundaries are not
selected properly for the crosshatching procedure.

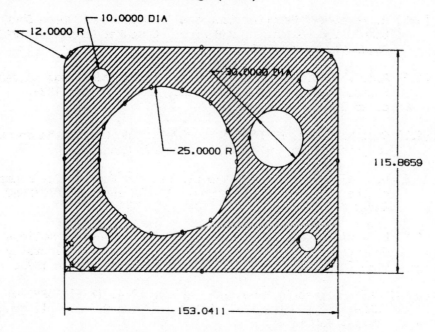

Step 11. The plotted image using a correct boundary
description. (Note that the NC program generation
of this part is described in the next Chapter.)

References and further reading

[4.1] Joan E. Scott: Introduction to Interactive Computer
Graphics, Wiley-Interscience, New York, 1982.

[4.2] Richard Stover: An Analysis of CAD/CAM Applications with
an introduction to CIM, Prentice-Hall Inc. Englewood
Cliffs, 1984. 290 p.

[4.3] Max Schindler: Better Graphics opens new windows on CAE
stations, Electronic Design, 20 January, 1983. p. 77–86.

[4.4] Apollo Domain Architecture, Apollo Computer, North
Billerica, MA, February, 1981.

[4.5] R. Weinberg: Computer Graphics in Support of Space Shuttle
Simulation, SIGGRAPH 1978 Proc., Computer Graphics, 12(3),
August 1978. p. 82-86.

[4.6] A.A.G. Requicha and H.B. Volker: Solid Modelling: Current
Status and Research Dirtections, IEEE, October 1983.
p. 25-37.

[4.7] Carl Machover and Robert E. Blauth: The CAD/CAM handbook,
Computervision Corporation, Bedford, Mass., 1980.

[4.8] B. Gaal and T. Varady: Experiences and Further Development
of the FFS (Free Form Shapes) CAD/CAM system, IFIP/IFAC
World Congress, Budapest 1984. p. 107-111.

[4.9] R.F. Riesenfeld:Applications of B-Spline Approximations to
Geometric problems of Computer Aided Design, University of
Utah Comp. Science Dept., UTEC CSc-73-126, March 1973.

[4.10] CAM-I Geometric Modelling Project 1984. (PR-83-ASPP-01.2)
CAM-I Inc., Suite 1107, 611 Ryan Plaza Drive, Arlington,
Texas, 76011.

[4.11] Paul G. Ránky and C.Y. Ho: Robot Modelling, Control and
Applications with Software, IFS(Publications) Ltd. UK,
and Springer Verlag, New York, 1985.

[4.12] E.A. Maxwell: General Homogenous Coordinates in Space of
Three dimensions, Cambridge University Press, Cambridge,
1951.

[4.13] W.M. Newman and R.F. Sproull: Principles of Interactive
Computer Graphics, 2nd ed., McGraw Hill, New York, 1979.

[4.14] James D. Foley and Andries Van Dam: Fundamentals of
Interactive Computer Graphics, Addison-Wesley Publ. Co.,
1982. 664 p.

[4.15] ISO: Graphical Kernel System (GKS) Functional Description ISO/DIS 7942

[4.16] R.W. Simons: Minimal GKS, Computer Graphics, 17, 3, 1983.
p. 183-189.

[4.17] F.R.A. Hopgood at all.: Introduction to the Graphical
Kernel System (GKS), Academic Press, London, 1983.

162   Computer Aided Design (CAD)

[4.18] BS 6390: British Standard Specification for a Set of Functions for Computer Graphics Programming, the Graphical Kernel System (GKS), 1983.

[4.19] ISO: FORTRAN interface of GKS 7.2, ISO TC97/SC5/WG2 N214, 1983.

[4.20] ISO (International Standards Organization: Graphical Kernel System (GKS), Version 6.6, May 1981.

[4.21] IGES Version 2.0, National Technical Information Service, 5285 Port Royal Road, Springfield, Virginia 20161, USA. In the UK: Microinfo Ltd., P O Box 3, Alton, Hampshire GU34 2PG

[4.22] AECMA Report of Geometry Data Exchange Study Group, March 1984. (Obtainable from H.Mason, British Aerospace, Kingston Division, UK.)

[4.23] NEDO Books, Computer Aided Design interchange of data: Guidelines for the use of IGES, July 1984. (Obtainable from: National Economic Development Office, Millbank Tower, Millbank, London SW1P 4QX.)

[4.24] Michael H. Liewald and Philip R. Kennicott: Intersystem Data Transfer via IGES, IEEE, May 1982, p. 55-63.

[4.25] W.W. Braithwaite: Boeings Approach to Interfacing Unlike CAD/CAM systems, Proc. 17th Numerical Control Society Ann. Meeting and Technical Conf. 27 April 1980.

[4.26] J.W. Lewis: Interchanging Spline curves using IGES,Proc. AUTOFACT West, CAD/CAM VIII, 17 Nov. 1980, p. 327-344.

[4.27] A.P. Armit and A.P. Forest: Interactive Surface Design, Computer Graphics 1970, Brunel University, April 1970.

[4.28] 3D Modelling for CAD/CAM, Engineering December 1983. Technical file No. 120.

[4.29] Edward J. Farrell: Colour Display and Interactive Interpretation of Three-dimensional Data, IBM J. Res. and Dev. Vol. 27, No. 4, July 1983.

[4.30] Ware Myers: An Industrial Perspective on Solid Modelling IEEE, March 1982. p. 86-97.

[4.31] McAuto Unigraphics: CAD Basic Unigraphics Manual,
MCDONNELL Douglas Corp. Box 516, St. Louis, MO 63166.

[4.32] McAuto Unigraphics: GRAF User Manual,
MCDONNELL Douglas Corp. Box 516, St. Louis, MO 63166.

[4.33] McAuto Unigraphics: GRIP User Manual,
MCDONNELL Douglas Corp. Box 516, St. Louis, MO 63166.

[4.34] McAuto Unigraphics: Unigraphics Operational Description,
MCDONNELL Douglas Corp. Box 516, St. Louis, MO 63166.

[4.35] UCSD-p. system manuals, TDI, Bristol, UK. 1984.

[4.36] Randy Clark and Stephan Koehler: The UCSD-Pascal
Handbook, Prentice-Hall Inc., Englewood Cliffs,
1982. 356 p.

CHAPTER FIVE

Computer Aided
Manufacture (CAM)

The term "Computer Aided Manufacture" is used in many
different ways to describe data processing assistance in preparing programs and production plans for numerically controlled and computer controlled (NC and CNC) machines, robots, Co-
ordinate Measuring Machines (CMM) and other programmable production equipment.

CAM often includes the DNC (Direct Numerical Control)
program, capable of editing, downloading the manufacturing
programs to NC equipment and receiving feedback data from such
machines, and also the preparation of schedules, batches, tool
and fixture requirement files and production plans for a much
shorter period of time than as with the Master Scheduler, or
with MRP etc. packages of the business system. In a computer
integrated factory CAM creates a bridge between CAD, the
business system and the flexible manufacturing facilities,
known as FMS (and discussed in the subsequent Chapters).

In this text besides the important role of generating
part programs based on the output of CAD, or on human input,
we underline this "bridge building" role of CAM, by illustrating and explaining with case studies:

* The main mechanical, electronic and software system
  components of NC and CNC equipment and how they work.

* Different NC and CNC part programming methods and
  systems.

* High level NC programming and a COMPACT II macro
  library, which is general enough to be implemented

using other than the COMPACT II language.

* Off-line robot program generation, using graphics to design the layout of the job and a "fill-in-the-screen" Manual Data Input technique to provide the required additional data enabling the computer to generate the requested robot program.

* Integrated CAD/CAM systems by showing the way a component and its part program can be created in integrated CAD/CAM systems.

As before, the main emphasis in this Chapter is put on understanding concepts by a series of self documented practical case studies and a short introduction of the topic they relate to. It is not our aim to replace programming manuals or trying to teach any of the applied languages or the CAD/CAM systems which have been used preparing the case studies. However references indicate where further reading is advisable for those who wish to learn more about the applied packages and systems (the CNC macro languages, the COMPACT II language, Unimation's VAL robot language, McAuto's Unigraphics CAD/CAM system, etc.).

5.1 The system architecture of CAM

With the integration of the design, manufacture and test procedures and by linking computer controlled machines to distributed data processing networks and databases of the manufacturing company the entire shape of industry changes. The important new aspect of CIM is that by utilizing advanced computer communication facilities in the factory, it brings the technical and business goals of design, the shop floor reality and the economics of tasks to be solved at the CAM workstation to real-time closeness. In other words, when utilizing CIM technology, each decision is supported by accessing relevant data from "above" and/or "below" to help the decision making process (Figure 5.1).

For example the FMS part programmer preparing a series of NC programs at a CAD/CAM workstation needs to communicate with graphics models created by the CAD system, which will provide him with the necessary geometrical description of the desired component, together with data on its material, the tolerance requirements, bill of material for requirements planning pur-

poses, etc.   He must select which surfaces and/or components
he will have to machine, grind, weld, assemble, etc. and
design the sequence of the operations. With the help of databases, tools and fixturing requirements must also be determined, selected and/or designed for each operation.

Ideally, all above data should be accessed from the CAD/
CAM system where the entire computer model of the part and the
workpiece-holding devices could be created together with the
part programs. This model could also be used later for assembly program generation, robot programming, tool-path simulation and process planning purposes.

If available, the FMS part programmer should also be
provided by the CAM system with machining data during his part
program writing phase, or any important data regarding the
programmed process, e.g. assembly tolerances, gripping forces
and possible gripping surfaces, advisable and compulsory test
procedures, etc. Because of the complexity of such processes
this is an area where expert systems could help a lot in
advising the CAD/CAM programmer and/or operator.

The operator interface is also an important component of
the data processing system. A communications system having
distributed intelligence and an associated man-machine,
machine-to-machine communication facility is essential in
order to achieve flexibility. The consideration of multiple
access points in a network to either a centralized or distributed database describing the state of the control system for
management, for the production engineers, for maintenance
supervisors etc., presents considerable database management
problems.

Finally it should be underlined that if the part programmer has to prepare part programs for a partly or totally
unmanned system of machine tools, Co-ordinate Measuring
Machines, assembly and inspection robots, a flexible welding
system, etc., rather than just concentrating on one or more
standalone CNC machines, or industrial robots then the task
also requires thinking in terms of a manufacturing system
consisting of a variety of processing cells, rather than just
a series of individual machines. (Note that this concept is
discussed further in the next Chapters.)

5.2 The main components of computer controlled machines

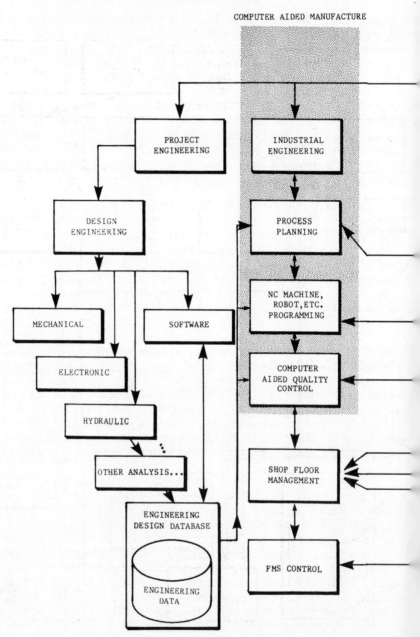

Figure 5.1 An overall CIM model indicating CAM (Computer Aided Manufacture) and its links to the rest of the system.

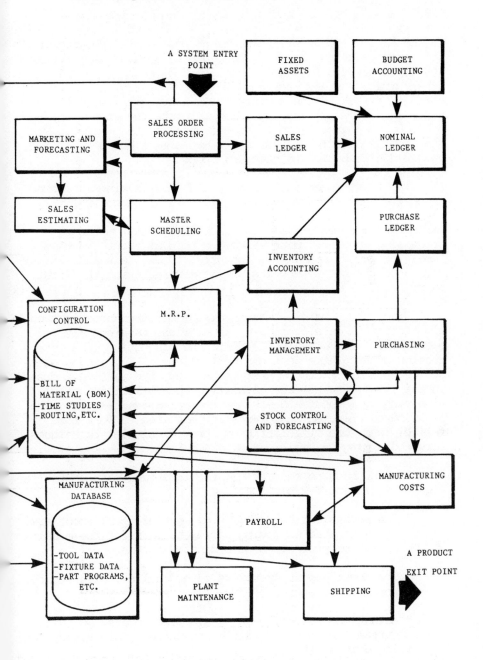

5.2.1 Fundamentals of computer controlled machines

The concept of "controlling machines by numbers" emerged in the early 1950s in the USA. The concept is much simpler than its implementation: "talk to machines" by means of numbers. In other words, express each desired function (e.g. switch on the spindle, switch on the coolant, take a new tool, move the table along the "x" axis by 200 mm, rotate the wrist of a robot by 30 degrees, etc.) and describe motions in set, or programmed co-ordinate systems by means of alphanumeric (i.e. combined alphabetic and numeric) codes.

The first Numerically Controlled (NC) machines utilized expensive and relatively large size hard-wired controllers, capable of reading the coded part programs from paper tapes (or magnetic tapes or punched cards borrowed from the data processing industry which was being born at about the same time) prepared manually on converted typewriters and similar devices or taught-by-showing whilst manufacturing the first part.

Since then, mainly because of the major changes in the electronic and data processing industries, machines have become more and more "intelligent" and productive. The introduction of Programmable Logic Controllers (PLC), microprocessor technology, microcomputer technology, the massive increase in the local data storage capacity, the availability of real-time sensing methods and the new communication features with distributed computer networks allow machines to work more reliably and more efficiently and provide simpler techniques for programming as well.

The CNC machine consists of the following main parts (see Figures 5.2/a and 5.2/b):

* The mechanical hardware (e.g. machine tool, robot arm, etc.) capable of moving different parts of the machine in set or programmed co-ordinate systems.

* The power supply system (e.g. hydraulics, electricity, pneumatics, etc.).

* Actuation devices, such as stepping motors, DC (Direct Current) motors, AC (Alternate-Current motors), hydraulic drives, etc.

Figure 5.2/a CNC precision milling center equipped with
          automated tool changing and pallet changing, and
          a powerful Siemens System 8 CNC controller. The
          machine can be integrated into FMS as a cell.
          (Csepel Machine Tool Company, Budapest, Hungary.)

* Power amplification systems.

* Feedback devices, such as encoders, resolvers, inductosyns, tachometers, etc.

* The computer control unit, consisting of the machine control system, and machine control software.

According to the method by which NC and CNC machines can perform different motions computer controlled machines (including machine tools, industrial robots, Co-ordinate Measuring Machines, etc.) can be classified into two broad groups:

1. Machines which need no functional relationship between their co-ordinate axis. These machines are the point-to-point and the straight-line controlled machines.

Figure 5.2/b CNC grinding machine. (Courtesy of Jones and Shipman Ltd., Leicester, England.)

Figure 5.3 CNC sheet metal punching and nibbling machine.

2. In the other group there are machines which need to follow a mathematically described and/or taught (often three dimensional) path,thus there must be a functional relationship between the movements along the different co-ordinate axis. These machines are also known as continuous path controlled machines.

In the case of the point-to-point controlled machines the tool is never in contact with the workpiece while any of the slides are in motion and the control system merely moves the selected slide to a discrete co-ordinate point. Typical examples include the NC co-ordinate drilling machine having an NC controlled xy table, NC jig borers, spot welding robots, NC punching, nibbling, laser cutting machines, printed circuit assembly devices, etc. (Figures 5.3 to 5.6).

The straight-line control system allows the tool to be in contact with the workpiece while the slide is moved, but the movements are always parallel to the axes of the machine and extend from a specified starting point to a determined finishing point. (A typical example is a simple NC milling machine with a traverse table.)

When guiding a tool, or a robot arm along a mathematically described path, at least two, but often as many as six axes must be controlled simultaneously while the tool is continuously in contact with the workpiece. Examples include NC profile milling machines and lathes, machining centres, assembly, deburring, glueing and spray painting robots, etc. (See again Figure 5.2 and refer to Figures 5.7 to 5.9).

Each axis of the NC machine, (or robot, or CMM, etc.) must have the information on where to move the selected slide and also must have the data of the current position to be able to decide whether it has reached the desired location or not (Figure 5.10).

The schematic structure and the basic elements of one numerically controlled axis are shown in Figure 5.11. In this closed loop control system the comparison of measured position data is done in the comparator. (Open loop control systems are also used in NC and robotics, because of the low cost. These machines are not as accurate as the closed loop controlled systems and are mainly equipped with stepper motors, rather than DC motors.)

5.2.2 Measuring systems, sensors and their integration into the control architecture of the machine

Machine integrated position measurement devices and sensors are very important, because they increase the reliability of

Figure 5.4 Point-to-point controlled welding robot.

Figure 5.5 Fanuc part loading robot.

Figure 5.6 Cincinnati Milacron machine loading/unloading robot.

Figure 5.7 CNC Laser cutting machine.

Figure 5.8 Fanuc deburring robot in action.

the machine, as well as facilitating unmanned manufacturing.
The simplest method of measuring distances by digital methods
is to divide the distance moved into equal small increments.
Such increments are also called as "distance quanta".

Most NC systems are capable of measuring distance and
working in absolute or in incremental modes (Figure 5.12).
The schematic arrangement of the incremental and the absolute
measuring system is shown in Figures 5.13 and 5.14.

Supposing that the scales are scanned by a device, called
the encoder, and that the signal obtained is amplified and fed
into the comparator, the machine will be able to sense its
current position as well as being able to measure any movement
along the slide by adding together the sensed electric pulses.

The important difference between the two systems is that
in the case of the relative (or incremental) measuring system
only displacements are measured and each position is determined by the displacement of its predecessor, whereas the
absolute system will calculate every distance of movement from
a fixed datum. Absolute systems are more widely used, if the
maximum length to be measured allows this arrangement, because
in the incremental system there is a real danger that any
error during the scanning or counting operations will affect

Figure 5.9 Dual arm DEA Pragma A 3000 assembly robot.

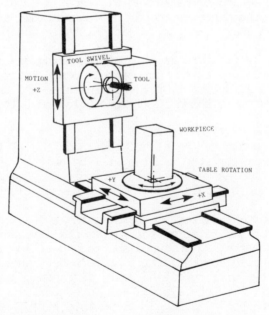

Figure 5.10 The machine co-ordinate system of a five axis CNC milling center.

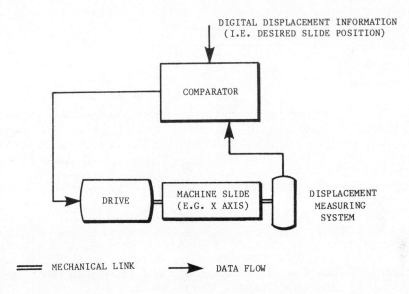

Figure 5.11 Elements of a closed loop numerically controlled axis.

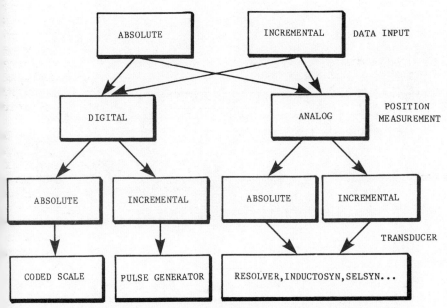

Figure 5.12 The traditional "data input and position measurement chart" for NC and CNC machines.

CIM-G

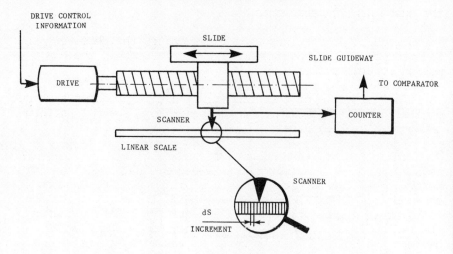

Figure 5.13 Relative, or incremental position measurement principle using a linear scale.

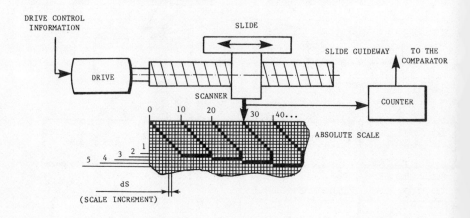

Figure 5.14 Absolute displacement measuring principle using a coded scale.

the subsequent measurements.

Feedback devices, such as the encoder, the resolver, the inductosyn and the tachometers are key devices in the NC motion control system.

The principle of the encoder is that it produces a unique digital output signal which corresponds to a special pattern put onto a disk or linear scale.

The resolver, mounted as a shaft position measuring device, works similarly to a small AC motor. It consists of a rotor and a stator, both having two windings at 90 degrees to one another. If an AC voltage is applied to one of the stator coils, the maximum voltage will appear at a rotor coil when the two coils are face to face. If they are aligned at 90 degrees the voltage will be zero. As the shaft turns, the sinusodial signal can be digitized and used for position measurement purposes.

The inductosyn is a very accurate (and relatively expensive) measuring system. Its principle is very similar to the resolver, incorporating often over a hundred stator poles.

It is also important to mention the all above discussed devices are manufactured both in rotary and in linear shapes.

Machine intelligence and safety largely depends on the source and speed of information the controller can obtain from its real-time sensors. Machine integrated sensors can be used for identifying components, pallets, part orientations and dimensions, etc., for monitoring tool wear, force, torque, deflection, vibration, temperature, etc., and for self diagnosis purposes (Figures 5.15 to 5.19). At its most advanced level, sensory feedback data can be used in Adaptive Control (AC) to react in real-time according to changes in the previously programmed conditions and also in generating data for building rules for the knowledge base of expert systems.

Sensors are not only employed to measure the position of machine table slides, or the position of robot arms, but also to provide information about:

* The work environment.

* State of the control system.

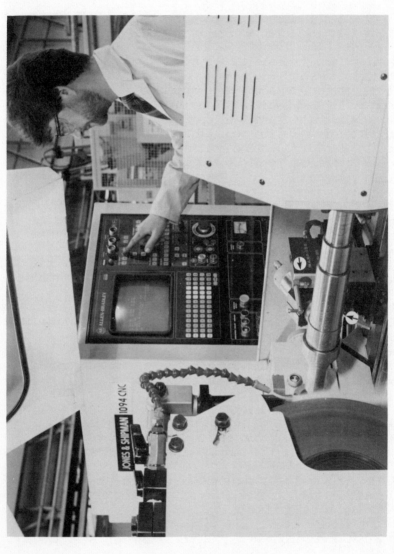

Figure 5.15 In-process gauging system integrated into the CNC controller provides real-time data feedback on part dimensions. (Courtesy of Jones and Shipman, Leicester, UK.)

* The condition and availability of tools and grippers to be used and/or currently being used in the process.

* The condition, size and availability of the components to be processed and/or currently being processed (see Figures 5.16 and 5.18),

* or the condition of the machine itself (e.g. overheated gearbox, vibration, overloaded robot or machine tool).

In measuring systems an electrical signal, or a series of signals are generated and sent to the processing computer, which then records the actual position of the machine, either in motion or else stopped before having the position sensors along the axis read. Sensors producing this electrical signal are basically divided into two large groups, those sensors working along contact and those along non-contact principles.

The most commonly used contact sensor is the electro-mechanical touch trigger probe (see Figure 5.19). It gives a direction dependent electronic signal, in most cases in 3D. This system allows a light and flexible mechanism to carry the probe and the measuring unit, which may be remote from the indicating instrument. The probe basically consists of a kinematic stylus-holder location, which is electrically connected to give a signal on stylus deflection. Touch trigger probes are widely used not only on Co-ordinate Measuring Machines, but are also integrated into machine tools for tool wear monitoring, and in-cycle and in-process workpart gauging purposes (see again Figure 5.15).

The economic and technical motivation for using noncontact sensing, vision and image processing techniques is:

* To increase productivity.

* To increase operation safety and to help self-recovery particularly in the case of unmanned operation.

* To gain a full 3D image of the environment for collision avoidance.

* To "learn" the environment for generative task planning purposes.

* To inspect tool wear and part condition, to detect tool break.

Figure 5.16 Approximity switches are often used as an integrated part of the robot tool in order to detect part loss, or to avoid collision.

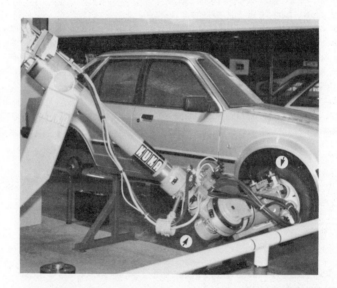

Figure 5.17 Force and torque sensors are required in robotised assembly to ensure reliable task execution, to avoid collision and difficulties relating to compliance.

* For part inspection purposes.

Non-contact sensing principles employ optical, capacitive and inductive probes, laser beam, infra-red light and other techniques. There are sensors which can detect variation in pressure, vibration, force, torque, heat, colour. Many different types of image processing techniques and systems which can transfer and store digitized images of the silhouette and shape of the measured part at very high speed are also used in unmanned manufacturing.

5.2.3 Automated Tool Changing (ATC) and Pallet Changing (APC) mechanisms

To ensure the shortest setup times and unmanned operation, Automated Tool Changing (ATC) devices, tool magazines and pallet magazines have been developed by most machine tool and tool manufacturers as well as slowly beginning to be developed by robot manufacturers who wish to see their robots working in flexible production systems.

Time spent on tool changing, setting up, gauging the workpiece and the tool for establishing the tool length correction data can be lengthy, unless solved with modular preset tooling systems and by means of in-cycle gauging, or both.

The principle of preset tooling is that the shank and the cutting part of the tool are separated and can be recoupled by means of a suitable coupling interface set to known (tool) coordinate values. The cutting units (i.e. blocks) can thus regarded as the operative units, enabling the storage of a large number of tools (i.e. usually 60 to 240 per tool magazine) at a lower cost.

In the Sandvik-Coromant system several hundred block tool cutting units can be stored in the machine's block tool magazine(s). (This tooling system is illustrated in Figure 6.5, in Chapter 6.)

In the Krupp Multiflex tooling system the tool head is put onto a cylindrical spigot from the front and is pulled back onto a seating by a central drawbar while driving keys transmit the required torque.

The Hertel Flexible Tooling System enables recoupling of

Figure 5.18 Non-contact part and pallet detection sensors in a chain type pallet magazine of an industrial assembly robot.

Figure 5.19 Contact measurement principle as applied in this Zeiss Co-ordinate Measuring Machine (CMM.)

tools by means of a face fitting Hirth-gear-tooth design and a collet. The Hirth coupling provides good torque transmission with high accuracy for both static and rotating tools. (Note that other systems do not provide both static and rotating tools within the same system). Each tool carries its own coding, enabling different machine integrated sensors to identify and check details on different tools in the magazine or in action. The code is carried on the tool-head on a ceramic storage plug, and can be read directly by a sensor in the tool transfer gripping device (Figure 5.20).

A further major development step is when a microprocessor is integrated into the block tool head for gathering and conditioning real-time cutting data. Since tool life measurements are expensive and there is insufficient amount of machining data available, this solution enables:

* Adaptive control systems to gain real-time data.

* Expert systems to learn directly about the cutting process by evaluating cutting forces, tool deflection, heat and torque in real-time during "normal" operation at no additional cost.

* As a short term benefit, with sufficient control software this microchip would be able to protect tools against break.

Pallet magazines and Automated Pallet Changing (APC) devices are similar in principle to the earlier developed tool magazines and tool changing systems. The purpose of utilizing a pallet store, or pallet pool usually offering 6 to 12 stations, is to automate part loading for nonconform parts, often arriving at the machine in random order. In a typical configuration pallets are automatically loaded onto the machine's table by a transfer arm, in most cases operated hydraulically, from the pallet magazine and they are accurately located by means of curvic coupling, or precise shot pins.

Pallet magazines vary. They are often chain type, vertical or horizontal, or are arranged into a line as a series of fixed position buffer stores. By employing work changing systems (e.g. pallet shuttles, mobile robots, workchanger caroussels, etc.) not only are savings substantial, but these techniques also enable the cell, or workstation to be integrated into FMS and work unmanned. (See Figures 5.21 and 5.22

CIM-G*

Figure 5.20 The Karl Hertel FTS (Flexible Tooling System) and the drum type of tool magazine. (Courtesy of Karl Hertel (UK) Ltd., Warwickshire, UK.)

and refer to Chapter 6, where FMS cells are discussed.)

5.2.4 Introduction to the control architecture of CNC machines

The control system of the machine tool, robot, inspection machine, or any other machine using CNC principles is the most rapidly developing field in the manufacturing industry.

There is an interesting contrast between the increasingly user-friendly and intelligent local machine and cell controllers, and the DNC (Direct Numerical Control, or sometimes Distributed Numerical Control) concept, in which the power and programming intelligence is concentrated in the DNC host computer rather than at individual machine level.

Very likely both the powerful CNC offering high level programming facilities, 3D tool path animation with graphics, real-time tool database, automated cutting parameter check and/or preselection, etc. on the shop floor, as well as the DNC/FMS integration concept will develop and prosper together mainly because of economic reasons. Many companies cannot afford FMS to start with, but wish to install machines, or cells, capable of being linked into FMS in the future and for them this is the most suitable solution.

The up-to-date machine controller is a kind of "work-station" in a LAN (Local Area Network), or alternatively in a distributed processing system. Obviously the functional requirements are different, the programming languages used are different, but principally they are powerful 16 to 32 bit microcomputers, often equipped with parallel processing capabilities and with a variety of input/output data handling features. They offer large read/write (bubble) memory extension which can be replaced similarly to a cartridge program in a microcomputer and they are capable of communicating with other machine controllers.

Figure 5.23 illustrates the general system architecture of an up-to-date CNC controller. As with any powerful microcomputer this architecture integrates four major modules, these being:

* The CPU (Central Processing Unit), which executes all the available software in the system, controls machine axis functions, and generally acts as a supervisor of

Figure 5.21 Chain type pallet magazine and tool magazine of the
Makino MC40 FMS cell.

Figure 5.22 Vertical part and tool magazines from Traub.

the system as a whole.

* The bus system, which is a fast bidirectional data
  transmission link between different parts of the CNC
  system, (usually consisting of the system bus, on which
  there is the CPU, the memory cards, the axis controller
  and the input/output unit; and the local input/output
  bus handling additional memory and digital I/O).

* The memory, usually providing bubble memory for the
  storage of system software (e.g. the monitor, or
  operating system, part program and macro libraries,
  utilities, etc.) and a faster dynamic RAM (Random
  Access Memory) into which programs are copied from the
  bubble memory before execution.(The important fact to
  remember is that bubble memory is protected against
  power faults or system shutdown, RAM is usually not.)

* The input/output system providing communications
  interfaces to:

  (a) different outside peripherals and/or
      computers by means of standard serial (e.g.
      RS232c, or the faster RS-422, etc.) or parallel
      (IEEE, or 8-bit parallel output for a paper tape
      punch, etc.) interfaces, or

  (b) to the machine itself, including: servo
      connections to the axis (axis can be driven by
      analog or by digital outputs, and position
      feedback signals can be analog or digital as
      well), spindle control, digital input/output
      connections, additional memory connections, etc.

When such a system "boots-up" the CPU executes the
monitor, or in more advanced systems the operating system
programs from the system memory, by means of an IPL (Initial
Program Loader) routine, or hardware, or firmware. Any new CNC
part program executed after this is an input via one of the
Input/Output control system's interfaces (ports) or the keyboard. The part program is executed by the CPU, via the axis
controller and the input/output controller. (The axis instructions of the part program go via the axis controller, the
others via the digital I/O). After successful program execution the "new part program" becomes an "old" or debugged program and can be saved in the non-volatile, or bubble memory
from where it can be transferred for execution into the system

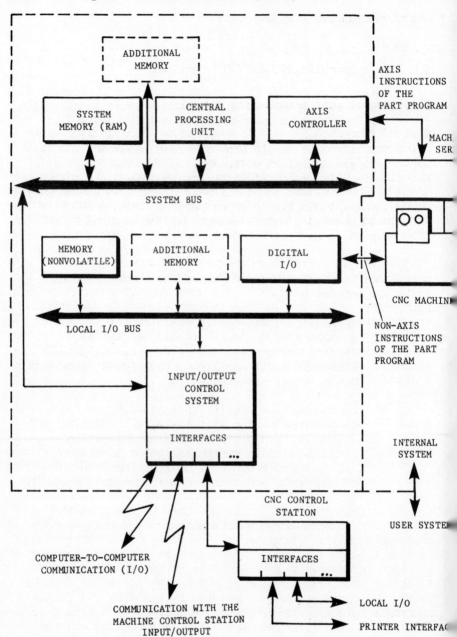

Figure 5.23 General system architecture of an up-to-date CNC
control system.

RAM when required the next time.

5.3 Part programming NC and CNC machines

5.3.1 Manual part programming methods

Most NC (Numerical Control) programming work is done today using some level of computer assistance (Computer Numerical Control, off-line programming, high level NC programming, Manual Data Input (MDI) NC programming, etc.), but it is all based on, or compiled to, or interpreted a large extent to the conventional, standard NC part program format, a brief introduction to which is thus necessary to understand the whole process.

Manual part programming of NC machines involves the manual selection of the required tools, clamping devices and fixtures, the description of the cutting parameters, the tool path geometry and the cutting sequence based on the drawing of the component. Manual programming was widely used in a variety of different forms mainly for milling, turning, boring and drilling operations from the mid 60's to the end of the 70's, when microprocessor based CNC controllers became available on a large scale.

5.3.2 NC tape formats

The NC tape format (word address, tab sequential, or variable block) is the general order in which data, or instructions appear on the input medium (e.g. magnetic tape, paper tape, etc.) or in the protocol transmitted via a DNC line.

Most systems follow the word address input data format, for example:

N015 X03240 Z01280 F180 S265 EOB

where N015 is the block number,

X and Z are position co-ordinates in a cartesian co-ordinate system,

F180 represents a feedrate code,

S265 a spindle speed code and

EOB is the "End-Of-Block", usually [LF-RETURN] character.

There are other less popular methods as well, these being the tab sequential format, where individual words are identified by their position in the block, for example:

N015 TAB TAB 03240 TAB 01280 TAB 180 TAB 265 EOB

and the interchangeable variable block format, representing different less important variations of the above listed methods, for example:

N015 TAB TAB X03240 TAB Z01280 TAB F180 TAB S265 EOB

As can be seen from the first example above, the word address format identifies each NC function by a word and the word is defined by a letter where:

N identifies a set of simultaneously executable instructions; it is the basic building block of an NC part program and is called block number,

G are preparatory words, discussed in detail below,

X, Y and Z are the co-ordinate axes identifiers, with the additional U, V and W if more than three axes are controlled,

I, J and K are special axis words, used in circular interpolation, thread cutting, etc.

F stands for the feedrate code,

S is the spindle speed code,

Q is the spindle orient word, enabling the programmer to orient the spindle to a programmed angular position,

R is the radius word, used to program the cutting or the

arc radius,

T is a tool identifier in the tool magazine of the
machine incorporating tool turret or magazine station
code and the tool offset values required for compensating tool wear, tool length deviations and tool
deflections, and finally

M are the miscellaneous words, discussed in more detail
below.

Because of their importance let us discuss the "G" and
the "M" words in more detail.

"G" codes represent preparatory functions for establishing different modes of operation, which can be modal (i.e.
once executed active until a cancelling G code is executed),
or non-modal (i.e. active only for the block in which it is
issued) such as for example:

G00 Positioning/Rapid,

G01 linear interpolation,

G02 Clock Wise (CW) circular interpolation,

G03 Counter Clockwise (CCW) circular interpolation,

G33 to 35   thread cutting,

G40 to G45 tool-tip-radius compensation, (see discussed
     later in section 5.3.4),

G53 to G55 part position offsets,

G70 metric format,

G71 inch format,

G74 to G76 full circle programming,

G80 ...canned cycles, macros and subroutines,

G90 absolute programming, (i.e. programming every
     displacement form the same, "absolute" origin),

G91 incremental, or relative programming, when for

certain geometry different co-ordinate system is defined by the programmer,

G96 constant surface speed programming, allowing to program part surface speed with respect to the tool tip,

G97 direct RPM (Revolution per Minute) programming, etc., and

"M" codes are in most cases modal words and represent miscellaneous functions for establishing machine control conditions, including:

M00 program stop,

M01 optional program stop,

M02 end of program,

M03 spindle on, clockwise,

M04 spindle on, counter clockwise,

M05 stop the spindle,

M06 change tool,

M08 coolant on,

M09 coolant off,

M30 end of program, etc.

Note that although word formats are standerdized by ISO, they are unfortunately not 100 % interchangeable in the case of different controllers, mainly due to the differences in which:

* Canned cycles, macros, or sub-programs (in other words predefined routines which generate multiple tool movements from a single parametric block) and their parameters are defined and called.

* Words are interpreted with different type of machines, e.g. CNC grinding machines often use two axes CNC lathe controllers, but there are obvious differences in the

processes thus some canned cycles and codes must be different as well.

* Words affect different functions in combination with other words already issued.

Because of even minor differences, before writing a part program one should consult carefully the machine's programming manual.

To give a simple example of an NC part program section consider a 10.00 mm thick plate into which we wish to drill a hole at a location described by x = 125.50 and y = 56.00 co-ordinates in the part co-ordinate system (note that z = 0 is fixed on the upper surface of the plate and that comments in part programs are not necessarily used the way we have used them):

N050   T0201
       (* Select tool *)

N051   G00 G90 X125.50 Y56.00
       (* Move rapid, in the absolute co-ordinate system
         to the given x and y co-ordinate values *)

N052   Z10.00
       (* Rapid until 10.00 above surface *)

N053   S250 M03 M08
       (* Select spindle speed code, switch on the
         spindle CW and the coolant *)

N054   G01 Z-14.00 F03
       (* Downto Z=-14.00, which means 4.00 mm overshoot
         at the set feedrate *)

N055   Z10.00 F12
       (* Up 10.00 mm above surface at a faster feedrate
         than in N054 *)

N056... (* Ready for the next operation *)

etc.

Note that because the above described tasks are often required, normally this job would have been programmed using the G81 canned cycle for drilling. This canned cycle requires the X and Y position co-ordinates, an R value describing a safety zone, or reference surface until which rapid movement is used, a Z drilling depth, a programmed feedrate and a spindle speed after which the job is done automatically.

Canned cycles are very useful in the case where the same operation must be repeated many times at different locations or orientations on the part, when the task is complex and tested sub-programs are essential, or when in FMS alternative routes must be taken, because of dynamic scheduling. (For further details regarding this aspect refer to the next Chapter.)

5.3.3 NC tape codes

Besides the format, the physical code in which the data is punched on paper tape, or recorded on disk, or downloaded via a DNC line is important too. There are three widely used codes, these being:

* The ISO (International Standards Organization), or DIN (Deutsche Industrie Normen) 66024 code, based on but not fully equal to the ASCII (American Standard Code for Information Interchange) and is an even parity 7 bits of information code, with a parity bit punched in track 8.

* The ASCII code, which is an even parity 7 bit code for information and one parity bit punched in track 8.

* The EIA code, which is an odd parity code for 6 bits of information. The parity bit is punched in track 5 and the EOB in track 8. (Note that more information on these codes can be found in many NC books dealing with numerical control, including [5.1].)

5.3.4 Tool-tip-radius compensation

Tool-tip radius compensation is required to enable the programmer to instruct the path of a nil tip radius tool. In other words the programmed point of the tool is a theoretical point, whereas the real tool tip is never a geometrical point, but a known radius (Figure 5.24).

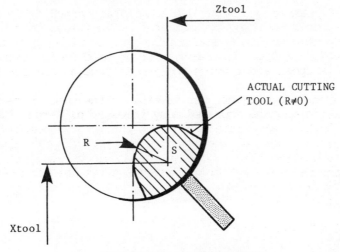

S = CUTTER RADIUS CENTER POINT

R = TOOL TIP RADIUS

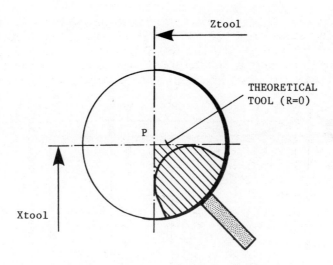

P = PROGRAMMED (I.E. THEORETICAL POINT)

Figure 5.24 The difference between the programmed (i.e. theoretical) point and the actual cutting tool when
programming NC machines.

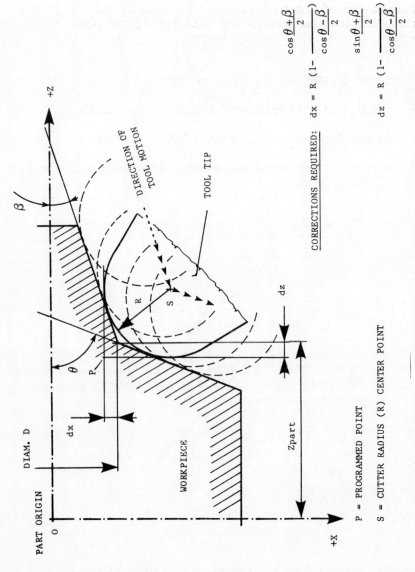

P = PROGRAMMED POINT

S = CUTTER RADIUS (R) CENTER POINT

CORRECTIONS REQUIRED: $dx = R\left(1 - \dfrac{\cos\dfrac{\theta + \beta}{2}}{\cos\dfrac{\theta - \beta}{2}}\right)$

$dz = R\left(1 - \dfrac{\sin\dfrac{\theta + \beta}{2}}{\cos\dfrac{\theta - \beta}{2}}\right)$

Figure 5.25 Offset value calculation with tool tip radius correction.

In the case of manual programming these calculations must be done by the programmer, otherwise the programmed and manufactured parts will be different. (For a general case refer to Figure 5.25, where the calculation of the dx and dz offset values are given for turning). In the case of CNC all trigonometric calculations and the necessary adjustments in the axis feed movements of the tool path can be done by the controller, thus simplifying the process and eliminating the possibility of errors.

5.4 Computer assisted part programming methods

Computer assisted part programming can be implemented:

* In the increasingly powerful CNC controller,

* or in an other computer, remote from the machine tool, robot, etc.

Because of the rapid development of microcomputers most up-to-date CNC controllers offer powerful computing facilities. The trend is that CNC systems offer an increasingly sophisticated software interface for the systems engineer, dealing with macro programming, interfacing, implementing Adaptive Control features and sensory feedback processing loops, etc., and the extremely user-friendly high level for Manual Data Input type part programming and operation control, often with 3D color graphics animation.

It is also clear that sophisticated CNC machine programming is becoming more and more a computer programming job and requires the additional understanding of the involved processes, this should be done by production engineers who have computing skills. (Note what has been said in Chapter 1, regarding the required skills.) The available low cost and large memories (i.e. 1 Mbyte RAM and 2-4 Mbyte bubble) enable manufacturers to make their machines more and more "intelligent" and user-friendly by providing some level of interactive, or dialog-type part programming software with many new programming and graphics features, an interface software for communication purposes and cell control functions and software representing expert knowledge. (See again the CNC control system architecture presented in Figure 5.23.)

Figure 5.26 illustrates some advanced CNC programming features implemented in the portable FANUC FAPT (FANUC-APT)

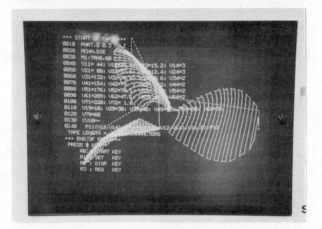

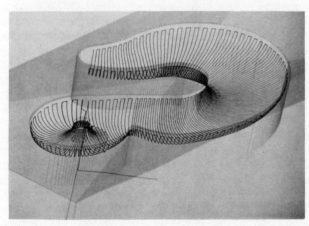

Figure 5.26 Advanced five axis milling programming and tool path generation using the FANUC/FAPT language.

programming station. The FAPT language is capable of supporting up-to five axes milling work, incorporates graphics and handles different peripherals, including plotter, digitizer, disk storage, printer, paper tape punch, etc.

Different CNCs use different software depending on the level of access required (e.g. high level user, system programming, machine diagnostics, etc.) and the programmed processes involved, (e.g. milling, turning, grinding, flame cutting, welding, etc.) but there is a core of software modules representing the practically unlimited new possibilities of using powerful micros in this industry.

The most important modules of this core system are integrated in an operating system offering the following system, application, diagnostics and other support:

* The system software is very similar to the operating system of a micro-, or minicomputer. It is independent of the applications the controller is involved in and generally acts as a supervisor controlling the total system, consisting of the hardware and the modular software sub-system. (The programming language of this level is mostly assembler and machine code).

* The machine control software provides the machine interface to the CNC unit. Standard machine activities, such as tool changing, tool magazine organization, spindle speed control and codes, feedrate control and codes, pallet magazine control, etc., are organized by this module. Often real-time sensors, adaptive control features, tool file data structures, etc. can be user-defined and programmed by using this system level interface. (Since this level very often involves programming more than one processor and PLCs (Programmable Logic Controllers), for example one PLC controlling the tool magazine, the second the pallet pool, a third one used for interfacing, etc., the programming language of this level is mostly assembler and machine code.)

* Because the CNC system itself, as well as the number and variety of controlled processes and devices become more and more complicated to understand and maintain within a short period of time, the machine diagnostics module is an increasingly important part of the CNC system software in order to avoid unnecessary machine downtime.

According to many reports it is not the electronics which tends to break down, but rather the mechanical or electro-mechanical components controlled by the CNC, thus the controller should be able to provide extended diagnostics of all controlled devices.

Fault diagnosis is an area where the growth rate of expert systems seems to be the biggest.

* The machine interface module allows the system programmer to interface the controller with the outside world using standard digital (serial and parallel) interfaces. These activities are crucial when integrating the CNC into a distributed computer network, when linking machine controllers together and also when developing adaptive control and real-time sensory feedback control systems.

This interface also handles the available peripherals, such as the keyboard, the function keys, the display (CRT or graphics), paper tape reader/punch, printer, disk drives (if provided) and all other possible peripherals presented to the machine tool operator.

The programming language of this level is often Assembler, but there is a trend to use structured languages such as Pascal, C or ADA. Keeping in mind that the average lifetime of a CNC control system is less than for computers and is in the region of 10 years, one should consider the implications of software support and maintenance, which is a good deal simpler if the code is developed and documented according to structured programming principles as described in Chapter 2 (see again Figure 2.1).

As part of the applications, or user interface module, most CNC controllers provide a Manual Data Input (MDI) or dialog type, high-level and user-friendly part programming environment. This means that during the part programming procedure the user needs to fill-in screens, answer questions raised by the controller, press function keys, etc. rather than use a conventional, or formal symbolic language.

The other trend is that part of the CNC application software is becoming similar to a high level programming language, such as BASIC and that both a high level and a limited system level programming interface (e.g. for designing macros, changing conditions of operation, etc.) are provided.

The most important extensions in the CNC system compared to "conventional" NC features include:

* More advanced canned cycles and tested macros, or sub-programs and a powerful NC programming and program development environment (e.g. part program editing, file handling, tool file editing, real-time input/output programmimg and other operating system utilities).

* The possibility of arithmetic calculations and logical decision making, allowing safe and efficient part program creation, verification and execution.

* Branch-on-condition, program control transfer, IF-THEN-ELSE loops, etc. allowing execution sequence control similarly to computer programming languages.

* Parametric programming using subroutines, or macros written by the vendor and/or the users.

  (A subroutine is a set of part program blocks which are executed by calling a block in the host, or active program. When the execution is completed in the called routine, control is transferred to the next block of the host, or active program. The most powerful sub-routines are parameterized, thus can be used for a whole "family of parts". Most advanced systems allow the transfer of parameters not only by value, but also by variable. Obviously these are not new features in computing, but are very powerful tools in CNC programming.)

* Manual Data Input and high level part programming, meaning that the operator must select from different menus such as material, machining unit, cutting pattern, geometry, tooling, cutting parameters, etc. and must answer questions set by the controller during the part program data input procedure.

  This high level operator assistance and support also means storing optimum cutting data and advising the user on selecting appropriate tools to specified material, optimizing tool layout in the magazine, checking programmed geometry, offering the controller

for programming whilest the machine is cutting,
measuring tool offset values automatically, storing
cutting paths and cutting conditions, displaying the
machined contours in 3D often using color, etc. (This
is an area again where expert systems integrated at CNC
level can have an enormous impact on machine programming, performance and diagnostics).

* Closed loop machining and in-process gauging means
the possibility of measuring part dimensions and
feeding the measured values back into the CNC controller
for decision making purposes. Measurements, using
special purpose, mainly contact probes, are capable of
detecting deviations so that corrections can be made
automatically in the tool offset file. Such probes are
also useful in detecting tool wear, avoiding tool collision, identifying automatically loaded parts for
example by robots into the hydraulically operated chuck
of the lathe and when setting up tool length and diameter correction values. (See again Figure 5.15 and
refer also to Figures 6.15 and 6.16 in Chapter 6.)

* Sensor programming, offering the possibility of incorporating user designed analog and digital sensory
feedback loops (e.g. force and torque measurement at
tool tip, vibration sensing, etc.) into the control
architecture.

* Input/output communications and interfacing.

* Part manufacturing time monitoring; useful for
example when scheduling parts in FMS operation.

* Tool life monitoring; important to avoid tool breakage.

Figure 5.27 lists a portion of a system macro written
for a CNC grinding machine using the Allen Bradly Paramacro
(TM) language. This figure partly illustrates the complex
tasks macro programs can deal with, and in this sense under-
lines what has been said above, but it also demonstrates
that despite the many powerful facilities available in such
languages there is still much to be done to decrease the gap
between the way up-to-date general purpose computers and
"machine controller computers" are programmed.

Without trying to explain all details but concentrating
on methods and program structures only, when analyzing the

Figure 5.27 Sample CNC grinding program segment written for the Jones and Shipman CNC grinding machine. (See the machine again in Figure 5.2/b.)

SAMPLE CNC GRINDING MACRO PROGRAM SEGMENT

+--+

```
(DM,G88)          Note that this is a macro

                  program line, and this

    *** DEFINE MACRO NAME = G88, PLUNGE GRIND MACRO

                  is a comment line

(G88 REVISION CODE, DATE...)

    *** DEFINE MACRO RELEASE NUMBER, REVISION DATE, ETC.

N10(CM,INIT)

    *** CALL COMMON INITIALIZATION AND CONDITION PRE-CHECK

        MACRO TO MAKE SURE THAT THE MACHINE STATUS IS OK.

N20 M59(AP,P1=94,P2=PZ+P119,Q3=0)

        (IFT,G95,P1=95)

        (IFT,PQ,NE,0,P2=PZ+P134,Q3=1)

    *** INITIALIZE AND ASSIGN PARAMETERS ASSUMING MM/MIN

        PROGRAMMING AND THAT THE WHEEL IS ON THE LEFT SIDE

        OF THE SHOULDER. IF MM/REV PROGRAMMING IS ACTIVE,

        THEN (IF_THEN;IFT) MODIFY P1 GLOBAL SYSTEM PARAMETER.

        IF SHOULDER TO THE RIGHT OF WHEEL, THEN MODIFY

        FOR WHEEL WIDTH.

N30(CM,CHK)

    *** CALL CO-ORDINATE CHECK MACRO

N40 G90 G21 F0 X=P100+P117

    *** MOVE TO WORKPIECE CLEARANCE DIAMETER AT RAPID

N50 Z=P2

    *** MOVE Z TO START POINT AT RAPID.
```

given listing one can realize that:

* it is very similar to a CNC language, with the difference, that

* there are many system parameters used (e.g. P1, P2, P100, P117, etc.), some of them being global, some local,

* values can be assigned to system parameters by means of the AP (Assign Parameter) instruction,

* conditions can be evaluated between parametric expressions, using Boolean operators, e.g. EQ = equal, NE = not equal, LT = less than, etc.

* program control (i.e. execution order) can be transferred by IFT (IF-THEN...) instructions, and

* other macros can be called (note: not defined) within a macro definition.

The Paramacro (TM) language offers many more features than those listed above including arithmetic calculations, mathematical functions, macro nesting, GO TO program control transfer, etc. This is impressive for a CNC controller at the time this text was written, but there is nothing new among these features which has not been available during the past 10 years in terms of general computing techniques. (With this comment we wish to emphasize the need for further development in this area of CNC control.)

Just to raise a few questions, for example:

* Why should a system use system mnemonics by fixed name (e.g. P112, Q, etc.) which are hard to remember, and which are in most cases application dependent anyway, rather than offering the possibility of properly declaring local and global variables, as for example in ALGOL or Pascal?

* Why is the "ELSE" part missing in the IF-THEN instruction? (This could make programs more structured, thus better documented, and, as a consequence, safer.)

* For example the CASE-OF type multiple branch statement

could be really useful, when evaluating different
parameters handled run-time, read from input ports,
etc.

* Why couldn't one specify files which relate to
  different tasks, and taking the idea further: volumes
  which could be assigned to different physical ports
  simplifying input/output data handling, real-time
  sensory input data processing and similar tasks?

To summarize: why are such commonly known features and
programming techniques in computing not implemented in most if
not all machine controllers? In the "old days" there wasn't
enough memory and local computing power available, but today
there is plenty of memory and processing power available.

When developing CNC control system software tools one
should realize that the code to be written using such development tools must be tested, debugged and well documented
otherwise the maintenance of such software systems is going to
cost a lot, and become obsolete in 6 to 10 years time, when
the controller will still likely be in use.

5.5 Case study: A COMPACT II macro library

Having discussed the basic components of numerically controlled machines, the way they work and are programmed using their
own controllers, let us concentrate on "high level, or
symbolic" programming languages, among them COMPACT II.

The purpose of this case study is to introduce an APT
like symbolic language and to demonstrate the way a macro
library can be designed and implemented. Having discussed CNC
macros before, it also provides the opportunity to compare the
CNC level and the APT level NC programming approaches.

The library was designed by the author and was developed
and implemented by the author and Mr Ken Walker at Trent
Polytechnic, Nottingham during a period of 1980 to 1984. The
Library contains profiling, contouring and area cleaning
macros of simple shapes, e.g. circular and square/rectangular,
and the geometric definition of cam profiles, of which some
macros are shown in this text. The Library is in industrial
use.

As in previous cases our aim is not to teach the

language, but to introduce concepts, demonstrate different
solutions and analyze different structures, thus only a brief
overview is given of COMPACT II. (For further information on
the language refer to the COMPACT II system and user manuals.)

5.5.1 Overview on APT and COMPACT II

The universal symbolic NC programming languages are all based
on APT (Automatically Programmed Tools) developed in the USA
in the sixties at the Industrial Research Center of Illinois
Institute of Technology in Chicago. APT is suitable for two
and three dimensional part programming, but it considers tool
path geometry only. In other words APT does not deal with the
selection of cutting parameters, or with any other components
of the Machine Tool-Workpart-Tool-Fixture and clamping device
system. It must be emphasized that despite the updates,
because it was written in FORTRAN more than twenty years ago
it is in several ways out-of-date. However the way the
geometry was defined is still valid and followed by several
"APT like" languages, such as ADAPT, MINIAPT, EXAPT, NELAPT,
COMPACT II, UNIAPT, etc. as well as CNC level languages.

APT and APT-like languages use a geometrical input
language, which is processed and then post-processed to
generate the required NC tape format. Most systems are capable
of handling some sort of a tool file, but are limited in
programming real-time sensors, probing cycles and in general
have many features which are out-of-date and hard to use.
Despite the above discussed drawbacks, APT languages are still
important since most integrated CAD/CAM systems generate some
kind of APT, or APT-compatible code for NC programming.

COMPACT II is a commercially available NC programming
system for turning, milling, drilling, boring, punching and
nibbling, flame cutting, EDM, etc., from Manufacturing Data
Systems International (MDSI). COMPACT II converts the source
in a single interaction into manual NC code, rather than using
the more conventional two stage processing and post-processing
approach. COMPACT II post-processors are called "linkers"
and there are currently over 2000 of them available to
interface the output of COMPACT II and the different machine
controllers.

5.5.2 The architecture of COMPACT II

Although the vocabulary of the language consists of more than 150 words, part programs can be written using around 40.

There are two type of words used, these being:

1. MAJOR words, describing the operation to be performed.

2. MINOR words, describing the location of the operation, the machining sequence, tool and machine functions.

The part program is built up of statements, or instructions. Each statement begins with a major word, and consists of a major word and a set of logically associated minor words. Statements can be:

* Control statements, providing information to the COMPACT II system for program initialization, co-ordinate system location, tool specification and regulation and program termination.

* Geometry statements, establishing the workpiece contour to be machined, containing points, lines, circles, arcs, mathematically calculated and interpolated functions, patterns, etc..

* Motion statements, describing tool motion around the workpiece along a line or an arc, or a segment of a line or arc.

The basic structure of the COMPACT II part program consists of the following groups of instructions:

1. Initialization, MACHIN, IDENT, SETUP, BASE, etc.

2. Geometry description, DPTn (define point) DLNn (define line), DCIRn (define circle), etc.

3. Tool change commands ATCHG (automated tool change), MTCHG (manual tool change).

4. Tool motion statements, MOVE, CUT, CONT (contour), ICON (inside contour), OCON (outside contour), etc..

5. Program termination statement, END.

The COMPACT II system consists of:

* The Executive System, offering log-on and entry to the next module, which is the

* Numerical Control System (NCS). This is the major processing module of the system with the ability to execute part programs.

* The Editor (Quick Editor, QED) offers program correction facilities and source code input, and

* the Run Time Command Dispatcher (RTCD), for handling interrupts, whenever an error is detected during processing and

* Linkers, or "post-processors" to different machines.

5.5.3 The COMPACT II macro library

COMPACT II as well as other symbolic NC programming languages provide some level of possible software solution, or software tools (which are unfortunately quite limited) for building up macro libraries, for storing pre-tested programs and files which can be parameterized and called by other "host" programs. Particularly if unmanned operation is the aim as in FMS, or CIM, macros play a very important role, because they

* are safe to use, since they are debugged and tested,

* are efficient from programming point of view, because one needs to fill in parameters only, rather than writing often several pages full of code,

* are well documented, thus can be understood by all users and maintained easily,

* offer solutions for dynamic scheduling and for "variable route part programming", (discussed in Chapters 6 and 7) and

* because they are consistent with "good programming style, or practice" and can be relatively easily introduced, followed and maintained by using them.

COMPACT II has a feature known as RECALL, allowing a user to store part program modules, or macros in the system and recalling these macros automatically during program processing. The macro may contain variables, or parameters whose values are set when the statements are recalled, thus operations occurring many times, such as contouring, pocketing, cam milling, thread cutting, etc. can be programmed using parameters, rather then fixed values.

There are many limitations in building macros, more than is feasible, probably because the roots of problems are in Fortran. (For example COMPACT II does not allow parameter transfer via the macro head by variable, but only by value; macros cannot be directly written in the part program; the number of variables (local) are too limited; it does not allow the use of simple expressive mnemonic names because of strict format specifications, etc. For all limita-tions refer to the COMPACT II manuals.)

Before designing the macro library a simple macro classification method was drawn up (Figure 5.28) to provide a frame for the modular development.

According to this all macros are:

* GEO, geometrical,

* MO, motion, or

* CUT, cutting or

* a combination of the above, e.g. GEO+MO, MO+CUT, etc.

The groupings arise from the simple principle that a part program consists of the geometry of the workpiece, the tool motion commands and the cutting parameters (or in general the process parameters) of the tool.

The most simple way of building up macros is to write GEO macros. Only geometrical descriptions are used to describe the workpiece contour and elements to be machined and put in a macro, combined with user written control and tool motion statements. If chained GEO macros are used it is most important that the same geometric symbols are not used for different geometric elements because part programs will fail due to the system assuming that the element has been defined more than once. (Unfortunately the real problem is that the system is not capable of properly handling local and global parameters, as in structured high level computing languages.)

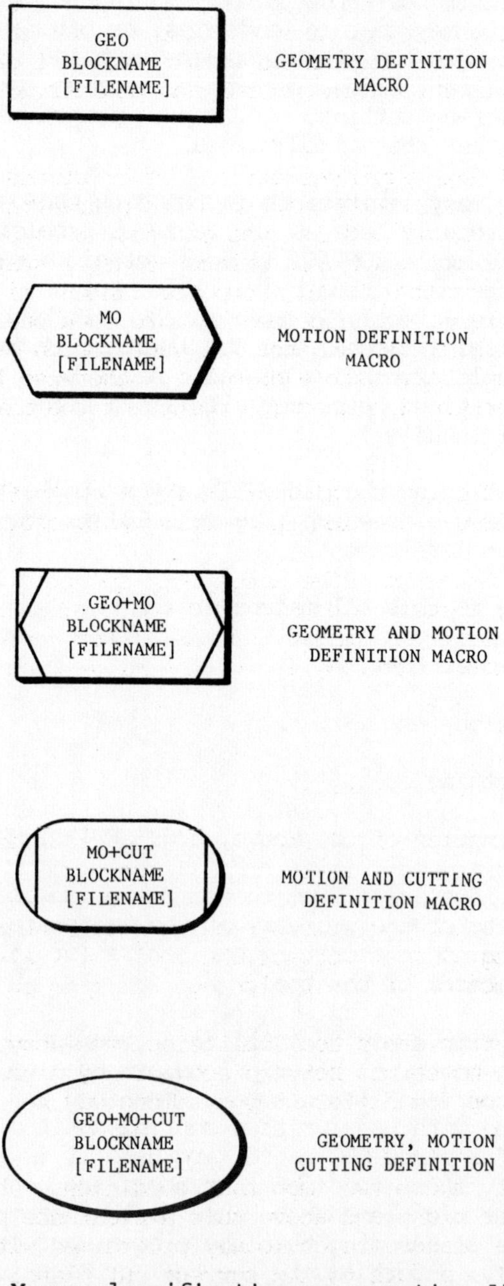

```
    GEO
  BLOCKNAME          GEOMETRY DEFINITION
 [FILENAME]                 MACRO
```

```
     MO
  BLOCKNAME          MOTION DEFINITION
 [FILENAME]                MACRO
```

```
  GEO+MO
  BLOCKNAME          GEOMETRY AND MOTION
 [FILENAME]           DEFINITION MACRO
```

```
   MO+CUT
  BLOCKNAME          MOTION AND CUTTING
 [FILENAME]           DEFINITION MACRO
```

```
  GEO+MO+CUT
  BLOCKNAME          GEOMETRY, MOTION AND
 [FILENAME]         CUTTING DEFINITION MACRO
```

Figure 5.28 Macro classification representing different combinations of geometry, motion and cutting definition macros as utilized in the COMPACT II macro library

Motion macros (MO) are put into the part program to describe the tool motion required to carry out the necessary machining. MO macros are called within the part program, combined with user written control and geometry definition instructions.

In the GEO+MO macro geometrical descriptions and the associated tool motion descriptions are concentrated and developed as one macro. This type of macro when used in a part program is combined with user written control statements. When for example using the GEO+MO macro for contour milling of a cam and the contour is described by more than one macro the geometrical description and the relevant motion description have to be compatible with each other, otherwise the surface will not be smooth at connection points of, the different contour segments.

The MO+CUT macro is similar in function to the MO macro, but includes tool selection and cutting parameter statements.

GEO+MO+CUT macros offer a complete solution to the manufacturing of a surface element, or a contour, or a whole part. This type of macro is limited in general use, unless designed for manufacturing part families with similar machining requirements.

5.5.4 Machining philosophy

The most important general aspects of the machining philosophy implemented in the Library is as follows:

* If different operations require different tooling then the program must retract the tool to a safe plane before rapid motion to the toolchange position occurs.

* In order to avoid tool collision, the above applies also to movements between operations, since the workpiece top for different operations may be on a different plane.

To satisfy these two important conditions, above the cut plane a retract plane and above this a clearance plane are defined. These planes are obviously programmed with parameters and are relative always to the current cut plane. After completion of the required operation the tool retracts at a program-

med feedrate to the retract plane and then with rapid motion
to the clearance plane, and then it can move safely to a tool
change location, or to the next operation. The three planes
divide the rapid and feedrate motions as follows:

* Rapid and/or toolchange are allowed only above the
  retract plane or on the clearance plane, or above.

* Between operations only the clearance plane (or above)
  can be used.

* The retract plane is a safe zone, but programmed (i.e.
  slower than rapid) feedrate should be used below it.

5.5.5 Macro documentation

Since macros are applied in industry it is essential to
provide clear and error-free documentation. Badly documented
NC (and other) programs are not only dangerous, but useless in
industry because nobody will use them. To ensure a general
format a documentation method was designed and put on the
computer (see Figures 5.29/a, 5.30/a and 5.31/a.)

The documentation of any macro in this Library consists
of the following items:

1. Macro name.

2. Version, Release and Modification number.

3. Date.

4. Macro classification (as discussed above).

5. Functional description, which is a brief description
   of the function of the macro and the operation(s) it
   performs.

6. Macro format, explaining exactly how the macro should
   be used in a part program.

7. The parameter description explaining and defining the
   parameters (local) used in the macro. (It must be
   mentioned that macros do not have to posess parameters.)

8. Source file listing with the necessary amount of comments enabling the user to follow the way the macro works.

9. Sample run, preferably with graphics plot.

Having discussed the most important principles of the language as well as the macro library, let us demonstrate them and the way they can be used in a part program.

5.5.6 Automatic toolchange macro

The automatic toolchange macro is a motion + cutting (MO + CUT) macro for the automatic toolchange of milling cutters. (Figure 5.29). It has been put into this section of the text because other macros also use it.

5.5.7 Internal profile milling macros

Internal profile milling macros are very useful since there are many cases when circular milling operations are required. The full documentation of these macros and an application including a plot is given in Figure 5.30/a-f.

5.5.8 Rectangular and square profiling macros

Rectangular and square profiling is a common procedure mainly when machining large gear-box castings. The full documentation and an application of these macros, including a plot is given in Figure 5.31/a-f.

5.6 Principles of robot programming

There are several different ways industrial manipulators and robots can be programmed. Within this text our purpose is only to give a brief summary of some available methods and in the next section to give the listing of an off line generated robot program, illustrating the fact that programming of different computer controlled machines and powerful industrial robots are very similar in principle. (Note that robot programming and VAL are discussed in much more detail in references [5.21] and [5.22]).

Figure 5.29/a Automatic toolchange macro (TOOLDEMO.)

```
TOOLDEMO (AUTOMATIC TOOLCHANGE MACRO USING "TOOLLIB") DOCU.
+-----------------------------------------------------------------


                    **************************
                    * MACRO NAME: TOOLDEMO *
                    **************************

                       Ver.00 Rel.01 Mod.12
                          14/March/1984
MACRO CLASSIFICATION:
********************

MO-CUT

FUNCTINAL DESCRIPTION:
*********************

MOTION-CUTTING MACRO PROGRAM FOR THE AUTOMATIC TOOLCHANGE
OF MILLING CUTTERS. A CUTTER OF DEFINED FORM IS SELECTED
BY THE USER DEFINING THE LIBRARY NO. OF THE TOOL AND THE
MAGAZINE LOCATION. THE USER ALSO DEFINES THE SPINDLE SPEED
AND THE CUTTING FEEDRATE DESIRED.

MACRO FORMAT:
*************

USE(TOOLDEMO),FILE(/TOOLLIB/),VR(#1,#2,#3,#4)

PARAMETER (VARIABLE) DESCRIPTION:
********************************

VR1=LIBRARY NO. OF TOOL SELECTED.
VR2=ATCHG MAGAZINE NO. OF SELECTED TOOL.
VR3=SPINDLE SPEED REQUIRED (RPM).
VR4=CUTTING FEEDRATE REQUIRED (MMPR).

TYPICAL APPLICATION:
*******************

USE(TOOLDEMO),FILE(/TOOLLIB/),VR(26,1,2000,0.1)

THIS STATEMENT INITIATES THE AUTOMATIC TOOLCHANGE OF
CUTTER NO. 26 FROM THE MAGAZINE. THIS TOOL IS AN END MILL
OF 8 MM DIA. AND 66.5 MM GAUGE LENGTH. THE SPINDLE SPEED
AND CUTTING FEEDRATE PROGRAMMED ARE 2000 RPM AND 0.1 MMPR
RESPECTIVELY.
```

Figure 5.29/b COMPACT II toolfile as used in the macro
        library.

```
           COMPACT II TOOL FILE USED IN THE MACRO LIBRARY
                                                  Page No: 1
+-------------------------------------------------------------------+




$ LIBRARY(/TOOLLIB/),BLOCKNAME(TOOLDEMO),

$ TOOL LIBRARY FILES

$ VR1= LIBRARY NO. OF TOOL SELECTED.

$ VR2= ATCHG MAGAZINE NO. OF SELECTED TOOL.

$ VR3=SPINDLE SPEED (RPM).

$ VR4=FEEDRATE (MMPM).

GOTO49

$ *** ROUGHING CUTTERS. ***

$ *** CYLINDRICAL FLAT BOTTOM CUTTERS WITH C/R. ***

$ ===================================================

$ *** 6 MM DIA. X 2 C/R X 60.5 GL. ***

<1>ATCHG,TOOL#2,60.5GL,6TD,TLCR2,#3RPM,#4MMPR

$ ------------------------------------

$ *** 8 MM DIA. X 2 C/R X 63.5 GL. ***

<2>ATCHG,TOOL#2,63.5GL,8TD,TLCR2,#3RPM,#4MMPR

$ ------------------------------------

$ *** 12 MM DIA. X 3 C/R X 70 MM GL. ***

<3>ATCHG,TOOL#2,70GL,12TD,TLCR3,#3RPM,#4MMPR

$ ------------------------------------

$ *** 16 MM DIA. X 4 C/R X 77 GL. ***

<4>ATCHG,TOOL#2,77GL,16TD,TLCR4,#3RPM,#4MMPR

$ ------------------------------------

$ *** 25 MM DIA. X 4 C/R X 101.5 GL. ***
```

CIM-H*

```
        COMPACT II TOOL FILE USED IN THE MACRO LIBRARY
                                              Page No: 2
+-----------------------------------------------------------------+


<5>ATCHG,TOOL#2,101.5GL,25TD,TLCR4,#3RPM,#4MMPR

$ -------------------------------------

$ *** 40 MM DIA. X 5 C/R X 105 GL. ***

<6>ATCHG,TOOL#2,105GL,40TD,TLCR5,#3RPM,#4MMPR

$ *** BALL END CUTTERS. ***

$ ------------------------------

$ *** 6 MM DIA. X 60.5 GL. ***

<11>ATCHG,TOOL#2,60.5GL,6TD,TLCR3,#3RPM,#4MMPR

$ ------------------------------

$ *** 8 MM DIA. X 63.5 GL. ***

<12>ATCHG,TOOL#2,63.5GL,8TD,TLCR4,#3RPM,#4MMPR

$ ------------------------------

$ *** 10 MM DIA. X 66.5 GL. ***

<13>ATCHG,TOOL#2,66.5GL,10TD,TLCR5,#3RPM,#4MMPR

$ ------------------------------

$ *** 12 MM DIA. X 70 GL. ***

<14>ATCHG,TOOL#2,70GL,12TD,TLCR6,#3RPM,#4MMPR

$ ------------------------------

$ *** 20 MM DIA. X 77 GL. ***

<15>ATCHG,TOOL#2,77GL,20TD,TLCR10,#3RPM,#4MMPR

$ ------------------------------

$ ***F 30 MM DIA. X 93.5 GL. ***

<16>ATCHG,TOOL#2,93.5GL,30TD,TLCR15,#3RPM,#4MMPR

$ *** END MILLS ***
```

```
COMPACT II TOOL FILE USED IN THE MACRO LIBRARY
                                    Page No: 3
+------------------------------------------------------------+


$ *** CYLINDRICAL FLAT BOTTOM CUTTERS. ***

$ =========================================

$ *** 3 MM DIA. X S/C X 54 GL. ***

<21>ATCHG,TOOL#2,54GL,3TD,#3RPM,#4MMPR

$ -----------------------------------

$ *** 4 MM DIA. X S/C X 57 GL. ***

<22>ATCHG,TOOL#2,57GL,4TD,#3RPM,#4MMPR

$ -----------------------------------

$ *** 5 MM DIA. X S/C X 60.5 GL. ***

<23>ATCHG,TOOL#2,60.5GL,5TD,#3RPM,#4MMPR

$ -----------------------------------

$ *** 6 MM DIA. X S/C X 60.5 GL. ***

<24>ATCHG,TOOL#2,60.5GL,6TD,#3RPM,#4MMPR

$ -----------------------------------

$ *** 7 MM DIA. X S/C X 60.5 GL. ***

<25>ATCHG,TOOL#2,60.5GL,7TD,#3RPM,#4MMPR

$ -----------------------------------

$ *** 8 MM DIA. X S/C X 63.5 GL. ***

<26>ATCHG,TOOL#2,63.5GL,8TD,#3RPM,#4MMPR

$ -----------------------------------

$ *** 9 MM DIA. X S/C X 66.5 GL. ***

<27>ATCHG,TOOL#2,66.5GL,9TD,#3RPM,#4MMPR

$ -----------------------------------

$ *** 10 MM DIA. X S/C X 66.5 GL. ***
```

COMPACT II TOOL FILE USED IN THE MACRO LIBRARY
                                              Page No: 4
+--+

```
<28>ATCHG,TOOL#2,66.5GL,10TD,#3RPM,#4MMPR

$ -----------------------------------

$ *** 12 MM DIA. X S/C X 70 GL. ***

<29>ATCHG,TOOL#2,70GL,12TD,#3RPM,#4MMPR

$ -----------------------------------

$ *** 16 MM DIA. X S/C X 77 GL. ***

<30>ATCHG,TOOL#2,77GL,16TD,#3RPM,#4MMPR

$ -------------------------------------

$ *** 20 MM DIA. X S/C X 83.5 GL. ***

<31>ATCHG,TOOL#2,83.5GL,20TD,#3RPM,#4MMPR

$ -----------------------------------

$ *** 25 MM DIA. X S/C X 101.5 GL. ***

<32>ATCHG,TOOL#2,101.5GL,25TD,#3RPM,#4MMPR

$ -----------------------------------

$ *** 40 MM DIA. X S/C X 117.5 GL. ***

<33>ATCHG,TOOL#2,117.5GL,40TD,#3RPM,#4MMPR

$ -----------------------------------

$ *** TOOL DEFINITIONS COMPLETE. ***

$ *** GOTO REQUIRED TOOL. ***

<49>DO#1

$ *** THIS MACRO IS COMPLETE. ***
```

Figure 5.30/a Internal profile milling macro (ICIR1) documentation.

```
ICIR1 (INTERNAL PROFILE MILLING MACRO) DOCUMENTATION
                                                    Page 1
+------------------------------------------------------------------+

              *************************
              * MACRO NAME: ICIR1     *
              *************************

                  Ver.00 Rel.01 Mod.10
                    12/March/1984

MACRO CLASSIFICATION:
*********************
GEO-MO

FUNCTINAL DESCRIPTION:
**********************

MACHINING OF INTERNAL PROFILE OF CIRCULAR POCKETS AND BORES
USING MACRO DEFINED GEOMETRY.

GEOMETRY - THE FOLLOWING GEOMETRIC ELEMENTS ARE
            DEFINED IN THIS MACRO:
            (1) CIRCLE CENTER POINT.
            (2) PROFILE CIRCLE.

 MOTION  - THE FOLLOWING TOOL MOTION IS GENERATED:
            (1) RETRACTION IN Z TO CLEARANCE-PLANE.
            (2) RAPID TRAVERSE TO CENTER POINT.
            (3) RAPID TRAVERSE IN Z TO RETRACT-PLANE.
            (4) FEED IN Z TO CUT-PLANE.
            (5) INTERNAL CONTOUR CLOCKWISE AROUND CIRCLE.
            (6) FEED TO CENTER POINT.
            (7) FEED IN Z TO RETACT-PLANE.
            (8) RAPID TRAVERSE IN Z TO CLEARANCE-PLANE.
            (9) RAPID TRAVERSE TO HOME POSITION ON
                CLEARANCE-PLANE.
            (10)RAPID TRAVERSE IN Z TO CLRP.

NOTE:
*****

COMPACT II DOES NOT ALLOW THE NESTED DEFINITION OF MORE
THAN ONE UNLABELED  GEOMETRIC ELEMENT. THE CENTER POINT
OF THE CIRCLE MUST THEREFORE BE DEFINED WITH AN IDENTITY
GIVEN IN THE CALL-UP OF THE MACRO PROGRAM. THIS POINT
IDENTITY MUST NOT BE USED ELSEWHERE IN THE PART PROGRAM.

MACRO FORMAT:
*************

USE(ICIR1),FILE(/MACLIB/),VR(#1,#2,#3,#4,#5,#6,#7)
```

ICIR1 (INTERNAL PROFILE MILLING MACRO) DOCUMENTATION

Page 2

+--

PARAMETER (VARIABLE) DESCRIPTION:
****

VR1=X COORDINATE OF THE CENTER OF THE CIRCLE(XB).
VR2=Y COORDINATE OF THE CENTER OF THE CIRCLE(YB).
VR3=RADIUS OF POCKET OR BORE.
VR4=HEIGHT OF CUT-PLANE(ZB).
VR5=HEIGHT OF RETRACT-PLANE(ZB).
VR6=HEIGHT OF CLEARANCE-PLANE(ZB).
VR7=POINT IDENTITY OF CENTER POINT REQUIRED.

Figure 5.30/b Internal profile milling macro (ICIR2) documentation.

```
      ICIR2 (INTERNAL PROFILE MILLING MACRO) DOCUMENTATION

+------------------------------------------------------------------+



                      *************************
                      * MACRO NAME: ICIR2   *
                      *************************

                          Ver.00 Rel.01 Mod.10
                            12/March/1984

MACRO CLASSIFICATION:
*********************
MO

FUNCTIONAL DESCRIPTION:
***********************

MACHINING OF INTERNAL PROFILE OF POCKETS AND BORES UTILIZING
PREVIOUSLY PART PROGRAM DEFINED CIRCLE AND CENTER POINT.

MOTION - THE FOLLOWING TOOL MOTION IS GENERATED:
         (1) RETRACTION IN Z TO CLEARANCE-PLANE.
         (2) RAPID TRAVERSE TO CENTER POINT.
         (3) RAPID TRAVERSE IN Z TO RETACT-PLANE.
         (4) FEED IN Z TO CUT-PLANE.
         (5) INTERNAL CONTOUR CLOCKWISE AROUND CIRCLE.
         (6) FEED TO CENTER POINT.
         (7) FEED IN Z TO RETRACT-PLANE.
         (8) RAPID TRAVERSE IN Z TO CLEARANCE-PLANE.
         (9) RAPID TRAVERSE TO HOME POSITION ON
             CLEARANCE-PLANE.
         (10)RAPID TRAVERSE IN Z TO CLRP.


MACRO FORMAT:
************

USE(ICIR2),FILE(/MACLIB/),VR(#1,#2,#3,#4,#5)

PARAMETER (VARIABLE) DESCRIPTION:
*********************************

VR1=HEIGHT OF CUT-PLANE(ZB).
VR2=HEIGHT OF RETRACT-PLANE(ZB).
VR3=HEIGHT OF CLEARANCE-PLANE(ZB).
VR4=POINT IDENTITY OF CIRCLE CENTER.
VR5=CIRCLE IDENTITY.
```

Figure 5.30/c Flowchart of the internal profile milling
demonstration program using the ICIR1 and ICIR2
macros and the TOOLIB.

Figure 5.30/d COMPACT II listings of the internal profile
milling macros.

```
INTERNAL PROFILE MILLING MACROS (ICIR1 AND ICIR2)
                                         Page No: 1
+----------------------------------------------------------+






$ LIBRARY(/MACLIB/),BLOCKNAME(ICIR1),INTERNAL CIRCULAR

$ POCKET & BORE PROFILING USING MACRO DEFINED GEOMETRY

$ VR1=X COORD. OF CNTR. OF POCKET(XB).

$ VR2=Y COORD. OF CNTR. OF POCKET(YB).

$ VR3=BORE RADIUS.

$ VR4=HEIGHT OF CUT-PLANE(ZB).

$ VR5=HEIGHT OF RETRACT-PLANE(ZB).

$ VR6=HEIGHT OF CLEARANCE-PLANE(ZB).

$ VR7=POINT IDENTITY OF CIRCLE CENTER REQUIRED.

$ *** GEOMETRY DEFINITION STATEMENT. ***

$ *** DEFINE CIRCLE CENTRE POINT. ***

DPT#7,#1XB,#2YB,#4ZB

$ *** GEOMETRY DEFINITION COMPLETE. ***

$ *** MOTION STATEMENTS. ***

$ *** MOVE IN Z TO CLEARANCE-PLANE. ***

MOVE,#6ZB

$ *** MOVE TO CNTR. OF POCKET ON CLEARANCE-PLANE. ***

MOVE,PT#7,NOZ

$ *** MOVE IN Z TO RETRACT-PLANE. ***

MOVE,#5ZB

$ *** FEED IN Z TO CUT-PLANE. ***

CUT,#4ZB

$ *** CONTOUR INSIDE OF POCKET. ***

ICON,CIR(PT#7,#3R),CW,S(0),F(360),#4ZB
```

INTERNAL PROFILE MILLING MACROS (ICIR1 AND ICIR2)

+--+

```
$ *** FEED TO CNTR. OF POCKET. ***

CUT,PT#7,NOZ

$ *** FEED IN Z TO RETRACT-PLANE. ***

CUT,#5ZB

$ *** RETRACT IN Z TO CLEARANCE-PLANE. ***

MOVE,#6ZB

$ *** MOVE TO HOME POSITION ON CLEARANCE-PLANE. ***

HOME,NOZ

$ *** MOVE IN Z TO HOME POSITION. ***

HOME

$ *** THIS MACRO IS COMPLETE. ***


- - - - - - - - ICIR2 listing starts here - - - - - - - - -


$ LIBRARY(/MACLIB/),BLOCKNAME(ICIR2),INTERNAL CIRCULAR

$ POCKET & BORE PROFILING USING PART PROGRAM DEFINED

$ GEOMETRY

$ VR1=HEIGHT OF CUT-PLANE(ZB).

$ VR2=HEIGHT OF RETRACT-PLANE(ZB).

$ VR3=HEIGHT OF CLEARANCE-PLANE(ZB).

$ VR4=POINT IDENTITY OF CIRCLE CNTR.

$ VR5=CIRCLE IDENTITY.

$ ***¯MOTION STATEMENTS. ***

$ *** MOVE IN Z TO CLEARANCE-PLANE. ***
```

INTERNAL PROFILE MILLING MACROS (ICIR1 AND ICIR2)
                                    Page No: 3
+---+

MOVE,#3ZB

$ *** MOVE TO CNTR. OF POCKET ON CLEARANCE-PLANE. ***

MOVE,PT#4,NOZ

$ *** MOVE IN Z TO RETRACT-PLANE. ***

MOVE,#2ZB

$ *** FEED IN Z TO CUT-PLANE. ***

CUT,#1ZB

$ *** CONTOUR INSIDE OF POCKET. ***

ICON,CIR#5,CW,S(0),F(360),#4ZB

$ *** FEED TO CNTR. OF POCKET. ***

CUT,PT#4,NOZ

$ *** FEED IN Z TO RETRACT-PLANE. ***

CUT,#2ZB

$ *** RETRACT IN Z TO CLEARANCE-PLANE. ***

MOVE,#3ZB

$ MOVE TO HOME POSITION ON CLEARANCE-PLANE. ***

HOME,NOZ

$ MOVE IN Z TO HOME POSITION. ***

HOME

*** THIS MACRO IS COMPLETE. ***

230 Computer Aided Manufacture (CAM)

Figure 5.30/e Demonstration program using the internal profile milling macros.

```
        DEMONSTRATION PROGRAM USING ICIR1 AND ICIR2 MACROS
                                          Page No: 1
+-------------------------------------------------------------------+



MACHIN,MILL

IDENT,ICIR1 & ICIR2 MACRO DEMONSTRATION

$ *** METRIC INPUT & OUTPUT. ***

INIT,METRIC/IN,METRIC/OUT

$ *** HOME POSITION. ***

SETUP,275LX,200LY,250LZ

$ *** BASE POSITIONED ON MACHINE AXES. ***

BASE,XA,YA,ZA

$ *** PLOT SCALE & POSITIONING STATEMENT. ***

DRAW,SCALE1,PT(-60XA,-50YA,-50ZA)

$ *** GEOMETRY DEFINITION FOR ICIR2 MACRO PROGRAM. ***

$ *** POINT DEFINITION. ***

DPT1,100XB,150YB,45ZB

$ *** CIRCLE DEFINITION. ***

DCIR1,PT1,50R

$ *** GEOMETRY DEFINITION COMPLETE. ***

$ *** AUTOMATIC TOOLCHANGE USING TOOLDEMO MACRO. ***

$ *** CUTTER NO. 25. ***

$ *** 8 MM DIA. S/C CYLINDRICAL CUTTER-60.5 MM G.L. ***

$ *** SPINDLE SPEED=2000 RPM ,FEEDRATE=0.1 MMPR. ***

USE(TOOLDEMO),FILE(/TOOLLIB/),VR(25,1,2000.0.1)

$ *** USE ICIR1 MACRO TO MACHINE CIRCULAR PROFILE 1. ***

USE(ICIR1),FILE(/MACLIB/),VR(200,75,60,50,75,100,2)
```

DEMONSTRATION PROGRAM USING ICIR1 AND ICIR2 MACROS
                                        Page No: 2
+--+

$ *** AUTOMATIC TOOLCHANGE USING TOOLDEMO MACRO. ***

$ *** CUTTER NO. 13. ***

$ *** 10 MM DIA. BALL END CUTTER-66.5 MM G.L. ***

$ *** SPINDLE SPEED=2000 RPM ,FEEDRATE=0.05 MMPR. ***

USE(TOOLDEMO),FILE(/TOOLLIB/),VR(13,2,2000,0.05)

$ *** USE ICIR2 MACRO TO MACHINE CIRCULAR PROFILE 2. ***

USE(ICIR2),FILE(/MACLIB/),VR(30,60,100,1,1)

$ *** PLOT IDENTIFICATION STATEMENT. ***

DRAW,PEN4

DRAW,LTR,TITLE

END

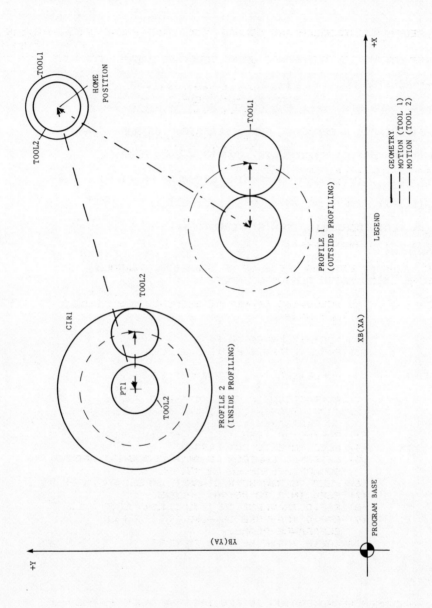

Figure 5.30/f Demonstration plot of the internal profile milling macros.

Figure 5.31/a COMPACT II rectangular and square profiling macro
(ORECT1) documentation.

```
ORECT1 (RECTANGULAR AND SQUARE PROFILING) MACRO DOCUMENTATION
                                                          Page 1
+---------------------------------------------------------------+


                    ***********************
                    * MACRO NAME: ORECT1  *
                    ***********************

                       Ver.00 Rel.01 Mod.10
                         14/March/1984

MACRO CLASSIFICATION:
*********************

GEO-MO

FUNCTIONAL DESCRIPTION:
**********************

MACHINING OF EXTERNAL PROFILE OF SQUARE/RECTANGULAR
SHAPES USING MACRO DEFINED GEOMETRY.

GEOMETRY - THE FOLLOWING GEOMETRIC ELEMENTS ARE DEFINED
           IN THIS MACRO:

           (1) POCKET CENTER POINT.
           (2) POCKET PART BOUNDARY (CLOSED).

MOTION    - THE FOLLOWING TOOL MOTION IS GENERATED:

           (1) RETRACTION IN Z TO CLEARANCE-PLANE.
           (2) RAPID TRAVERSE TO SQUARE/RECTANGLE OFFSET
               ON CLEARANCE-PLANE.
           (3) RAPID TRAVERSE IN Z TO RETRACT-PLANE.
           (4) FEED IN Z TO CUT-PLANE.
           (5) EXTERNAL CONTOUR COUNTER-CLOCKWISE AROUND
               SQUARE/RECTANGLE ON CUT-PLANE.
           (6) FEED TO SQUARE/RECTANGLE OFFSET ON CUT-PLANE.
           (7) FEED IN Z TO RETRACT-PLANE.
           (8) RAPID TRAVERSE IN Z TO CLEARANCE-PLANE.
           (9) RAPID TRAVERSE TO HOME POSITION ON
               CLEARANCE-PLANE.
           (10)RAPID TRAVERSE IN Z TO CLRP.

NOTE:
*****

IF NO CORNER BLEND RADIUS IS REQUIRED VARIABLE #8 MUST BE
DEFINED AS 0.

MACRO FORMAT:
*************
```

ORECT1 (RECTANGULAR AND SQUARE PROFILING) MACRO DOCUMENTATION
+---

USE(ORECT1),FILE(/MACLIB/),VR(#1,#2,#3,#4,#5,#6,#7,#8,#9.#10,#11

PARAMETER (VARIABLE) DESCRIPTION:
****

VR1=X COORDINATE OF LOWER RIGHT-HAND CORNER OF SHAPE(XB).
VR2=Y COORDINATE OF LOWER RIGHT-HAND CORNER OF SHAPE(YB).
VR3=X COORDINATE OF UPPER LEFT-HAND CORNER OF SHAPE(XB).
VR4=Y COORDINATE OF UPPER LEFT-HAND CORNER OF SHAPE(YB).
VR5=HEIGHT OF CUT-PLANE(ZB).
VR6=HEIGHT OF RETRACT-PLANE(ZB).
VR7=HEIGHT OF CLEARANCE-PLANE(ZB).
VR8=SQUARE/RECTANGLE CORNER BLEND RADIUS.
VR9=POINT IDENTITY OF SQUARE/RECTANGLE CENTRE REQUIRED.
VR10=SQUARE/RECTANGLE PART BOUNDARY REQUIRED.
VR11=OFFSET DISTANCE FROM SQUARE/RECTANGLE FOR INITIAL
     & FINAL MOVE.

Figure 5.31/b COMPACT II rectangular and square profiling macro (ORECT2) documentation.

```
ORECT2 (RECTANGULAR AND SQUARE PROFILING) MACRO DOCUMENTATION
                                                         Page 1
+------------------------------------------------------------------+



                    ************************
                    * MACRO NAME: ORECT2   *
                    ************************

                       Ver.00 Rel.01 Mod.9
                          14/March/1984

MACRO CLASSIFICATION:
*********************

MO

FUNCTIONAL DESCRIPTION:
***********************

MACHINING OF EXTERNAL PROFILE OF SQUARE/RECTANGULAR
SHAPES USING PREVIOUSLY PART PROGRAM DEFINED GEOMETRY.

MOTION   - THE FOLLOWING TOOL MOTION IS GENERATED:

              (1) RETRACTION IN Z TO CLEARANCE-PLANE.
              (2) RAPID TRAVERSE TO SQUARE/RECTANGLE OFFSET
                  ON CLEARANCE-PLANE.
              (3) RAPID TRAVERSE IN Z TO RETRACT-PLANE.
              (4) FEED IN Z TO CUT-PLANE.
              (5) EXTERNAL CONTOUR COUNTER-CLOCKWISE AROUND
                  SQUARE/RECTANGLE ON CUT-PLANE.
              (6) FEED TO SQUARE/RECTANGLE OFFSET ON CUT-PLANE.
              (7) FEED IN Z TO RETRACT-PLANE.
              (8) RAPID TRAVERSE IN Z TO CLEARANCE-PLANE.
              (9) RAPID TRAVERSE TO HOME POSITION ON
                  CLEARANCE-PLANE.
              (10)RAPID TRAVERSE IN Z TO CLRP.

NOTE:
*****

IF NO CORNER BLEND RADIUS IS REQUIRED VARIABLE #9 MUST BE
DEFINED AS 0.

MACRO FORMAT:
************

USE(ORECT2),FILE(/MACLIB/),VR(#1,#2,#3,#4,#5,#6,#7,#8,#9.#10,#11)

PARAMETER (VARIABLE) DESCRIPTION:
*********************************
```

ORECT2 (RECTANGULAR AND SQUARE PROFILING) MACRO DOCUMENTATION
                                                              Page
+--

VR1=HEIGHT OF CUT-PLANE(ZB).
VR2=HEIGHT OF RETRACT-PLANE(ZB).
VR3=HEIGHT OF CLEARANCE-PLANE(ZB).
VR4=IDENTITY OF SQUARE/RECTANGLE CENTER POINT.
VR5=IDENTITY OF LOWER BOUNDARY LINE.
VR6=IDENTITY OF LEFT BOUNDARY LINE.
VR7=IDENTITY OF UPPER BOUNDARY LINE.
VR8=IDENTITY OF RIGHT BOUNDARY LINE.
VR9=SQUARE/RECTANGLE CORNER BLEND RADIUS.
VR10=SQUARE/RECTANGLE PART BOUNDARY REQUIRED.
VR11=OFFSET DISTANCE FROM SQUARE/RECTANGLE FOR INITIAL
     & FINAL MOVE.

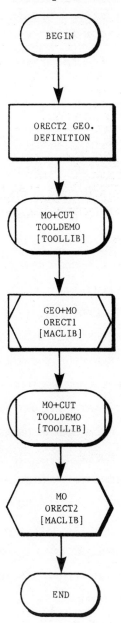

Figure 5.31/c The flowchart of the external rectangular and
square profiling macro demonstration program.

Figure 5.31/d COMPACT II source code listing of the external
      rectangular and square profiling macros.

```
              EXTERNAL RECTANGULAR AND SQUARE PROFILING MACROS
                                                    Page No: 1
+-----------------------------------------------------------------+


  $ LIBRARY(/MACLIB/),BLOCKNAME(ORECT1),EXTERNAL RECTANGULAR

  $ & SQUARE PROFILING USING MACRO DEFINED GEOMETRY

  $ VR1=X COORD. OF LOWER R/H CORNER(XB).

  $ VR2=Y COORD. OF LOWER R/H CORNER(YB).

  $ VR3=X COORD. OF UPPER L/H CORNER(XB).

  $ VR4=Y COORD. OF UPPER L/H CORNER(YB).

  $ VR5=HEIGHT OF CUT-PLANE(ZB).

  $ VR6=HEIGHT OF RETRACT-PLANE(ZB).

  $ VR7=HEIGHT OF CLEARANCE-PLANE(ZB).

  $ VR8=SQUARE/RECTANGLE CORNER BLEND RADIUS.

  $ VR9=POINT IDENTITY OF SQUARE/RECTANGLE CENTRE REQUIRED.

  $ VR10=SQUARE/RECTANGLE PART BOUNDARY IDENTITY REQUIRED.

  $ VR11=OFFSET DISTANCE FROM SQUARE/RECTANGLE FOR

  $        INITIAL & FINAL MOVE.

  $ *** GEOMETRY DEFINITION STATEMENTS. ***

  $ *** DEFINITION OF CNTR. POINT OF SQUARE/RECTANGLE. ***

  DPT#9,(#3+(#1-#3)/2)XB,(#4+(#2-#4)/2)YB,#5ZB

  $ *** DEFINITION OF SQUARE/RECTANGLE PART BOUNDARY. ***

  DPB#10,S(LN(PT#9,PARY)),LN(#2YB);LN(#3XB);LN(#4YB);LN(#1XB);

  LN(#2YB),F(LN(PT#9,PARY)),NOMORE

  $ *** GEOMETRY DEFINITION COMPLETE. ***

  $ *** MOTION STATEMENTS. ***

  $ *** MOVE IN Z TO CLEARANCE-PLANE. ***

  MOVE,#7ZB
```

EXTERNAL RECTANGULAR AND SQUARE PROFILING MACROS

Page No: 2

+--+

$ *** MOVE TO SQUARE/RECTANGLE OFFSET ON CLEARANCE-PLANE. ***

MOVE,PT#9,(#2-#11)YB,NOZ

$ *** MOVE IN Z TO RETRACT-PLANE. ***

MOVE,#6ZB

$ *** FEED IN Z TO CUT-PLANE. ***

CUT,#5ZB

$ *** CONTOUR OUTSIDE PROFILE OF SQUARE/RECTANGLE WITH ***

$ *** SPECIFIED BLEND CORNER RAD. ON CUT-PLANE. ***

CUT,PB#10/CL,#8R,#5ZB

$ *** FEED TO SQUARE/RECTANGLE OFFSET ON CUT-PLANE. ***

CUT,PT#9,(#2-#11)YB,NOZ

$ *** FEED IN Z TO RETRACT-PLANE. ***

CUT,#6ZB

$ *** RETRACT IN Z TO CLEARANCE-PLANE. ***

MOVE,#7ZB

$ *** MOVE TO HOME POSITION ON CLEARANCE-PLANE. ***

HOME,NOZ

$ *** MOVE IN Z TO HOME POSITION. ***

HOME

$ *** THIS MACRO IS COMPLETE. ***

- - - - - - - - - - - OREC2 listing starts here - - - - - -

$ LIBRARY(/MACLIB/),BLOCKNAME(ORECT2),EXTERNAL RECTANGULAR

```
          EXTERNAL RECTANGULAR AND SQUARE PROFILING MACROS
                                             Page No: 3
+-------------------------------------------------------------+



$ & SQUARE PROFILING USING PART PROGRAM DEFINED GEOMETRY

$ VR1=HEIGHT OF CUT-PLANE(ZB).

$ VR2=HEIGHT OF RETRACT-PLANE(ZB).

$ VR3=HEIGHT OF CLEARANCE-PLANE(ZB).

$ VR4=IDENTITY OF SQUARE/RECTANGLE CENTRE POINT.

$ VR5=IDENTITY OF LOWER BOUNDARY LINE.

$ VR6=IDENTITY OF LEFT BOUNDARY LINE.

$ VR7=IDENTITY OF UPPER BOUNDARY LINE.

$ VR8=IDENTITY OF RIGHT BOUNDARY LINE.

$ VR9=SQUARE/RECTANGLE CORNER BLEND RADIUS.

$ VR10=SQUARE/RECTANGLE PART BOUNDARY IDENTITY REQUIRED.

$ VR11=OFFSET DISTANCE FROM SQUARE/RECTANGLE FOR

$        INITIAL & FINAL MOVE.

$ *** GEOMETRY DEFINITION STATEMENT. ***

$ *** DEFINITION OF SQUARE/RECTANGLE PART BOUNDARY. ***

DPB#10,S(LN(PT#4,PARY)),LN#5;LN#;LN#6;LN#7;LN#8,

F(LN(PT#4,PARY),NOMORE

$ *** GEOMETRY DEFINITION COMPLETE. ***

$ *** MOTION STATEMENTS. ***

$ *** MOVE IN Z TO CLEARANCE-PLANE. ***

MOVE,#3ZB

S *** CALCULATE DISTANCE BETWEEN CENTER POINT & LOWER LINE ***

DVR75,PT#4,LN#5,DIST

$ *** MOVE TO SQUARE/RECTANGLE OFFSET ON CLEARANCE-PLANE. ***
```

EXTERNAL RECTANGULAR AND SQUARE PROFILING MACROS
                                        Page No: 4
+--+

```
MOVE,PT#4,-(#75+#11)Y,NOZ

$ *** MOVE IN Z TO RETRACT-PLANE. ***

MOVE,#2ZB

$ *** FEED IN Z TO CUT-PLANE. ***

CUT,#1ZB

$ *** CONTOUR OUTSIDE PROFILE OF SQUARE/RECTANGLE WITH  ***

$ *** SPECIFIED BLEND RADIUS ON CUT-PLANE. ***

CUT,PB#10,CL,#9R,#1ZB

$ *** FEED TO SQUARE/RECTANGLE OFFSET ON CUT-PLANE. ***

CUT,PT#4,-(#75+#11)Y,NOZ

$ *** FEED IN Z TO RETRACT-PLANE. ***

CUT,#2ZB

$ *** RETRACT IN Z TO CLEARANCE-PLANE. ***

MOVE,#3ZB

$ *** MOVE TO HOME POSITION ON CLEARANCE-PLANE. ***

HOME,NOZ

$ *** MOVE IN Z TO HOME POSITION. ***

HOME

$ *** THIS MACRO IS COMPLETE. ***
```

Figure 5.31/e Demonstration program using the external rectangular and square profiling macros.

```
DEMONSTRATION PROGRAM CALLING THE RECTANG. PROFILING MACROS
                                                   Page No: 1
+--------------------------------------------------------------+


MACHIN,MILL

IDENT,ORECT1 & ORECT2 DEMONSTRATION

$ *** METRIC INPUT & OUTPUT. ***

INIT,METRIC/IN,METRIC/OUT

$ *** HOME POSITION. ***

SETUP,275LX,175LY,250LZ

$ *** BASE POSITIONED ON MACHINE AXES. ***

BASE,XA,YA,ZA

$ *** PLOT SCALE & POSITIONING STATEMENT. ***

DRAW,SCALE1,PT(-60XA,-30YA,-50ZA)

$ *** GEOMETRY DEFINITION FOR IRECT2 MACRO PROGRAM. ***

$ *** POINT DEFINITION. ***

DPT1,100XB,125YB,45ZB

$ *** LINE DEFINITIONS. ***

DLN1,100YB

DLN2,75XB

DLN3,150YB

DLN4,125XB

$ *** GEOMETRY DEFINITION COMPLETE. ***

$ *** AUTOMATIC TOOLCHANGE USING TOOLDEMO MACRO. ***

$ *** CUTTER NO. 31. ***

$ *** 20 MM DIA. S/C CYLINDRICAL CUTTER-83.5 MM GL. ***

$ *** SPINDLE SPEED=2000 RPM ,FEEDRATE=0.1 MMPR. ***

USE(TOOLDEMO),FILE(/TOOLLIB/),VR(31,1,2000,0.1)
```

DEMONSTRATION PROGRAM CALLING THE RECTANG. PROFILING MACROS
```
                                              Page No: 2
+------------------------------------------------------------------+
```

```
$ *** USE ORECT1 MACRO TO MACHINE INSIDE OF RECTANGULAR ***

$ *** PROFILE 1. ***

$ *** X COORD. LOWER R/H CORNER=250XB. ***

$ *** Y COORD. LOWER R/H CORNER=30YB. ***

$ *** X COORD. UPPER L/H CORNER=150XB. ***

$ *** Y COORD. UPPER L/H CORNER=60YB. ***

$ *** CUT-PLANE=50ZB. ***

$ *** RETRACT-PLANE=80ZB. ***

$ *** CLEARANCE-PLANE=100ZB. ***

$ *** CORNER BLEND RADIUS=10. ***

$ *** POINT IDENTITY=2. ***

$ *** PART BOUNDARY IDENTITY=1. ***

$ *** RECTANGLE OFFSET=10. ***

USE(ORECT1),FILE(/MACLIB/),VR(250,30,150,60,50,80,100,10,2,1,10)

$ *** AUTOMATIC TOOLCHANGE USING TOOLDEMO MACRO. ***

$ *** CUTTER NO. 32. ***

$ *** 25 MM DIA. S/C CYLINDRICAL CUTTER-101.5 GL. ***

$ *** SDPINDLE SPEED=2000 RPM ,FEEDRATE=0.15 MMPR. ***

USE(TOOLDEMO),FILE(/TOOLLIB/),VR(32,2,2000,0.15)

$ *** USE ORECT2 MACRO TO MACHINE EXTERNAL RECTANGULAR ***

$ *** PROFILE 2. ***

$ *** CUT-PLANE=40ZB. ***

$ *** RETRACT-PLANE=60ZB. ***

$ *** CLEARANCE-PLANE=100ZB. ***
```

CIM-I

```
DEMONSTRATION PROGRAM CALLING THE RECTANG. PROFILING MACROS
                                              Page No: 3
+-----------------------------------------------------------------+



$ *** POINT IDENTITY=1. ***

$ *** LINE IDENTITIES=1,2,3,4. ***

$ *** CORNER BLEND RADIUS=20. ***

$ *** PART BOUNDARY IDENTITY=2. ***

$ *** RECTANGLE OFFSET=15. ***

USE(ORECT2),FILE(/MACLIB/),VR(40,60,100,1,1,2,3,4,20,2,15)

$ *** PLOT IDENTIFICATION STATEMENT. ***

DRAW,PEN4

DRAW,LTR,TITLE

END
```

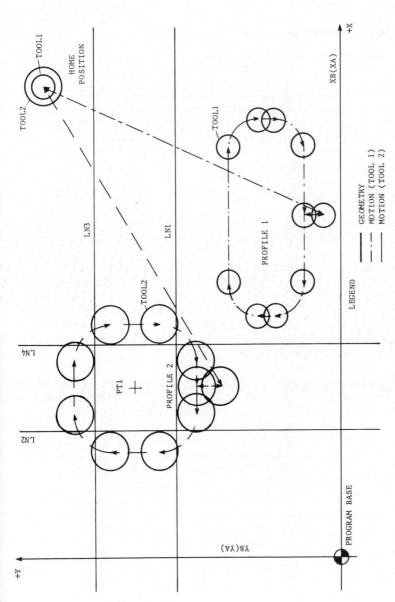

Figure 5.31/f Sample plot of the demo program using the external rectangular and square profiling macros.

When programming industrial robots it is important to think in terms of a system, consisting of several different subsystems in which the robot arm itself is only one component. Generally such a system consists of the robot arm itself, with its controller, the necessary grippers and/or robot tools, and the part feeding, locating, orientating and other mechanical, electronic and sensory based (vision, force and torque sensing, etc.) devices which are interfaced with it.

If the robot cell is integrated into a larger system a further important feature is the necessary provision of interfaces and communication facilities between the robot controller and the "outside world". It is important to realize that distributed, sensory feedback processing makes robots more intelligent, more reliable and more flexible.

Robot and manipulator programming methods include:

* Physical setup programming, in which the operator sets up programs by fixing limit switches, stops, etc.,

* Teach mode programming, when the operator leads the robot through the desired locations by means of a teach pendant which is interfaced with the robot controller. The taught locations and the approach vectors are recorded in digital format and are used as co-ordinate values and approach angles to generate the trajectory for the robot during operation.

* Teach-by-showing programming, the most commonly known method of programming paint spraying, welding and other robots involving the operator teaching the robot by using a detachable grip handle, which causes the position of the arm to be remembered by the controller. Auxiliary commands, such as spray gun On/Off, I/O handling, gripper "in and out", motion speed, etc. can be programmed from the teach-pendant.

* Off-line programming, representing the most important method for CIM applications, involves the use of a high level robot programming language. This allows writing and editing programs in a language which is closer to the operator's language than to the machine's. In some cases such programs can also be prepared remote, or off-line from the actual robot, or generated in a remote computer on the basis of the CAD/CAM database and

downloaded into the robot controller via a communications line.

Off-line programming can be based on an explicit programming language, or on a "world-modeling", (i.e. implicit or "model based") language. Explicit programming deals with motion-oriented commands, whereas "world-modeling" attempts to provide the programmer with a software environment capable of describing a handling task by giving a problem-oriented description only, and letting the robot discover by means of its sensors what kind of actual motion and action statements have to be executed.

The major benefit of the latter method is that the parts to be handled have to be described and stored as models in the database only once, and this "world-model" does not then have to be altered even if the production process changes. It should be emphasized that such systems are currently under research and only partly operational, but they will have a great impact on the assembly industry in particular when they eventually come onto the commercial market.

One must realize that as with the high-level, off-line programming languages of NC machines none of these robot programming methods provide a universal or an optimum solution in every case, thus combined programming methods must be used.

5.7 Case study: Off-line robot program generation with graphics

The off-line robot generation program shows the way "fill-in the screen" type, task oriented packages can help the robot (and other machine) programmer to solve a relatively complex task in a matter of a few minutes.

The program allows design of the pallet arrangement on the screen and is capable of generating Unimation VAL code for a general three dimensional pallet arrangement, representing the most general case. The design process is supported by graphics and the input values are checked (both the syntax as well as the semantics). The method can be applied in programming other robots.

The purpose of this case study is to illustrate the way

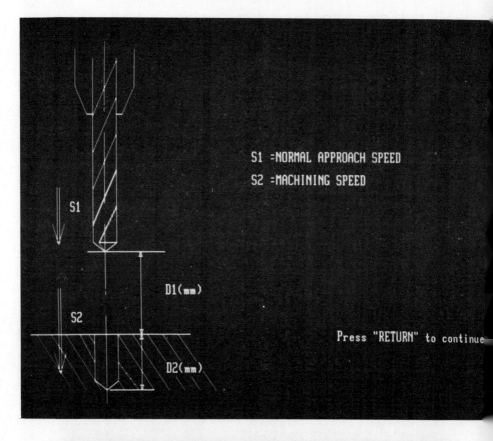

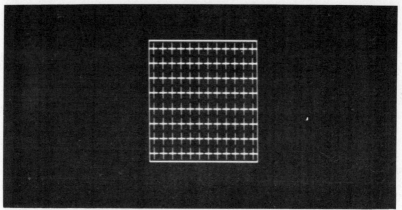

Figure 5.32 Graphics output from the robot program generator, showing the drill and the designed pattern to be drilled with the robot.

simple graphics design and off-line robot programming can be combined. The extension of this case study is to store the generated program and download it in DNC to the robot. (The further benefit is that the source can be edited, further code can be added to it etc. as required) This method not only saves programming and machine time, etc. but also improves programming reliability, which is very important in unmanned operations.

The sequence of the input data process and some of the graphics and the generated output are given in Figure 5.32. The input data to this run was as follows:

The number of holes in the X direction = 12 and 8 in the Y direction.

The hole center distance in the X direction = 5.8 mm and 6.55 mm in the Y direction.

The drilling height above the surface is 3.6 mm, and the depth of the drilled hole is 6.5 mm.

The program speed is 50 and the drilling feedrate (expressed in robot motion speed) is 40% of normal.

The generated drilling co-ordinate points and drilling order are shown in Figure 5.33 and the generated VAL code in Figure 5.34.

(Note that because of the lack of space Unimation's VAL language is not described in this text. For those who are not familiar with this language it is advisable to read [5.21] and [5.22].)

5.8 Case study: integrated CAD/CAM system. Prismatic workpiece design and manufacture using the McAuto Unigraphics CAD/CAM system

This case study is concerned with the NC programming of the same part shown in the CAD case study in the previous Chapter.

The demonstration is organized into steps, "similarly to a slide show", indicating the most important operator actions and system responses.

Figure 5.33 The generated X, Y co-ordinate points and the drilling order.

```
SUMMARY OF INPUT DATA
*********************
```

DRILLING ORDER	THESE ARE THE INTERSECTION POINTS	
	X-COORDINATES	Y-COORDINATES
1(0, 0)	0.00	0.00
2(0, 1)	0.00	6.55
3(0, 2)	0.00	13.10
4(0, 3)	0.00	19.65
5(0, 4)	0.00	26.20
6(0, 5)	0.00	32.75
7(0, 6)	0.00	39.30
8(0, 7)	0.00	45.85
9(1, 0)	5.80	0.00
10(1, 1)	5.80	6.55
11(1, 2)	5.80	13.10
12(1, 3)	5.80	19.65
13(1, 4)	5.80	26.20
14(1, 5)	5.80	32.75
15(1, 6)	5.80	39.30
16(1, 7)	5.80	45.85
17(2, 0)	11.60	0.00
18(2, 1)	11.60	6.55
19(2, 2)	11.60·	13.10
20(2, 3)	11.60	19.65
21(2, 4)	11.60	26.20
22(2, 5)	11.60	32.75
23(2, 6)	11.60	39.30
24(2, 7)	11.60	45.85
25(3, 0)	17.40	0.00
26(3, 1)	17.40	6.55
27(3, 2)	17.40	13.10
28(3, 3)	17.40	19.65
29(3, 4)	17.40	26.20
30(3, 5)	17.40	32.75
31(3, 6)	17.40	39.30
32(3, 7)	17.40	45.85
33(4, 0)	23.20	0.00
34(4, 1)	23.20	6.55
35(4, 2)	23.20	13.10
36(4, 3)	23.20	19.65
37(4, 4)	23.20	26.20
38(4, 5)	23.20	32.75
39(4, 6)	23.20	39.30
40(4, 7)	23.20	45.85
41(5, 0)	29.00	0.00
42(5, 1)	29.00	6.55
43(5, 2)	29.00	13.10
44(5, 3)	29.00	19.65
45(5, 4)	29.00	26.20
46(5, 5)	29.00	32.75
47(5, 6)	29.00	39.30
48(5, 7)	29.00	45.85
49(6, 0)	34.80	0.00
50(6, 1)	34.80	6.55
51(6, 2)	34.80	13.10
52(6, 3)	34.80	19.65

53(6, 4)	34.80	26.20
54(6, 5)	34.80	32.75
55(6, 6)	34.80	39.30
56(6, 7)	34.80	45.85
57(7, 0)	40.60	0.00
58(7, 1)	40.60	6.55
59(7, 2)	40.60	13.10
60(7, 3)	40.60	19.65
61(7, 4)	40.60	26.20
62(7, 5)	40.60	32.75
63(7, 6)	40.60	39.30
64(7, 7)	40.60	45.85
65(8, 0)	46.40	0.00
66(8, 1)	46.40	6.55
67(8, 2)	46.40	13.10
68(8, 3)	46.40	19.65
69(8, 4)	46.40	26.20
70(8, 5)	46.40	32.75
71(8, 6)	46.40	39.30
72(8, 7)	46.40	45.85
73(9, 0)	52.20	0.00
74(9, 1)	52.20	6.55
75(9, 2)	52.20	13.10
76(9, 3)	52.20	19.65
77(9, 4)	52.20	26.20
78(9, 5)	52.20	32.75
79(9, 6)	52.20	39.30
80(9, 7)	52.20	45.85
81(10, 0)	58.00	0.00
82(10, 1)	58.00	6.55
83(10, 2)	58.00	13.10
84(10, 3)	58.00	19.65
85(10, 4)	58.00	26.20
86(10, 5)	58.00	32.75
87(10, 6)	58.00	39.30
88(10, 7)	58.00	45.85
89(11, 0)	63.80	0.00
90(11, 1)	63.80	6.55
91(11, 2)	63.80	13.10
92(11, 3)	63.80	19.65
93(11, 4)	63.80	26.20
94(11, 5)	63.80	32.75
95(11, 6)	63.80	39.30
96(11, 7)	63.80	45.85

Figure 5.34 The generated VAL code.

```
PROGRAM GRID
REM **********************************************
REM * THIS VAL PROGRAM WAS CREATED BY THE        *
REM * MARTI OFF-LINE ROBOT PROGRAMMING SYSTEM    *
REM **********************************************
REM ***   THIS PROGRAM USES CENTER HOLE DIMENSIONS AS INPUT DATA
REM ***   BEFORE OPERATION THE WORK FRAME MUST BE DEFINED
GOSUB PLANE
REM ***   THIS SUBROUTINE IS USED TO DEFINE THE WORKING PLANE
REM ***   THE ROBOT IS TO EXECUTE THE MACHINING OPERATIONS
SPEED 50 ALWAYS
MOVE F1
REM ***   THIS IS THE FIRST CORNER HOLE TO BE DRILLED
SETI ROW=0
SETI COL=0
REM ***   THE X/Y COUNTERS SET TO ZERO
GOSUB DRILL
REM ***   VERTICAL OPERATION OF THE ARM FOR DRILLING
10 SETI ROW=ROW+1
IF ROW GT 7 THEN 20
REM ***   WHEN LOGIC IS TRUE GOTO LINE 20
REM ***   ELSE
TYPEI ROW
TYPEI COL
REM ***   IF ROW VALUE NOT GREATER THAN 7 THEN
REM ***   INCREMENT IN THE Y-DIRECTION BY 1.
SHIFT DOWN BY 0, 6.55 ,0
SET DRILLING=GRID:DOWN
MOVE DRILLING
DELAY 3
GOSUB DRILL
REM ***   VERTICAL OPERATION OF THE ARM FOR DRILLING
REM ***   THIS PROCEDURE NOW COMPLETED,RETURN TO LINE 10
GOTO 10
20 SETI COL=COL+1
REM ***   ROW NUMBER NOW EXCEDES INPUT,THEREFORE INCREMENT COL
IF COL GT 11 THEN 30
REM ***   IF COL VALUE NOT GREATER THAN 11 THEN
REM ***   INCREMENT THE X-DIRECTION BY 1
SHIFT DOWN BY   5.80,- 52.40,0.0
REM ***   ELSE GOTO LINE NUMBER 30
SET DRILLING=GRID:DOWN
MOVE DRILLING
GOSUB DRILL
REM ***   DRILL HOLE INCREMENTED IN THE X-DIRECTION
REM ***   RE-SET ROW COUNTER FOR RE-RUN OF PROCEDURE
SETI ROW=0
GOTO 10

30 TYPE ===== THIS IS THE END OF PROGRAM GRID =====



PROGRAM DRILL
REM **********************************************
REM * THIS SUBROUTINE IS CONCERNED WITH THE      *
REM * ACTUAL DRILLING OPERATION.                 *
REM **********************************************
REM
SHIFT DOWN BY 0,0,  3.60
REM *** MOVE   3.60MM IN THE Z-AXIS
SET DRILLING=GRID:DOWN
MOVE DRILLING
REM *** DOWNWARD MOVEMENT (ZD) BEFORE ACTUAL DRILLING.
REM *** REDUCE SPEED FOR ACTUAL DRILLING INTO MATERIAL
REM *** A DISTANCE  6.50
SHIFT DOWN BY 0,0,  6.50
```

```
SET DRILLING=GRID:DOWN
SPEED 40
REM *** SPEED NOW 40 % OF NORMAL
MOVE DRILLING
REM *** MOVE TO LOCATION
SPEED 100
REM *** SPEED BACK TO NORMAL(50).
DELAY 3
SHIFT DOWN BY 0,0, -10.10
REM *** MOVE ARM UPWARDS A DISTANCE  10.10
SET DRILLING=GRID:DOWN
MOVE DRILLING
RETURN 0
REM *** RETURN TO MAIN PROGRAM AT LINE AFTER
REM *** BRANCHING COMMAND


PROGRAM PLANE
REM *********************************************
REM * THIS SUBROUTINE IS USED TO DETERMINE THE *
REM * PLANE OF OPERATION FOR THE DRILLING      *
REM *********************************************
REM
TYPE POSITION THE ROBOT AT THE CORNER HOLE 0,0  (F1)
TYPE AT THE CORRECT HEIGHT (TOOL Z-AXIS)
GOSUB REPLY
HERE F1
TYPE POSITION THE ROBOT AT A POINT ON THE
TYPE FRAMES +VE X-AXIS(F2)
GOSUB REPLY
HERE F2
TYPE POSITION THE ROBOT AT A POINT IN THE
TYPE X/Y PLANE (F3)
GOSUB REPLY
HERE F3
FRAME GRID=F1,F2,F3
REM *** NEW FRAME DEFINED BY GRID
INVERSE DRILL=GRID
REM *** ROBOT IS NOW DEFINED WITH RESPECT TO FRAME GRID
REM *** NOW DEFINE DRILL START POINT IN GRID COORDINATES
SET HOLE=DRILL:F1
REM *** ORIGIN IS NOW SET
SET DOWN=HOLE
SET DRILLING=GRID:DOWN
REM *** NEW REFERENCE FRAME FULLY DEFINED
RETURN 0
REM *** RETURN TO MAIN PROGRAM AT LINE AFTER
REM *** BRANCHING COMMAND


PROGRAM REPLY
REM *************************************************
REM * THIS SUBROUTINE ALLOWS THE ROBOT TO BE MOVED  *
REM * TO THE DESIRED POSITION BEFORE DEFINING IT    *
REM *************************************************
REM
PAUSE *** PLEASE TYPE PROCEED [CR] TO CONTINUE
RETURN 0
```

Please follow the steps in the given order.

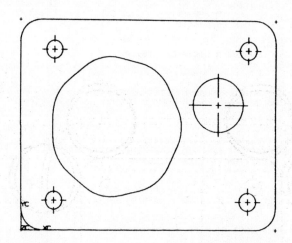

Step 1. In this case study we recall from the data base the same design we have saved in a previous session and demonstrated in the CAD Chapter.

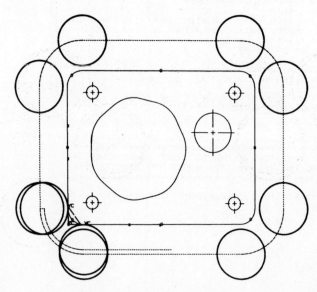

Step 2. Having described the required tool size as well as the cutting conditions the defined two roughing cuts are automatically generated for the selected boundary.

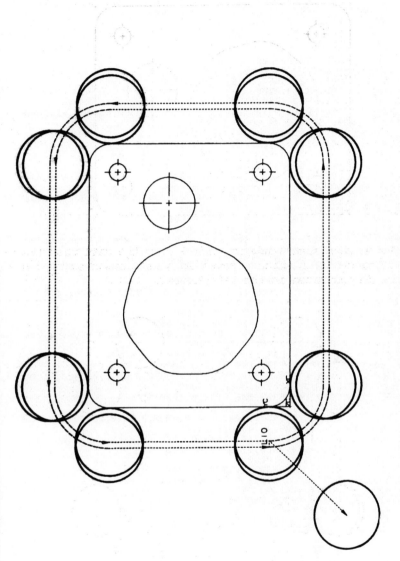

Step 3. The plotted tool path indicating two cuts around the outside contour of the part.

```
UNIGRAPHICS  CL-SOURCE FILE  3305PAULM       12--71--84


     10 * TOOL PATH/20.0000,0.5000,60.0000,P1
     20     MSYS/0.0000000,0.0000000,0.0000000,1.0000000,0.0000
000,0.0000000,0.0000000,1.0000000,0.0000000
     30     PAINT/TOOL,FULL,1
     40     PAINT/ARROW
     50     PAINT/LINNO
     60     PAINT/DASH
     70     FEDRAT/200.0000
     80     PAINT/COLOR,CYAN
     90     GOTO/-23.5580,11.9179,15.0000
    100     FEDRAT/10.0000
    110     GOTO/-23.5580,11.9179,0.0000
    120     CIRCLE/11.9416,12.0898,0.0000,0.0000,0.0000,-1.0000
,35.5000,0.0508,0.5000,40.0000,0.5000
    130     GOTO/12.2075,-23.4092,0.0000
    140     GOTO/140.7992,-22.4460,0.0000
    150     CIRCLE/140.5333,13.0530,0.0000,0.0000,0.0000,-1.000
0,35.5000,0.0508,0.5000,40.0000,0.5000
    160     GOTO/176.0333,13.0530,0.0000
    170     GOTO/176.0333,104.9192,0.0000
    180     CIRCLE/140.5333,104.9192,0.0000,0.0000,0.0000,-1.00
00,35.5000,0.0508,0.5000,40.0000,0.5000
    190     GOTO/140.5333,140.4192,0.0000
    200     GOTO/11.4921,140.4192,0.0000
    210     CIRCLE/11.4921,104.9192,0.0000,0.0000,0.0000,-1.000
0,35.5000,0.0508,0.5000,40.0000,0.5000
    220     GOTO/-24.0075,104.7473,0.0000
    230     GOTO/-23.5580,11.9179,0.0000
    240     GOTO/-20.0580,11.9348,0.0000
    250     CIRCLE/11.9416,12.0898,0.0000,0.0000,0.0000,-1.0000
,32.0000,0.0508,0.5000,40.0000,0.5000
    260     GOTO/12.1813,-19.9093,0.0000
    270     GOTO/140.7730,-18.9461,0.0000
    280     CIRCLE/140.5333,13.0530,0.0000,0.0000,0.0000,-1.000
0,32.0000,0.0508,0.5000,40.0000,0.5000
    290     GOTO/172.5333,13.0530,0.0000
    300     GOTO/172.5333,104.9192,0.0000
    310     CIRCLE/140.5333,104.9192,0.0000,0.0000,0.0000,-1.00
00,32.0000,0.0508,0.5000,40.0000,0.5000
    320     GOTO/140.5333,136.9192,0.0000
    330     GOTO/11.4921,136.9192,0.0000
    340     CIRCLE/11.4921,104.9192,0.0000,0.0000,0.0000,-1.000
0,32.0000,0.0508,0.5000,40.0000,0.5000
    350     GOTO/-20.5075,104.7643,0.0000
    360     GOTO/-20.0580,11.9348,0.0000
    370     FEDRAT/200.0000
    380     GOTO/-20.0580,11.9348,15.0000
    390     GOTO/-63.5802,-34.7915,0.0000
    400     PAINT/TOOL,NOMORE
    410     PAINT/ARROW,NOMORE
    420     PAINT/LINNO,NOMORE
    430     PAINT/DASH,NOMORE
    440 * END-OF-PATH
```

Step 4. Listing of the automatically generated UNIAPT program.
It is important to mention that some post-processor
words must be added to this program manually before
generating the NC tape file using a selected post-
processor.

UNIGRAPHICS CL-SOURCE FILE 3305PAULM   12--71--84

```
        1    PARTNO/PAULM
☞       2    MACHIN/50,OPTION,106,50,51
        4    FROM/-25,-25,50
       10  * TOOL PATH/20.0000,0.5000,60.0000,P1
       20      MSYS/0.0000000,0.0000000,0.0000000,1.0000000,0.0000
000,0.0000000,0.0000000,1.0000000,0.0000000
       30      PAINT/TOOL,FULL,1
       40      PAINT/ARROW
       50      PAINT/LINNO
       60      PAINT/DASH
       70      FEDRAT/200.0000
       80      PAINT/COLOR,CYAN
       90      GOTO/-23.5580,11.9179,15.0000
      100      FEDRAT/10.0000
      110      GOTO/-23.5580,11.9179,0.0000
      120      CIRCLE/11.9416,12.0898,0.0000,0.0000,0.0000,-1.0000
,35.5000,0.0508,0.5000,40.0000,0.5000
      130      GOTO/12.2075,-23.4092,0.0000
      140      GOTO/140.7992,-22.4460,0.0000
      150      CIRCLE/140.5333,13.0530,0.0000,0.0000,0.0000,-1.000
0,35.5000,0.0508,0.5000,40.0000,0.5000
      160      GOTO/176.0333,13.0530,0.0000
      170      GOTO/176.0333,104.9192,0.0000
      180      CIRCLE/140.5333,104.9192,0.0000,0.0000,0.0000,-1.00
00,35.5000,0.0508,0.5000,40.0000,0.5000
      190      GOTO/140.5333,140.4192,0.0000
      200      GOTO/11.4921,140.4192,0.0000
      210      CIRCLE/11.4921,104.9192,0.0000,0.0000,0.0000,-1.000
0,35.5000,0.0508,0.5000,40.0000,0.5000
      220      GOTO/-24.0075,104.7473,0.0000
      230      GOTO/-23.5580,11.9179,0.0000
      240      GOTO/-20.0580,11.9348,0.0000
      250      CIRCLE/11.9416,12.0898,0.0000,0.0000,0.0000,-1.0000
,32.0000,0.0508,0.5000,40.0000,0.5000
      260      GOTO/12.1813,-19.9093,0.0000
      270      GOTO/140.7730,-18.9461,0.0000
      280      CIRCLE/140.5333,13.0530,0.0000,0.0000,0.0000,-1.000
0,32.0000,0.0508,0.5000,40.0000,0.5000
      290      GOTO/172.5333,13.0530,0.0000
      300      GOTO/172.5333,104.9192,0.0000
      310      CIRCLE/140.5333,104.9192,0.0000,0.0000,0.0000,-1.00
00,32.0000,0.0508,0.5000,40.0000,0.5000
      320      GOTO/140.5333,136.9192,0.0000
      330      GOTO/11.4921,136.9192,0.0000
      340      CIRCLE/11.4921,104.9192,0.0000,0.0000,0.0000,-1.000
0,32.0000,0.0508,0.5000,40.0000,0.5000
      350      GOTO/-20.5075,104.7643,0.0000
      360      GOTO/-20.0580,11.9348,0.0000
      370      FEDRAT/200.0000
      380      GOTO/-20.0580,11.9348,15.0000
      390      GOTO/-63.5802,-34.7915,0.0000
      400      PAINT/TOOL,NOMORE
      410      PAINT/ARROW,NOMORE
      420      PAINT/LINNO,NOMORE
      430      PAINT/DASH,NOMORE
      440  * END-OF-PATH
☞     441    FINI
```

Step 5. The manually edited UNIAPT listing. Note the special
☞ marks, indicating where changes had to be made.

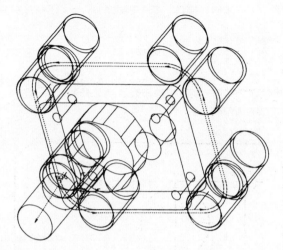

Step 6. Three dimensional tool path generation, using the manually modified UNIAPT program. Note the effect of the FROM statement.

```
L,PAULM


FILE - PAULM!LIB1

000010  GIGAA!FIOII'IIIIF!AAAAO'IOIII!!!!!''!!!!!!!!!!!!!!'!!!!!
        !!!!!!!!!''!!!!!!!!!!!!!!!!!!
000020  !!!!!!!!!!!!!!!!!!!!!!!!!!!!!!!!!'''''''''!!!!!!!!!!!!!'!!!!
        !!!!!!!!!!!!!!!!!!!!!!!!!!!!!
000030  !!!!!!!!!!!!!!!!!!!!!'X<$
000040  @PAULM<$
000050  N0010G71<$
000060  N0020G01X-23.558Y11.918Z15.F200.<$
000070  N0030Z.0F10.<$
000080  N0040G17<$
000090  N0050G03X12.208Y-23.409I35.5J.172<$
000100  N0060G01X140.799Y-22.446<$
000110  N0070G03X176.033Y13.053I.266J35.499<$
000120  N0080G01Y104.919<$
000130  N0090G03X140.533Y140.419I35.5J.0<$
000140  N0100G01X11.492<$
000150  N0110G03X-24.007Y104.747I.0J35.5<$
000160  N0120G01X-23.558Y11.918<$
000170  N0130X-20.058Y11.935<$
000180  N0140G03X12.181Y-19.909I32.J.155<$
000190  N0150G01X140.773Y-18.946<$
000200  N0160G03X172.533Y13.053I.24J31.999<$
000210  N0170G01Y104.919<$
000220  N0180G03X140.533Y136.919I32.J.0<$
000230  N0190G01X11.492<$
000240  N0200G03X-20.508Y104.764I.0J32.<$
000250  N0210G01X-20.058Y11.935<$
000260  N0220Z15.F200.<$
000270  N0230X-63.58Y-34.792Z.0<$
```

Step 7. The generated NC tape file for the outside contouring job. Note that the first three blocks contain unprintable paper tape header data and that the NC program termination "M30" code was not generated by this post-processor.

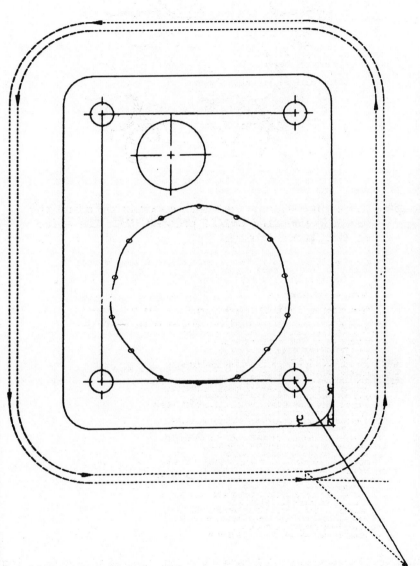

Step 8. The tool path of the drilling operations on the part.

```
UNIGRAPHICS  CL-SOURCE FILE  330SPAULM       12--71--84

    10    PARTNO/PAULM
    20    MACHIN/50,OPTION,106,50,51
    30    FROM/-25,-25,50
    40  * TOOL PATH/20.0000,0.5000,60.0000,P1
    50      MSYS/0.0000000,0.0000000,0.0000000,1.0000000,0.0000
000,0.0000000,0.0000000,1.0000000,0.0000000
    60    PAINT/TOOL,FULL,1
    70    PAINT/ARROW
    80    PAINT/LINNO
    90    PAINT/DASH
   100    FEDRAT/200.0000
   110    PAINT/COLOR,CYAN
   120    GOTO/-23.5580,11.9179,15.0000
   130    FEDRAT/10.0000
   140    GOTO/-23.5580,11.9179,0.0000
   150    CIRCLE/11.9416,12.0898,0.0000,0.0000,0.0000,-1.0000
,35.5000,0.0508,0.5000,40.0000,0.5000
   160    GOTO/12.2075,-23.4092,0.0000
   170    GOTO/140.7992,-22.4460,0.0000
   180    CIRCLE/140.5333,13.0530,0.0000,0.0000,0.0000,-1.000
0,35.5000,0.0508,0.5000,40.0000,0.5000
   190    GOTO/176.0333,13.0530,0.0000
   200    GOTO/176.0333,104.9192,0.0000
   210    CIRCLE/140.5333,104.9192,0.0000,0.0000,0.0000,-1.00
00,35.5000,0.0508,0.5000,40.0000,0.5000
   220    GOTO/140.5333,140.4192,0.0000
   230    GOTO/11.4921,140.4192,0.0000
   240    CIRCLE/11.4921,104.9192,0.0000,0.0000,0.0000,-1.000
0,35.5000,0.0508,0.5000,40.0000,0.5000
   250    GOTO/-24.0075,104.7473,0.0000
   260    GOTO/-23.5580,11.9179,0.0000
   270    GOTO/-20.0580,11.9348,0.0000
   280    CIRCLE/11.9416,12.0898,0.0000,0.0000,0.0000,-1.0000
,32.0000,0.0508,0.5000,40.0000,0.5000
   290    GOTO/12.1813,-19.9093,0.0000
   300    GOTO/140.7730,-18.9461,0.0000
   310    CIRCLE/140.5333,13.0530,0.0000,0.0000,0.0000,-1.000
0,32.0000,0.0508,0.5000,40.0000,0.5000
   320    GOTO/172.5333,13.0530,0.0000
   330    GOTO/172.5333,104.9192,0.0000
   340    CIRCLE/140.5333,104.9192,0.0000,0.0000,0.0000,-1.00
00,32.0000,0.0508,0.5000,40.0000,0.5000
   350    GOTO/140.5333,136.9192,0.0000
   360    GOTO/11.4921,136.9192,0.0000
   370    CIRCLE/11.4921,104.9192,0.0000,0.0000,0.0000,-1.000
0,32.0000,0.0508,0.5000,40.0000,0.5000
   380    GOTO/-20.5075,104.7643,0.0000
   390    GOTO/-20.0580,11.9348,0.0000
   400    FEDRAT/200.0000
   410    GOTO/-20.0580,11.9348,15.0000
   420    GOTO/-63.5802,-34.7915,0.0000
   430    PAINT/TOOL,NOMORE
   440    PAINT/ARROW,NOMORE
   450    PAINT/LINNO,NOMORE
   460    PAINT/DASH,NOMORE
   470  * END-OF-PATH
   480    FINI
   490  * TOOL PATH/5.0000,0.0000,45.0000,P2
   500      GOTO/19.6186,16.7572,0.0000
   510      GOTO/19.9994,100.5429,0.0000
   520      GOTO/136.1570,100.5429,0.0000
   530      GOTO/135.7761,16.7572,0.0000
   540  * END-OF-PATH
```

Step 9. The updated UNIAPT file, containing the milling
operations and the added drilling operations.

UNIGRAPHICS CL-SOURCE FILE 3305PAULM   12--71--84

```
        10   PARTNO/PAULM
        20   MACHIN/50,OPTION,106,50,51
        21    LOAD/TOOL,1
        30   FROM/-25,-25,50
        40 * TOOL PATH/20.0000,0.5000,60.0000,P1
        50     MSYS/0.0000000,0.0000000,0.0000000,1.0000000,0.0000
000,0.0000000,0.0000000,1.0000000,0.0000000
        60     PAINT/TOOL,FULL,1
        70     PAINT/ARROW
        80     PAINT/LINNO
        90     PAINT/DASH
       100   FEDRAT/200.0000
       110   PAINT/COLOR,CYAN
       120   GOTO/-23.5580,11.9179,15.0000
       130   FEDRAT/10.0000
       140   GOTO/-23.5580,11.9179,0.0000
       150   CIRCLE/11.9416,12.0898,0.0000,0.0000,0.0000,-1.0000
,35.5000,0.0508,0.5000,40.0000,0.5000
       160   GOTO/12.2075,-23.4092,0.0000
       170   GOTO/140.7992,-22.4460,0.0000
       180   CIRCLE/140.5333,13.0530,0.0000,0.0000,0.0000,-1.000
0,35.5000,0.0508,0.5000,40.0000,0.5000
       190   GOTO/176.0333,13.0530,0.0000
       200   GOTO/176.0333,104.9192,0.0000
       210   CIRCLE/140.5333,104.9192,0.0000,0.0000,0.0000,-1.00
00,35.5000,0.0508,0.5000,40.0000,0.5000
       220   GOTO/140.5333,140.4192,0.0000
       230   GOTO/11.4921,140.4192,0.0000
       240   CIRCLE/11.4921,104.9192,0.0000,0.0000,0.0000,-1.000
0,35.5000,0.0508,0.5000,40.0000,0.5000
       250   GOTO/-24.0075,104.7473,0.0000
       260   GOTO/-23.5580,11.9179,0.0000
       270   GOTO/-20.0580,11.9348,0.0000
       280   CIRCLE/11.9416,12.0898,0.0000,0.0000,0.0000,-1.0000
,32.0000,0.0508,0.5000,40.0000,0.5000
       290   GOTO/12.1813,-19.9093,0.0000
       300   GOTO/140.7730,-18.9461,0.0000
       310   CIRCLE/140.5333,13.0530,0.0000,0.0000,0.0000,-1.000
0,32.0000,0.0508,0.5000,40.0000,0.5000
       320   GOTO/172.5333,13.0530,0.0000
       330   GOTO/172.5333,104.9192,0.0000
       340   CIRCLE/140.5333,104.9192,0.0000,0.0000,0.0000,-1.00
00,32.0000,0.0508,0.5000,40.0000,0.5000
       350   GOTO/140.5333,136.9192,0.0000
       360   GOTO/11.4921,136.9192,0.0000
       370   CIRCLE/11.4921,104.9192,0.0000,0.0000,0.0000,-1.000
0,32.0000,0.0508,0.5000,40.0000,0.5000
       380   GOTO/-20.5075,104.7643,0.0000
       390   GOTO/-20.0580,11.9348,0.0000
       400   FEDRAT/200.0000
       410   GOTO/-20.0580,11.9348,15.0000
       420   GOTO/-63.5802,-34.7915,0.0000
       430   PAINT/TOOL,NOMORE
       440   PAINT/ARROW,NOMORE
       450   PAINT/LINNO,NOMORE
       460   PAINT/DASH,NCMORE
       470 * END-OF-PATH
       480   GOHOME
       482   LOAD/TOOL,2
       490 * TOOL PATH/5.0000,0.0000,45.0000,P2
       500   GOTO/19.6186,16.7572,0.0000
       510   GOTO/19.9994,100.5429,0.0000
       520   GOTO/136.1570,100.5429,0.0000
       530   GOTO/135.7761,16.7572,0.0000
       540 * END-OF-PATH
       550   GOHOME
       560   FINI
```

Step 10. The manually edited and corrected UNIAPT file. Note
        the post-processor words we had to add to the listing.

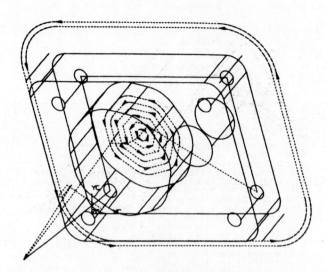

Step 11. The automatically generated three dimensional tool
path indicating the outside contouring, the drilling
and the inside contouring operations.

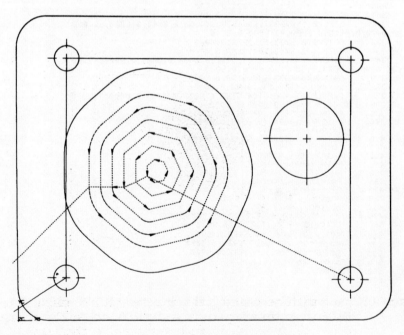

Step 12. Top view of the generated tool path.

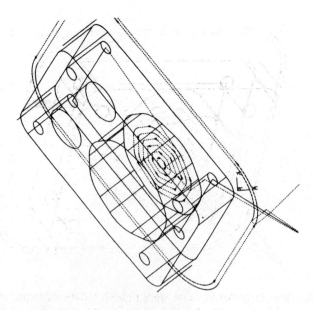

Step 13. A different three dimensional view of the generated
tool path.

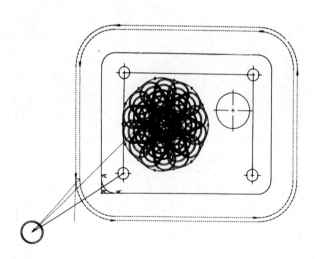

Step 14. Tool path generation showing the tool diameter.
Note that this tool path is generated on the screen
dynamically which is more informative than this
relatively overcrowded static image.

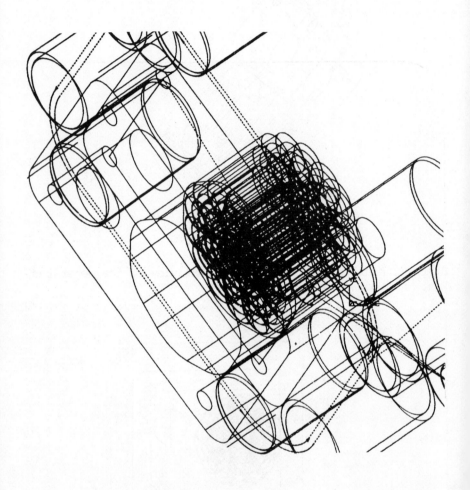

Step 15. Three dimensional view of the tool path showing the tool in 3D. Note that this tool path is generated on the screen dynamically which is more informative than this static image.

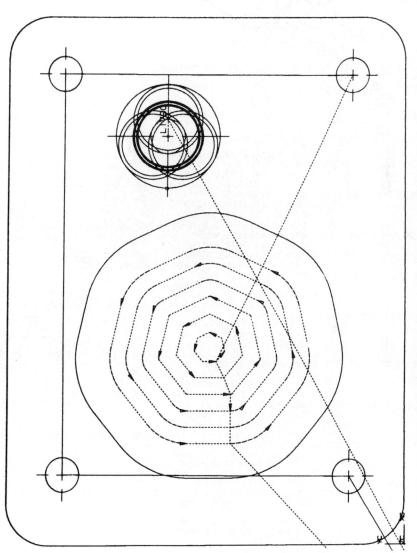

Step 16. This plot shows all operations perfomed on the part.

Step 17. Listing of the automatically generated UNIAPT program. Note that the system has added the remaining two operation programs to the current (i.e. manually edited) UNIAPT file.

UNIGRAPHICS CL-SOURCE FILE 3305PAULM 12--71--84

```
 10    PARTNO/PAULM
 20    MACHIN/50,OPTION,106,50,51
 21      LOAD/TOOL,1
 30    FROM/-25,-25,50
 40  * TOOL PATH/20.0000,0.5000,60.0000,P1
 50      MSYS/0.0000000,0.0000000,0.0000000,1.0000000,0.0000
000,0.0000000,0.0000000,1.0000000,0.0000000
 60    PAINT/TOOL,FULL,1
 70    PAINT/ARROW
 80    PAINT/LINNO
 90    PAINT/DASH
100    FEDRAT/200.0000
110    PAINT/COLOR,CYAN
120    GOTO/-23.5580,11.9179,15.0000
130    FEDRAT/10.0000
140    GOTO/-23.5580,11.9179,0.0000
150    CIRCLE/11.9416,12.0898,0.0000,0.0000,0.0000,-1.0000
,35.5000,0.0508,0.5000,40.0000,0.5000
160    GOTO/12.2075,-23.4092,0.0000
170    GOTO/140.7992,-22.4460,0.0000
180    CIRCLE/140.5333,13.0530,0.0000,0.0000,0.0000,-1.000
0,35.5000,0.0508,0.5000,40.0000,0.5000
190    GOTO/176.0333,13.0530,0.0000
200    GOTO/176.0333,104.9192,0.0000
210    CIRCLE/140.5333,104.9192,0.0000,0.0000,0.0000,-1.00
00,35.5000,0.0508,0.5000,40.0000,0.5000
220    GOTO/140.5333,140.4192,0.0000
230    GOTO/11.4921,140.4192,0.0000
240    CIRCLE/11.4921,104.9192,0.0000,0.0000,0.0000,-1.000
0,35.5000,0.0508,0.5000,40.0000,0.5000
250    GOTO/-24.0075,104.7473,0.0000
260    GOTO/-23.5580,11.9179,0.0000
270    GOTO/-20.0580,11.9348,0.0000
280    CIRCLE/11.9416,12.0898,0.0000,0.0000,0.0000,-1.0000
,32.0000,0.0508,0.5000,40.0000,0.5000
290    GOTO/12.1813,-19.9093,0.0000
300    GOTO/140.7730,-18.9461,0.0000
310    CIRCLE/140.5333,13.0530,0.0000,0.0000,0.0000,-1.000
0,32.0000,0.0508,0.5000,40.0000,0.5000
320    GOTO/172.5333,13.0530,0.0000
330    GOTO/172.5333,104.9192,0.0000
340    CIRCLE/140.5333,104.9192,0.0000,0.0000,0.0000,-1.00
00,32.0000,0.0508,0.5000,40.0000,0.5000
350    GOTO/140.5333,136.9192,0.0000
360    GOTO/11.4921,136.9192,0.0000
370    CIRCLE/11.4921,104.9192,0.0000,0.0000,0.0000,-1.000
0,32.0000,0.0508,0.5000,40.0000,0.5000
380    GOTO/-20.5075,104.7643,0.0000
390    GOTO/-20.0580,11.9348,0.0000
400    FEDRAT/200.0000
410    GOTO/-20.0580,11.9348,15.0000
420    GOTO/-63.5802,-34.7915,0.0000
430    PAINT/TOOL,NOMORE
440    PAINT/ARROW,NOMORE
450    PAINT/LINNO,NOMORE
460    PAINT/DASH,NOMORE
470  * END-OF-PATH
480    GOHOME
482      LOAD/TOOL,2
490  * TOOL PATH/5.0000,0.0000,45.0000,P2
500    GOTO/19.6186,16.7572,0.0000
```

```
510       GOTO/19.9994,100.5429,0.0000
520       GOTO/136.1570,100.5429,0.0000
530       GOTO/135.7761,16.7572,0.0000
540   * END-OF-PATH
550       GOHOME
560       FINI
570   * TOOL PATH/10.0000,1.0000,45.0000,P3
580       MSYS/0.0000000,0.0000000,0.0000000,1.0000000,0.0000
000,0.0000000,0.0000000,1.0000000,0.0000000
590       PAINT/TOOL,FULL,1
600       PAINT/ARROW
610       PAINT/LINNO
620       PAINT/DASH
630       FEDRAT/200.0000
640       PAINT/COLOR,CYAN
650       GOTO/52.9132,55.9485,15.0000
660       FEDRAT/10.0000
670       GOTO/52.9132,55.9485,0.0000
680       GOTO/55.9465,53.5295,0.0000
690       GOTO/59.7289,54.3928,0.0000
700       GOTO/61.4123,57.8884,0.0000
710       GOTO/59.7289,61.3839,0.0000
720       GOTO/55.9465,62.2472,0.0000
730       GOTO/52.9132,59.8282,0.0000
740       GOTO/52.9132,55.9485,0.0000
750       GOTO/48.1507,53.6550,0.0000
760       GOTO/54.7702,48.3760,0.0000
770       GOTO/63.0247,50.2601,0.0000
780       GOTO/66.6983,57.8884,0.0000
790       GOTO/63.0247,65.5166,0.0000
800       GOTO/54.7702,67.4007,0.0000
810       GOTO/48.1507,62.1217,0.0000
820       GOTO/48.1507,53.6550,0.0000
830       GOTO/43.3882,51.7046,0.0000
840       CIRCLE/44.1007,51.7046,0.0000,0.0000,0.0000,-1.0000
,0.7125,0.0508,0.5000,20.0000,1.0000
850       GOTO/43.6564,51.1475,0.0000
860       GOTO/53.3257,43.4365,0.0000
870       CIRCLE/53.7700,43.9936,0.0000,0.0000,0.0000,-1.0000
,0.7125,0.0508,0.5000,20.0000,1.0000
880       GOTO/53.9285,43.2990,0.0000
890       GOTO/65.9859,46.0510,0.0000
900       CIRCLE/65.8274,46.7456,0.0000,0.0000,0.0000,-1.0000
,0.7125,0.0508,0.5000,20.0000,1.0000
910       GOTO/66.4693,46.4365,0.0000
920       GOTO/71.8354,57.5792,0.0000
930       CIRCLE/71.1934,57.8884,0.0000,0.0000,0.0000,-1.0000
,0.7125,0.0508,0.5000,20.0000,1.0000
940       GOTO/71.8354,58.1975,0.0000
950       GOTO/66.4693,69.3402,0.0000
960       CIRCLE/65.8274,69.0311,0.0000,0.0000,0.0000,-1.0000
,0.7125,0.0508,0.5000,20.0000,1.0000
970       GOTO/65.9859,69.7257,0.0000
980       GOTO/53.9285,72.4778,0.0000
990       CIRCLE/53.7700,71.7831,0.0000,0.0000,0.0000,-1.0000
,0.7125,0.0508,0.5000,20.0000,1.0000
1000      GOTO/53.3257,72.3402,0.0000
1010      GOTO/43.6564,64.6292,0.0000
1020      CIRCLE/44.1007,64.0721,0.0000,0.0000,0.0000,-1.0000
,0.7125,0.0508,0.5000,20.0000,1.0000
1030      GOTO/43.3882,64.0721,0.0000
1040      GOTO/43.3882,51.7046,0.0000
1050      GOTO/38.6257,51.7046,0.0000
1060      CIRCLE/44.1007,51.7046,0.0000,0.0000,0.0000,-1.0000
```

```
,5.4750,0.0508,0.5000,20.0000,1.0000
      1070      GOTO/40.6870,47.4241,0.0000
      1080      GOTO/50.3564,39.7131,0.0000
      1090      CIRCLE/53.7700,43.9936,0.0000,0.0000,0.0000,-1.0000
,5.4750,0.0508,0.5000,20.0000,1.0000
      1100      GOTO/54.9883,38.6559,0.0000
      1110      GOTO/67.0457,41.4079,0.0000
      1120      CIRCLE/65.8274,46.7456,0.0000,0.0000,0.0000,-1.0000
,5.4750,0.0508,0.5000,20.0000,1.0000
      1130      GOTO/70.7602,44.3701,0.0000
      1140      GOTO/76.1262,55.5128,0.0000
      1150      CIRCLE/71.1934,57.8884,0.0000,0.0000,0.0000,-1.0000
,5.4750,0.0508,0.5000,20.0000,1.0000
      1160      GOTO/76.1262,60.2639,0.0000
      1170      GOTO/70.7602,71.4066,0.0000
      1180      CIRCLE/65.8274,69.0311,0.0000,0.0000,0.0000,-1.0000
,5.4750,0.0508,0.5000,20.0000,1.0000
      1190      GOTO/67.0457,74.3688,0.0000
      1200      GOTO/54.9883,77.1208,0.0000
      1210      CIRCLE/53.7700,71.7831,0.0000,0.0000,0.0000,-1.0000
,5.4750,0.0508,0.5000,20.0000,1.0000
      1220      GOTO/50.3564,76.0636,0.0000
      1230      GOTO/40.6870,68.3526,0.0000
      1240      CIRCLE/44.1007,64.0721,0.0000,0.0000,0.0000,-1.0000
,5.4750,0.0508,0.5000,20.0000,1.0000
      1250      GOTO/38.6257,64.0721,0.0000
      1260      GOTO/38.6257,51.7046,0.0000
      1270      GOTO/33.8632,51.7046,0.0000
      1280      CIRCLE/44.1007,51.7046,0.0000,0.0000,0.0000,-1.0000
,10.2375,0.0508,0.5000,20.0000,1.0000
      1290      GOTO/37.7177,43.7006,0.0000
      1300      GOTO/47.3870,35.9896,0.0000
      1310      CIRCLE/53.7700,43.9936,0.0000,0.0000,0.0000,-1.0000
,10.2375,0.0508,0.5000,20.0000,1.0000
      1320      GOTO/56.0480,34.0128,0.0000
      1330      GOTO/68.1054,36.7648,0.0000
      1340      CIRCLE/65.8274,46.7456,0.0000,0.0000,0.0000,-1.0000
,10.2375,0.0508,0.5000,20.0000,1.0000
      1350      GOTO/75.0511,42.3037,0.0000
      1360      GOTO/80.4171,53.4465,0.0000
      1370      CIRCLE/71.1934,57.8884,0.0000,0.0000,0.0000,-1.0000
,10.2375,0.0508,0.5000,20.0000,1.0000
      1380      GOTO/80.4171,62.3302,0.0000
      1390      GOTO/75.0511,73.4730,0.0000
      1400      CIRCLE/65.8274,69.0311,0.0000,0.0000,0.0000,-1.0000
,10.2375,0.0508,0.5000,20.0000,1.0000
      1410      GOTO/68.1054,79.0119,0.0000
      1420      GOTO/56.0480,81.7639,0.0000
      1430      CIRCLE/53.7700,71.7831,0.0000,0.0000,0.0000,-1.0000
,10.2375,0.0508,0.5000,20.0000,1.0000
      1440      GOTO/47.3870,79.7871,0.0000
      1450      GOTO/37.7177,72.0761,0.0000
      1460      CIRCLE/44.1007,64.0721,0.0000,0.0000,0.0000,-1.0000
,10.2375,0.0508,0.5000,20.0000,1.0000
      1470      GOTO/33.8632,64.0721,0.0000
      1480      GOTO/33.8632,51.7046,0.0000
      1490      GOTO/29.1007,51.7046,0.0000
      1500      CIRCLE/44.1007,51.7046,0.0000,0.0000,0.0000,-1.0000
,15.0000,0.0508,0.5000,20.0000,1.0000
      1510      GOTO/34.7483,39.9771,0.0000
      1520      GOTO/44.4176,32.2661,0.0000
      1530      CIRCLE/53.7700,43.9936,0.0000,0.0000,0.0000,-1.0000
,15.0000,0.0508,0.5000,20.0000,1.0000
      1540      GOTO/57.1078,29.3697,0.0000
```

```
1550      GOTO/69.1652,32.1217,0.0000
1560      CIRCLE/65.8274,46.7456,0.0000,0.0000,0.0000,-1.0000
,15.0000,0.0508,0.5000,20.0000,1.0000
1570      GOTO/79.3419,40.2374,0.0000
1580      GOTO/84.7080,51.3801,0.0000
1590      CIRCLE/71.1934,57.8884,0.0000,0.0000,0.0000,-1.0000
,15.0000,0.0508,0.5000,20.0000,1.0000
1600      GOTO/84.7080,64.3966,0.0000
1610      GOTO/79.3419,75.5393,0.0000
1620      CIRCLE/65.8274,69.0311,0.0000,0.0000,0.0000,-1.0000
,15.0000,0.0508,0.5000,20.0000,1.0000
1630      GOTO/69.1652,83.6550,0.0000
1640      GOTO/57.1078,86.4070,0.0000
1650      CIRCLE/53.7700,71.7831,0.0000,0.0000,0.0000,-1.0000
,15.0000,0.0508,0.5000,20.0000,1.0000
1660      GOTO/44.4176,83.5106,0.0000
1670      GOTO/34.7483,75.7996,0.0000
1680      CIRCLE/44.1007,64.0721,0.0000,0.0000,0.0000,-1.0000
,15.0000,0.0508,0.5000,20.0000,1.0000
1690      GOTO/29.1007,64.0721,0.0000
1700      GOTO/29.1007,51.7046,0.0000
1710      FEDRAT/200.0000
1720      GOTO/29.1007,51.7046,15.0000
1730      GOTO/-63.5802,-34.7915,0.0000
1740      PAINT/TOOL,NOMORE
1750      PAINT/ARROW,NOMORE
1760      PAINT/LINNO,NOMORE
1770      PAINT/DASH,NOMORE
1780 *  END-OF-PATH
1790 *  TOOL PATH/10.0000,1.0000,45.0000,P4
1800      MSYS/0.0000000,0.0000000,0.0000000,1.0000000,0.0000
000,0.0000000,0.0000000,1.0000000,0.0000000
1810      PAINT/TOOL,FULL,1
1820      PAINT/ARROW
1830      PAINT/LINNO
1840      PAINT/DASH
1850      FEDRAT/200.0000
1860      PAINT/COLOR,CYAN
1870      GOTO/118.4948,70.4562,15.0000
1880      FEDRAT/10.0000
1890      GOTO/118.4948,70.4562,0.0000
1900      CIRCLE/118.2573,70.4562,0.0000,0.0000,0.0000,-1.000
0,0.2375,0.0508,0.5000,20.0000,1.0000
1910      GOTO/118.1390,70.6621,0.0000
1920      CIRCLE/118.2573,70.4562,0.0000,0.0000,0.0000,-1.000
0,0.2375,0.0508,0.5000,20.0000,1.0000
1930      GOTO/118.1383,70.2507,0.0000
1940      CIRCLE/118.2573,70.4562,0.0000,0.0000,0.0000,-1.000
0,0.2375,0.0508,0.5000,20.0000,1.0000
1950      GOTO/118.4948,70.4562,0.0000
1960      GOTO/123.2573,70.4562,0.0000
1970      CIRCLE/118.2573,70.4562,0.0000,0.0000,0.0000,-1.000
0,5.0000,0.0508,0.5000,20.0000,1.0000
1980      GOTO/115.7664,74.7916,0.0000
1990      CIRCLE/118.2573,70.4562,0.0000,0.0000,0.0000,-1.000
0,5.0000,0.0508,0.5000,20.0000,1.0000
2000      GOTO/115.7528,66.1287,0.0000
2010      CIRCLE/118.2573,70.4562,0.0000,0.0000,0.0000,-1.000
0,5.0000,0.0508,0.5000,20.0000,1.0000
2020      GOTO/123.2573,70.4562,0.0000
2030      FEDRAT/200.0000
2040      GOTO/123.2573,70.4562,15.0000
2050      GOTO/-63.5802,-34.7915,0.0000
2060      PAINT/TOOL,NOMORE
2070      PAINT/ARROW,NOMORE
2080      PAINT/LINNO,NOMORE
2090      PAINT/DASH,NOMORE
2100 *  END-OF-PATH
```

Step 18. The manually edited UNIAPT program.

```
UNIGRAPHICS  CL-SOURCE FILE  3305PAULM      12--71--84

     10    PARTNO/PAULM
     20    MACHIN/50,OPTION,106,50,51
     21    LOAD/TOOL,1
     30    FROM/-25,-25,50
     40  * TOOL PATH/20.0000,0.5000,60.0000,P1
     50    MSYS/0.0000000,0.0000000,0.0000000,1.0000000,0.0000
000,0.0000000,0.0000000,1.0000000,0.0000000
     60    PAINT/TOOL,FULL,1
     70    PAINT/ARROW
     80    PAINT/LINNO
     90    PAINT/DASH
    100    FEDRAT/200.0000
    110    PAINT/COLOR,CYAN
    120    GOTO/-23.5580,11.9179,15.0000
    130    FEDRAT/10.0000
    140    GOTO/-23.5580,11.9179,0.0000
    150    CIRCLE/11.9416,12.0898,0.0000,0.0000,0.0000,-1.0000
,35.5000,0.0508,0.5000,40.0000,0.5000
    160    GOTO/12.2075,-23.4092,0.0000
    170    GOTO/140.7992,-22.4460,0.0000
    180    CIRCLE/140.5333,13.0530,0.0000,0.0000,0.0000,-1.000
0,35.5000,0.0508,0.5000,40.0000,0.5000
    190    GOTO/176.0333,13.0530,0.0000
    200    GOTO/176.0333,104.9192,0.0000
    210    CIRCLE/140.5333,104.9192,0.0000,0.0000,0.0000,-1.00
00,35.5000,0.0508,0.5000,40.0000,0.5000
    220    GOTO/140.5333,140.4192,0.0000
    230    GOTO/11.4921,140.4192,0.0000
    240    CIRCLE/11.4921,104.9192,0.0000,0.0000,0.0000,-1.000
0,35.5000,0.0508,0.5000,40.0000,0.5000
    250    GOTO/-24.0075,104.7473,0.0000
    260    GOTO/-23.5580,11.9179,0.0000
    270    GOTO/-20.0580,11.9348,0.0000
    280    CIRCLE/11.9416,12.0898,0.0000,0.0000,0.0000,-1.0000
,32.0000,0.0508,0.5000,40.0000,0.5000
    290    GOTO/12.1813,-19.9093,0.0000
    300    GOTO/140.7730,-18.9461,0.0000
    310    CIRCLE/140.5333,13.0530,0.0000,0.0000,0.0000,-1.000
0,32.0000,0.0508,0.5000,40.0000,0.5000
    320    GOTO/172.5333,13.0530,0.0000
    330    GOTO/172.5333,104.9192,0.0000
    340    CIRCLE/140.5333,104.9192,0.0000,0.0000,0.0000,-1.00
00,32.0000,0.0508,0.5000,40.0000,0.5000
    350    GOTO/140.5333,136.9192,0.0000
    360    GOTO/11.4921,136.9192,0.0000
    370    CIRCLE/11.4921,104.9192,0.0000,0.0000,0.0000,-1.000
0,32.0000,0.0508,0.5000,40.0000,0.5000
    380    GOTO/-20.5075,104.7643,0.0000
    390    GOTO/-20.0580,11.9348,0.0000
    400    FEDRAT/200.0000
    410    GOTO/-20.0580,11.9348,15.0000
    420    GOTO/-63.5802,-34.7915,0.0000
    430    PAINT/TOOL,NOMORE
    440    PAINT/ARROW,NOMORE
    450    PAINT/LINNO,NOMORE
    460    PAINT/DASH,NOMORE
    470  * END-OF-PATH
    420    GOHOME
    482    LOAD/TOOL,2
    490  * TOOL PATH/5.0000,0.0000,45.0000,P2
    500    GOTO/19.6186,16.7572,0.0000
```

```
510     GOTO/19.9994,100.5429,0.0000
520     GOTO/136.1570,100.5429,0.0000
530     GOTO/135.7761,16.7572,0.0000
540   * END-OF-PATH
550     GOHOME
551     LOAD/TOOL,4
570   * TOOL PATH/10.0000,1.0000,45.0000,P3
580     MSYS/0.0000000,0.0000000,0.0000000,1.0000000,0.0000
000,0.0000000,0.0000000,1.0000000,0.0000000
590     PAINT/TOOL,FULL,1
600     PAINT/ARROW
610     PAINT/LINNO
620     PAINT/DASH
630     FEDRAT/200.0000
640     PAINT/COLOR,CYAN
650     GOTO/52.9132,55.9485,15.0000
660     FEDRAT/10.0000
670     GOTO/52.9132,55.9485,0.0000
680     GOTO/55.9465,53.5295,0.0000
690     GOTO/59.7289,54.3928,0.0000
700     GOTO/61.4123,57.8884,0.0000
710     GOTO/59.7289,61.3839,0.0000
720     GOTO/55.9465,62.2472,0.0000
730     GOTO/52.9132,59.8282,0.0000
740     GOTO/52.9132,55.9485,0.0000
750     GOTO/48.1507,53.6550,0.0000
760     GOTO/54.7702,48.3760,0.0000
770     GOTO/63.0247,50.2601,0.0000
780     GOTO/66.6983,57.8884,0.0000
790     GOTO/63.0247,65.5166,0.0000
800     GOTO/54.7702,67.4007,0.0000
810     GOTO/48.1507,62.1217,0.0000
820     GOTO/48.1507,53.6550,0.0000
830     GOTO/43.3882,51.7046,0.0000
840     CIRCLE/44.1007,51.7046,0.0000,0.0000,0.0000,-1.0000
,0.7125,0.0508,0.5000,20.0000,1.0000
850     GOTO/43.6564,51.1475,0.0000
860     GOTO/53.3257,43.4365,0.0000
870     CIRCLE/53.7700,43.9936,0.0000,0.0000,0.0000,-1.0000
,0.7125,0.0508,0.5000,20.0000,1.0000
880     GOTO/53.9285,43.2990,0.0000
890     GOTO/65.9859,46.0510,0.0000
900     CIRCLE/65.8274,46.7456,0.0000,0.0000,0.0000,-1.0000
,0.7125,0.0508,0.5000,20.0000,1.0000
910     GOTO/66.4693,46.4365,0.0000
920     GOTO/71.8354,57.5792,0.0000
930     CIRCLE/71.1934,57.8884,0.0000,0.0000,0.0000,-1.0000
,0.7125,0.0508,0.5000,20.0000,1.0000
940     GOTO/71.8354,58.1975,0.0000
950     GOTO/66.4693,69.3402,0.0000
960     CIRCLE/65.8274,69.0311,0.0000,0.0000,0.0000,-1.0000
,0.7125,0.0508,0.5000,20.0000,1.0000
970     GOTO/65.9859,69.7257,0.0000
980     GOTO/53.9285,72.4778,0.0000
990     CIRCLE/53.7700,71.7831,0.0000,0.0000,0.0000,-1.0000
,0.7125,0.0508,0.5000,20.0000,1.0000
1000    GOTO/53.3257,72.3402,0.0000
1010    GOTO/43.6564,64.6292,0.0000
1020    CIRCLE/44.1007,64.0721,0.0000,0.0000,0.0000,-1.0000
,0.7125,0.0508,0.5000,20.0000,1.0000
1030    GOTO/43.3882,64.0721,0.0000
1040    GOTO/43.3882,51.7046,0.0000
1050    GOTO/38.6257,51.7046,0.0000
1060    CIRCLE/44.1007,51.7046,0.0000,0.0000,0.0000,-1.0000
```

```
,5.4750,0.0508,0.5000,20.0000,1.0000
     1070      GOTO/40.6870,47.4241,0.0000
     1080      GOTO/50.3564,39.7131,0.0000
     1090      CIRCLE/53.7700,43.9936,0.0000,0.0000,0.0000,-1.0000
,5.4750,0.0508,0.5000,20.0000,1.0000
     1100      GOTO/54.9883,38.6559,0.0000
     1110      GOTO/67.0457,41.4079,0.0000
     1120      CIRCLE/65.8274,46.7456,0.0000,0.0000,0.0000,-1.0000
,5.4750,0.0508,0.5000,20.0000,1.0000
     1130      GOTO/70.7602,44.3701,0.0000
     1140      GOTO/76.1262,55.5128,0.0000
     1150      CIRCLE/71.1934,57.8884,0.0000,0.0000,0.0000,-1.0000
,5.4750,0.0508,0.5000,20.0000,1.0000
     1160      GOTO/76.1262,60.2639,0.0000
     1170      GOTO/70.7602,71.4066,0.0000
     1180      CIRCLE/65.8274,69.0311,0.0000,0.0000,0.0000,-1.0000
,5.4750,0.0508,0.5000,20.0000,1.0000
     1190      GOTO/67.0457,74.3688,0.0000
     1200      GOTO/54.9883,77.1208,0.0000
     1210      CIRCLE/53.7700,71.7831,0.0000,0.0000,0.0000,-1.0000
,5.4750,0.0508,0.5000,20.0000,1.0000
     1220      GOTO/50.3564,76.0636,0.0000
     1230      GOTO/40.6870,68.3526,0.0000
     1240      CIRCLE/44.1007,64.0721,0.0000,0.0000,0.0000,-1.0000
,5.4750,0.0508,0.5000,20.0000,1.0000
     1250      GOTO/38.6257,64.0721,0.0000
     1260      GOTO/38.6257,51.7046,0.0000
     1270      GOTO/33.8632,51.7046,0.0000
     1280      CIRCLE/44.1007,51.7046,0.0000,0.0000,0.0000,-1.0000
,10.2375,0.0508,0.5000,20.0000,1.0000
     1290      GOTO/37.7177,43.7006,0.0000
     1300      GOTO/47.3870,35.9896,0.0000
     1310      CIRCLE/53.7700,43.9936,0.0000,0.0000,0.0000,-1.0000
,10.2375,0.0508,0.5000,20.0000,1.0000
     1320      GOTO/56.0480,34.0128,0.0000
     1330      GOTO/68.1054,36.7648,0.0000
     1340      CIRCLE/65.8274,46.7456,0.0000,0.0000,0.0000,-1.0000
,10.2375,0.0508,0.5000,20.0000,1.0000
     1350      GOTO/75.0511,42.3037,0.0000
     1360      GOTO/80.4171,53.4465,0.0000
     1370      CIRCLE/71.1934,57.8884,0.0000,0.0000,0.0000,-1.0000
,10.2375,0.0508,0.5000,20.0000,1.0000
     1380      GOTO/80.4171,62.3302,0.0000
     1390      GOTO/75.0511,73.4730,0.0000
     1400      CIRCLE/65.8274,69.0311,0.0000,0.0000,0.0000,-1.0000
,10.2375,0.0508,0.5000,20.0000,1.0000
     1410      GOTO/68.1054,79.0119,0.0000
     1420      GOTO/56.0480,81.7639,0.0000
     1430      CIRCLE/53.7700,71.7831,0.0000,0.0000,0.0000,-1.0000
,10.2375,0.0508,0.5000,20.0000,1.0000
     1440      GOTO/47.3870,79.7871,0.0000
     1450      GOTO/37.7177,72.0761,0.0000
     1460      CIRCLE/44.1007,64.0721,0.0000,0.0000,0.0000,-1.0000
,10.2375,0.0508,0.5000,20.0000,1.0000
     1470      GOTO/33.8632,64.0721,0.0000
     1480      GOTO/33.8632,51.7046,0.0000
     1490      GOTO/29.1007,51.7046,0.0000
     1500      CIRCLE/44.1007,51.7046,0.0000,0.0000,0.0000,-1.0000
,15.0000,0.0508,0.5000,20.0000,1.0000
     1510      GOTO/34.7483,39.9771,0.0000
     1520      GOTO/44.4176,32.2661,0.0000
     1530      CIRCLE/53.7700,43.9936,0.0000,0.0000,0.0000,-1.0000
,15.0000,0.0508,0.5000,20.0000,1.0000
     1540      GOTO/57.1073,29.3697,0.0000
```

```
1550        GOTO/69.1652,32.1217,0.0000
1560        CIRCLE/65.8274,46.7456,0.0000,0.0000,0.0000,-1.0000
,15.0000,0.0508,0.5000,20.0000,1.0000
1570        GOTO/79.3419,40.2374,0.0000
1580        GOTO/84.7080,51.3801,0.0000
1590        CIRCLE/71.1934,57.8884,0.0000,0.0000,0.0000,-1.0000
,15.0000,0.0508,0.5000,20.0000,1.0000
1600        GOTO/84.7080,64.3966,0.0000
1610        GOTO/79.3419,75.5393,0.0000
1620        CIRCLE/65.8274,69.0311,0.0000,0.0000,0.0000,-1.0000
,15.0000,0.0508,0.5000,20.0000,1.0000
1630        GOTO/69.1652,83.6550,0.0000
1640        GOTO/57.1078,86.4070,0.0000
1650        CIRCLE/53.7700,71.7831,0.0000,0.0000,0.0000,-1.0000
,15.0000,0.0508,0.5000,20.0000,1.0000
1660        GOTO/44.4176,83.5106,0.0000
1670        GOTO/34.7483,75.7996,0.0000
1680        CIRCLE/44.1007,64.0721,0.0000,0.0000,0.0000,-1.0000
,15.0000,0.0508,0.5000,20.0000,1.0000
1690        GOTO/29.1007,64.0721,0.0000
1700        GOTO/29.1007,51.7046,0.0000
1710        FEDRAT/200.0000
1720        GOTO/29.1007,51.7046,15.0000
1730        GOTO/-63.5802,-34.7915,0.0000
1740        PAINT/TOOL,NOMORE
1750        PAINT/ARROW,NOMORE
1760        PAINT/LINNO,NOMORE
1770        PAINT/DASH,NOMORE
1780 * END-OF-PATH
1790 * TOOL PATH/10.0000,1.0000,45.0000,P4
1800        MSYS/0.0000000,0.0000000,0.0000000,1.0000000,0.0000
000,0.0000000,0.0000000,1.0000000,0.0000000
1810        PAINT/TOOL,FULL,1
1820        PAINT/ARROW
1830        PAINT/LINNO
1840        PAINT/DASH
1850        FEDRAT/200.0000
1860        PAINT/COLOR,CYAN
1870        GOTO/118.4948,70.4562,15.0000
1880        FEDRAT/10.0000
1890        GOTO/118.4948,70.4562,0.0000
1900        CIRCLE/118.2573,70.4562,0.0000,0.0000,0.0000,-1.000
0,0.2375,0.0508,0.5000,20.0000,1.0000
1910        GOTO/118.1390,70.6621,0.0000
1920        CIRCLE/118.2573,70.4562,0.0000,0.0000,0.0000,-1.000
0,0.2375,0.0508,0.5000,20.0000,1.0000
1930        GOTO/118.1383,70.2507,0.0000
1940        CIRCLE/118.2573,70.4562,0.0000,0.0000,0.0000,-1.000
0,0.2375,0.0508,0.5000,20.0000,1.0000
1950        GOTO/118.4948,70.4562,0.0000
1960        GOTO/123.2573,70.4562,0.0000
1970        CIRCLE/118.2573,70.4562,0.0000,0.0000,0.0000,-1.000
0,5.0000,0.0508,0.5000,20.0000,1.0000
1980        GOTO/115.7664,74.7916,0.0000
1990        CIRCLE/118.2573,70.4562,0.0000,0.0000,0.0000,-1.000
0,5.0000,0.0508,0.5000,20.0000,1.0000
2000        GOTO/115.7528,66.1287,0.0000
2010        CIRCLE/118.2573,70.4562,0.0000,0.0000,0.0000,-1.000
0,5.0000,0.0508,0.5000,20.0000,1.0000
2020        GOTO/123.2573,70.4562,0.0000
2030        FEDRAT/200.0000
2040        GOTO/123.2573,70.4562,15.0000
2050        GOTO/-63.5802,-34.7915,0.0000
2060        PAINT/TOOL,NOMORE
2070        PAINT/ARROW,NOMORE
2080        PAINT/LINNO,NOMORE
2090        PAINT/DASH,NOMORE
2100 * END-OF-PATH
2101        GOHOME
2102        FINI
```

Step 19. Listing of the automatically generated NC tape file.


```
        Listing of the generated NC program (milling job)
                                          Page No: 1
+------------------------------------------------------------+

000010  GIGAA!FIOII!IIIIF!AAAAO!IOIII!!!!!!!!!!!!!!!!!!!!!!!

        !!!!!!!!!!!!!!!!!!!!!!!!!!!!!!

000020  !!!!!!!!!!!!!!!!!!!!!!!!!!!!!!!!!!!!!!!!!!!!!!!!!!!!!

        !!!!!!!!!!!!!!!!!!!!!!!!!!!!!!

000030  !!!!!!!!!!!!!!!!!!!!!!%<$

000040  @PAULM<$

000050  N0010G71<$

000060  N0020G90<$

000070  N0030G80SOOOOTO1M06<$

000080  N0040G00X-23.558Y11.918D00<$

000090  N0050Z15.M09<$

000100  N0060G01Z.OF10.<$

000110  N0070G17<$

000120  N0080G03X12.208Y-23.409I35.5J.172<$

000130  N0090G01X140.799Y-22.446<$

000140  N0100G03X176.033Y13.053I.266J35.499<$

000150  N0110G01Y104.919<$

000160  N0120G03X140.533Y140.419I35.5J.0<$

000170  N0130G01X11.492<$

000180  N0140G03X-24.007Y104.747I.0J35.5<$

000190  N0150G01X-23.558Y11.918<$

000200  N0160X-20.058Y11.935<$

000210  N0170G03X12.181Y-19.909I32.J.155<$

000220  N0180G01X140.773Y-18.946<$
```

Listing of the generated NC program (milling job)
                                            Page No: 2
+---+

000230   N0190G03X172.533Y13.053I.24J31.999<$

000240   N0200G01Y104.919<$

000250   N0210G03X140.533Y136.919I32.J.0<$

000260   N0220G01X11.492<$

000270   N0230G03X-20.508Y104.764I.0J32.<$

000280   N0240G01X-20.058Y11.935<$

000290   N0250Z15.F200.<$

000300   N0260X-63.58Y-34.792Z.0<$

000310   N0270G00Z50.<$

000320   N0280X-25.Y-25.<$

000330   N0290G90<$

000340   N0300G80S0000T02M06<$

000350   N0310G00X19.619Y16.757D00<$

000360   N0320Z.0M09<$

000370   N0330G01X19.999Y100.543F200.<$

000380   N0340X136.157<$

000390   N0350X135.776Y16.757<$

000400   N0360G00Z50.<$

000410   N0370X-25.Y-25.<$

000420   N0380G90<$

000430   N0390G80S0000T04M06<$

000440   N0400G00X52.913Y55.949D00<$

000450   N0410Z15.M09<$

000460   N0420G01Z.0F10.<$

000470   N0430X55.947Y53.53<$

000480   N0440X59.729Y54.393<$

000490   N0450X61.412Y57.888<$

CIM-J

Listing of the generated NC program (milling job)
                                                    Page No: 3
+-------------------------------- --+

000500   N0460X59.729Y61.384<$

000510   N0470X55.947Y62.247<$

000520   N0480X52.913Y59.828<$.

000530   N0490Y55.949<$

000540   N0500X48.151Y53.655<$

000550   N0510X54.77Y48.376<$

000560   N0520X63.025Y50.26<$

000570   N0530X66.698Y57.888<$

000580   N0540X63.025Y65.517<$

000590   N0550X54.77Y67.401<$

000600   N0560X48.151Y62.122<$

000610   N0570Y53.655<$

000620   N0580X43.388Y51.705<$

000630   N0590G03X43.656Y51.148I.713J.0<$

000640   N0600G01X53.326Y43.437<$

000650   N0610G03X53.77Y43.281I.444J.557<$

000660   N0620X53.929Y43.299I.0J.713<$

000670   N0630G01X65.986Y46.051<$

000680   N0640G03X66.469Y46.436I.159J.695<$

000690   N0650G01X71.835Y57.579<$

000700   N0660G03X71.906Y57.888I.642J.309<$

000710   N0670X71.835Y58.198I.713J.0<$

000720   N0680G01X66.469Y69.34<$

000730   N0690G03X65.986Y69.726I.642J.309<$

000740   N0700G01X53.929Y72.478<$

000750   N0710G03X53.77Y72.496I.159J.695<$

```
+------------------------------------------------------------+

000760   N0720X53.326Y72.34I.0J.713<$

000770   N0730G01X43.656Y64.629<$

000780   N0740G03X43.388Y64.072I.444J.557<$

000790   N0750G01Y51.705<$

000800   N0760X38.626<$

000810   N0770G03X40.687Y47.424I5.475J.0<$

000820   N0780G01X50.356Y39.713<$

000830   N0790G03X53.77Y38.519I3.414J4.281<$

000840   N0800X54.988Y38.656I.0J5.475<$

000850   N0810G01X67.046Y41.408<$

000860   N0820G03X70.76Y44.37I1.218J5.338<$

000870   N0830G01X76.126Y55.513<$

000880   N0840G03X76.668Y57.888I4.933J2.376<$

000890   N0850X76.126Y60.264I5.475J.0<$

000900   N0860G01X70.76Y71.407<$

000910   N0870G03X67.046Y74.369I4.933J2.376<$

000920   N0880G01X54.988Y77.121<$

000930   N0890G03X53.77Y77.258I1.218J5.338<$

000940   N0900X50.356Y76.064I.0J5.475<$

000950   N0910G01X40.687Y68.353<$

000960   N0920G03X38.626Y64.072I3.414J4.281<$

000970   N0930G01Y51.705<$

000980   N0940X33.863<$

000990   N0950G03X37.718Y43.701I10.238J.0<$

001000   N0960G01X47.387Y35.99<$

001010   N0970G03X53.77Y33.756I6.383J8.004<$
```

+--+

001020   N0980X56.048Y34.013I.0J10.238<$·

001030   N0990G01X68.105Y36.765<$

001040   N1000G03X75.051Y42.304I2.278J9.981<$

001050   N1010G01X80.417Y53.447<$

001060   N1020G03X81.431Y57.888I9.224J4.442<$

001070   N1030X80.417Y62.33I10.238J.0<$

001080   N1040G01X75.051Y73.473<$

001090   N1050G03X68.105Y79.012I9.224J4.442<$

001100   N1060G01X56.048Y81.764<$

001110   N1070G03X53.77Y82.021I2.278J9.981<$

001120   N1080X47.387Y79.787I.0J10.238<$

001130   N1090G01X37.718Y72.076<$

001140   N1100G03X33.863Y64.072I6.383J8.004<$

001150   N1110G01Y51.705<$

001160   N1120X29.101<$

001170   N1130G03X34.748Y39.977I15.J.0<$

001180   N1140G01X44.418Y32.266<$

001190   N1150G03X53.77Y28.994I9.352J11.728<$

001200   N1160X57.108Y29.37I.0J15.<$

001210   N1170G01X69.165Y32.122<$

001220   N1180G03X79.342Y40.237I3.338J14.624<$

001230   N1190G01X84.708Y51.38<$

001240   N1200G03X86.193Y57.888I13.515J6.508<$

001250   N1210X84.708Y64.397I15.J.0<$

001260   N1220G01X79.342Y75.539<$

001270   N1230G03X69.165Y83.655I13.515J6.508<$

001280   N1240G01X57.108Y86.407<$

+--+

001290   N1250G03X53.77Y86.783I3.338J14.624<$

001300   N1260X44.418Y83.511I.0J15.<$

001310   N1270G01X34.748Y75.8<$

001320   N1280G03X29.101Y64.072I9.352J11.728<$

001330   N1290G01Y51.705<$

001340   N1300Z15.F200.<$

001350   N1310X-63.58Y-34.792Z.0<$

001360   N1320X118.495Y70.456Z15.<$

001370   N1330Z.0F10.<$

001380   N1340G03X118.257Y70.694I.238J.0<$

001390   N1350X118.139Y70.662I.0J.238<$

001400   N1360X118.02Y70.456I.118J.206<$

001410   N1370X118.138Y70.251I.237J.0<$

001420   N1380X118.257Y70.219I.119J.206<$

001430   N1390X118.495Y70.456I.0J.237<$

001440   **N1400G01X123.257<$**

001450   N1410G03X118.257Y75.456I5.J.0<$

001460   N1420X115.766Y74.792I.0J5.<$

001470   N1430X113.257Y70.456I2.491J4.335<$

001480   N1440X115.753Y66.129I5.J.0<$

001490   N1450X118.257Y65.456I2.505J4.328<$

001500   N1460X123.257Y70.456I.0J5.<$

001510   N1470G01Z15.F200.<$

001520   N1480X-63.58Y-34.792Z.0<$

001530   **N1490G00Z50.<$**

001540   N1500X-25.Y-25.<$

5.9 Case study: integrated CAD/CAM system. Design and manufacture of a turned component using the McAuto Unigraphics CAD/CAM system

As with the milling case study this demonstration is organised into steps, "similarly to a slide show", indicating the most important operator actions and system responses, when designing and manufacturing a rotary shaped component.

Please follow the steps in the given order.

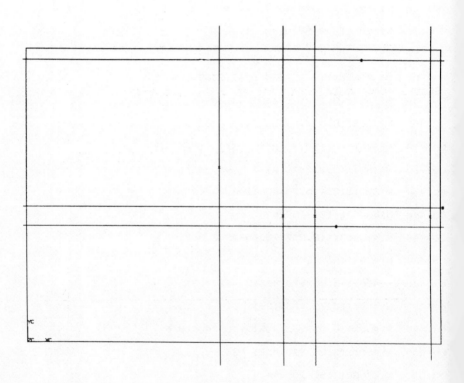

Step 1. In this case study the part design begins with the definition of construction lines.

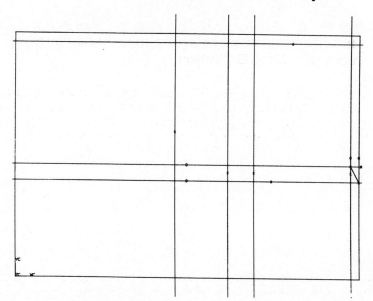

Step 2. Some more construction lines are added.

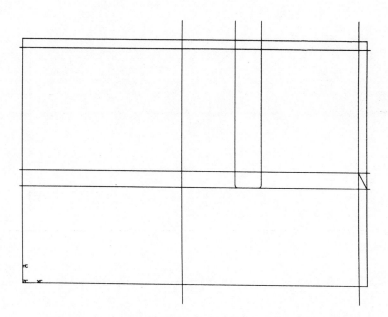

Step 3. Construction lines are added in order to design the
groove on the part.

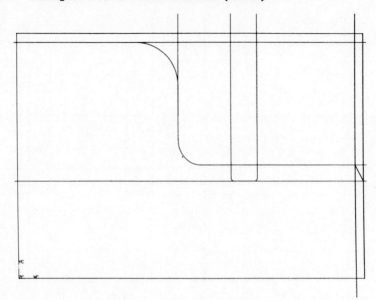

Step 4. Design of the outside contour.

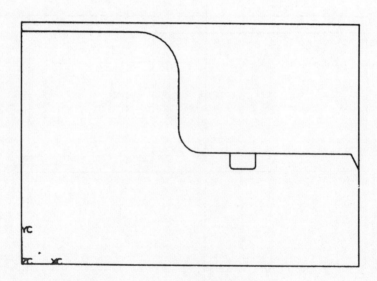

Step 5. The final design of the component.
       Construction lines have been removed, but a simplified
       shape has been kept for roughing.

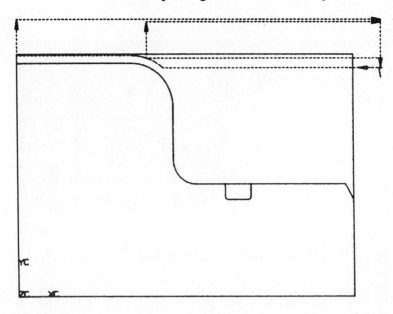

Step 6. Dynamic simulation of the tool path. First partial
plot.

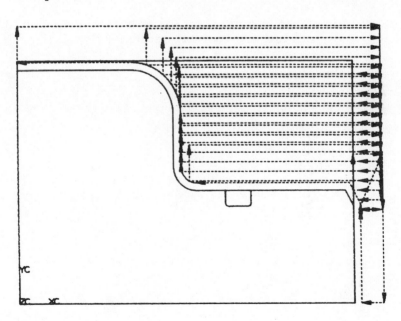

Step 7. Dynamic simulation of the tool path. Complete plot.

CIM-J*

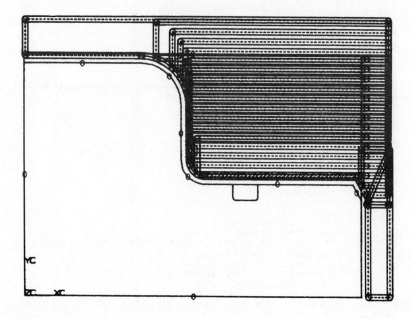

Step 8. Plot of the outside roughing operation indicating the tool path and the tool tip radius.

Step 9. The automatically generated UNIAPT code. This listing must be edited manually before post-processing too.

```
UNIGRAPHICS  CL-SOURCE FILE  3305PAULL       12--72--84


    10  * TOOL PATH/1.2000,0.0000,0.0000,P1
    20     MSYS/0.0000000,0.0000000,0.0000000,1.0000000,0.0000
000,0.0000000,0.0000000,1.0000000,0.0000000
    30     PAINT/TOOL
    40     PAINT/SPEED,1
    50     PAINT/ARROW
    60     PAINT/DASH
    70     TURRET/FACE,1,XOFF,100.0000,YOFF,100.0000
    80     SPINDL/SFM,220,CCLW,MAXRPM,2000,RANGE,MEDIUM
    90     FEDRAT/200.0000
   100     PAINT/COLOR,CYAN
   110     GOTO/162.4700,98.4700,0.0000
   120     FEDRAT/MMPR,0.7000
   130     GOTO/152.4700,98.4700,0.0000
   140     FEDRAT/MMPR,0.5000
   150     GOTO/58.3958,98.4700,0.0000
   160     FEDRAT/MMPR,0.8000
   170     CIRCLE/51.8551,78.3042,0.0000,0.0000,0.0000,-1.0000
,21.2000,0.0508,0.5000,2.4000,0.0000
   180     GOTO/51.8551,99.5042,0.0000
   190     GOTO/-0.0000,99.5042,0.0000
   200     FEDRAT/MMPR,5.0000
   210     GOTO/0.0000,114.5042,0.0000
   220     FEDRAT/200.0000
   230     GOTO/162.4700,114.5042,0.0000
   240     GOTO/162.4700,94.4700,0.0000
   250     FEDRAT/MMPR,0.7000
   260     GOTO/152.4700,94.4700,0.0000
   270     FEDRAT/MMPR,0.8000
   280     GOTO/65.5703,94.4700,0.0000
   290     CIRCLE/51.8551,78.3042,0.0000,0.0000,0.0000,-1.0000
,21.2000,0.0508,0.5000,2.4000,0.0000
   300     GOTO/58.3958,98.4700,0.0000
   310     FEDRAT/MMPR,5.0000
   320     GOTO/58.3958,113.4700,0.0000
   330     FEDRAT/200.0000
   340     GOTO/162.4700,113.4700,0.0000
   350     GOTO/162.4700,90.4700,0.0000
   360     FEDRAT/MMPR,0.7000
   370     GOTO/152.4700,90.4700,0.0000
   380     FEDRAT/MMPR,0.8000
   390     GOTO/69.2169,90.4700,0.0000
   400     CIRCLE/51.8551,78.3042,0.0000,0.0000,0.0000,-1.0000
,21.2000,0.0508,0.5000,2.4000,0.0000
   410     GOTO/65.5703,94.4700,0.0000
   420     FEDRAT/MMPR,5.0000
   430     GOTO/65.5703,109.4700,0.0000
   440     FEDRAT/200.0000
   450     GOTO/162.4700,109.4700,0.0000
   460     GOTO/162.4700,86.4700,0.0000
   470     FEDRAT/MMPR,0.7000
   480     GOTO/152.4700,86.4700,0.0000
   490     FEDRAT/MMPR,0.8000
   500     GOTO/71.4193,86.4700,0.0000
   510     CIRCLE/51.8551,78.3042,0.0000,0.0000,0.0000,-1.0000
,21.2000,0.0508,0.5000,2.4000,0.0000
   520     GOTO/69.2169,90.4700,0.0000
   530     FEDRAT/MMPR,5.0000
   540     GOTO/69.2169,105.4700,0.0000
   550     FEDRAT/200.0000
   560     GOTO/162.4700,105.4700,0.0000
```

```
570      GOTO/162.4700,82.4700,0.0000
580      FEDRAT/MMPR,0.7000
590      GOTO/152.4700,82.4700,0.0000
600      FEDRAT/MMPR,0.8000
610      GOTO/72.6418,82.4700,0.0000
620      CIRCLE/51.8551,78.3042,0.0000,0.0000,0.0000,-1.0000
,21.2000,0.0508,0.5000,2.4000,0.0000
630      GOTO/71.4193,86.4700,0.0000
640      FEDRAT/MMPR,5.0000
650      GOTO/71.4193,101.4700,0.0000
660      FEDRAT/200.0000
670      GOTO/162.4700,101.4700,0.0000
680      GOTO/162.4700,78.4700,0.0000
690      FEDRAT/MMPR,0.7000
700      GOTO/152.4700,78.4700,0.0000
710      FEDRAT/MMPR,0.8000
720      GOTO/73.0544,78.4700,0.0000
730      CIRCLE/51.8551,78.3042,0.0000,0.0000,0.0000,-1.0000
,21.2000,0.0508,0.5000,2.4000,0.0000
740      GOTO/72.6418,82.4700,0.0000
750      FEDRAT/MMPR,5.0000
760      GOTO/72.6418,97.4700,0.0000
770      FEDRAT/200.0000
780      GOTO/162.4700,97.4700,0.0000
790      GOTO/162.4700,74.4700,0.0000
800      FEDRAT/MMPR,0.7000
810      GOTO/152.4700,74.4700,0.0000
820      FEDRAT/MMPR,0.8000
830      GOTO/73.0551,74.4700,0.0000
840      GOTO/73.0551,78.3042,0.0000
850      CIRCLE/51.8551,78.3042,0.0000,0.0000,0.0000,-1.0000
,21.2000,0.0508,0.5000,2.4000,0.0000
860      GOTO/73.0544,78.4700,0.0000
870      FEDRAT/MMPR,5.0000
880      GOTO/73.0544,93.4700,0.0000
890      FEDRAT/200.0000
900      GOTO/162.4700,93.4700,0.0000
910      GOTO/162.4700,70.4700,0.0000
920      FEDRAT/MMPR,0.7000
930      GOTO/152.4700,70.4700,0.0000
940      FEDRAT/MMPR,0.8000
950      GOTO/73.0551,70.4700,0.0000
960      GOTO/73.0551,74.4700,0.0000
970      FEDRAT/MMPR,5.0000
980      GOTO/73.0551,89.4700,0.0000
990      FEDRAT/200.0000
1000     GOTO/162.4700,89.4700,0.0000
1010     GOTO/162.4700,66.4700,0.0000
1020     FEDRAT/MMPR,0.7000
1030     GOTO/152.4700,66.4700,0.0000
1040     FEDRAT/MMPR,0.8000
1050     GOTO/73.0551,66.4700,0.0000
1060     GOTO/73.0551,70.4700,0.0000
1070     FEDRAT/MMPR,5.0000
1080     GOTO/73.0551,85.4700,0.0000
1090     FEDRAT/200.0000
1100     GOTO/162.4700,85.4700,0.0000
1110     GOTO/162.4700,62.4700,0.0000
1120     FEDRAT/MMPR,0.7000
1130     GOTO/152.4700,62.4700,0.0000
1140     FEDRAT/MMPR,0.8000
1150     GOTO/73.0551,62.4700,0.0000
1160     GOTO/73.0551,66.4700,0.0000
1170     FEDRAT/MMPR,5.0000
```

```
1180      GOTO/73.0551,81.4700,0.0000
1190      FEDRAT/200.0000
1200      GOTO/162.4700,81.4700,0.0000
1210      GOTO/162.4700,58.4700,0.0000
1220      FEDRAT/MMPR,0.7000
1230      GOTO/152.4700,58.4700,0.0000
1240      FEDRAT/MMPR,0.8000
1250      GOTO/73.0551,58.4700,0.0000
1260      GOTO/73.0551,62.4700,0.0000
1270      FEDRAT/MMPR,5.0000
1280      GOTO/73.0551,77.4700,0.0000
1290      FEDRAT/200.0000
1300      GOTO/162.4700,77.4700,0.0000
1310      GOTO/162.4700,54.4700,0.0000
1320      FEDRAT/MMPR,0.7000
1330      GOTO/152.4700,54.4700,0.0000
1340      FEDRAT/MMPR,0.8000
1350      GOTO/73.3713,54.4700,0.0000
1360      CIRCLE/79.8551,56.5196,0.0000,0.0000,0.0000,1.0000,
6.8000,0.0508,0.5000,2.4000,0.0000
1370      GOTO/73.0551,56.5196,0.0000
1380      GOTO/73.0551,58.4700,0.0000
1390      FEDRAT/MMPR,5.0000
1400      GOTO/73.0551,73.4700,0.0000
1410      FEDRAT/200.0000
1420      GOTO/162.4700,73.4700,0.0000
1430      GOTO/162.4700,50.4700,0.0000
1440      FEDRAT/MMPR,0.7000
1450      GOTO/152.4700,50.4700,0.0000
1460      FEDRAT/MMPR,0.8000
1470      GOTO/76.7499,50.4700,0.0000
1480      CIRCLE/79.8551,56.5196,0.0000,0.0000,0.0000,1.0000,
6.8000,0.0508,0.5000,2.4000,0.0000
1490      GOTO/73.3713,54.4700,0.0000
1500      FEDRAT/MMPR,5.0000
1510      GOTO/73.3713,69.4700,0.0000
1520      FEDRAT/200.0000
1530      GOTO/162.4700,69.4700,0.0000
1540      GOTO/162.4700,46.4700,0.0000
1550      FEDRAT/MMPR,0.7000
1560      GOTO/152.4700,46.4700,0.0000
1570      FEDRAT/MMPR,0.8000
1580      GOTO/150.1023,46.4700,0.0000
1590      GOTO/148.3161,49.7196,0.0000
1600      GOTO/79.8551,49.7196,0.0000
1610      CIRCLE/79.8551,56.5196,0.0000,0.0000,0.0000,1.0000,
6.8000,0.0508,0.5000,2.4000,0.0000
1620      GOTO/76.7499,50.4700,0.0000
1630      FEDRAT/MMPR,5.0000
1640      GOTO/76.7499,65.4700,0.0000
1650      FEDRAT/200.0000
1660      GOTO/162.4700,65.4700,0.0000
1670      GOTO/162.4700,42.4700,0.0000
1680      FEDRAT/MMPR,0.7000
1690      GOTO/152.4700,42.4700,0.0000
1700      FEDRAT/MMPR,0.8000
1710      GOTO/152.3009,42.4700,0.0000
1720      GOTO/150.1023,46.4700,0.0000
1730      FEDRAT/MMPR,5.0000
1740      GOTO/150.1023,61.4700,0.0000
1750      FEDRAT/200.0000
1760      GOTO/163.2000,61.4700,0.0000
1770      GOTO/163.2000,-0.0000,0.0000
1780      FEDRAT/MMPR,0.7000
```

```
1790    GOTO/153.2000,0.0000,0.0000
1800    FEDRAT/MMPR,0.8000
1810    GOTO/153.2000,38.4700,0.0000
1820    FEDRAT/200.0000
1830    GOTO/162.4700,38.4700,0.0000
1840    FEDRAT/MMPR,0.7000
1850    GOTO/152.4700,38.4700,0.0000
1860    FEDRAT/200.0000
1870    GOTO/163.2000,61.4700,0.0000
1880    GOTO/163.2000,38.4700,0.0000
1890    FEDRAT/MMPR,0.7000
1900    GOTO/153.2000,38.4700,0.0000
1910    FEDRAT/MMPR,0.8000
1920    GOTO/153.2000,40.8343,0.0000
1930    GOTO/148.3161,49.7196,0.0000
1940    GOTO/79.8551,49.7196,0.0000
1950    CIRCLE/79.8551,56.5196,0.0000,0.0000,0.0000,1.0000,
6.8000,0.0508,0.5000,2.4000,0.0000
1960    GOTO/73.0551,56.5196,0.0000
1970    GOTO/73.0551,78.3042,0.0000
1980    CIRCLE/51.8551,78.3042,0.0000,0.0000,0.0000,-1.0000
,21.2000,0.0508,0.5000,2.4000,0.0000
1990    GOTO/51.8551,99.5042,0.0000
2000    GOTO/-0.0000,99.5042,0.0000
2010    FEDRAT/MMPR,5.0000
2020    GOTO/0.0000,114.5042,0.0000
2030    PAINT/TOOL,NOMORE
2040    PAINT/SPEED,10
2050    PAINT/ARROW,NOMORE
2060    PAINT/DASH,NOMORE
2070  * END-OF-PATH
```

Step 10. The manually edited UNIAPT program.

```
UNIGRAPHICS  CL-SOURCE FILE  3305PAULL        12--72--84


        10   PARTNO/PAULL
        20   MACHIN/40,OPTION,106,40,41
        30   FROM/1064,320,0
        40   $$ OTHER POST COMMANDS NOT INCLUDED
        50 * TOOL PATH/1.2000,0.0000,0.0000,P1
        60     MSYS/0.0000000,0.0000000,0.0000000,1.0000000,0.0000
000,0.0000000,0.0000000,1.0000000,0.0000000
        70     TURRET/FACE,1,XOFF,100.0000,YOFF,100.0000
        80     SPINDL/SFM,220,CCLW,MAXRPM,2000,RANGE,MEDIUM
        90     FEDRAT/200.0000
       100     GOTO/162.4700,98.4700,0.0000
       110     FEDRAT/MMPR,0.7000
       120     GOTO/152.4700,98.4700,0.0000
       130     FEDRAT/MMPR,0.5000
       140     GOTO/58.3958,98.4700,0.0000
       150     FEDRAT/MMPR,0.8000
       160     CIRCLE/51.8551,78.3042,0.0000,0.0000,0.0000,-1.0000
,21.2000,0.0508,0.5000,2.4000,0.0000
       170     GOTO/51.8551,99.5042,0.0000
       180     GOTO/-0.0000,99.5042,0.0000
       190     FEDRAT/MMPR,5.0000
       200     GOTO/0.0000,114.5042,0.0000
       210     FEDRAT/200.0000
       220     GOTO/162.4700,114.5042,0.0000
       230     GOTO/162.4700,94.4700,0.0000
       240     FEDRAT/MMPR,0.7000
       250     GOTO/152.4700,94.4700,0.0000
       260     FEDRAT/MMPR,0.8000
       270     GOTO/65.5703,94.4700,0.0000
       280     CIRCLE/51.8551,78.3042,0.0000,0.0000,0.0000,-1.0000
,21.2000,0.0508,0.5000,2.4000,0.0000
       290     GOTO/58.3958,98.4700,0.0000
       300     FEDRAT/MMPR,5.0000
       310     GOTO/58.3958,113.4700,0.0000
       320     FEDRAT/200.0000
       330     GOTO/162.4700,113.4700,0.0000
       340     GOTO/162.4700,90.4700,0.0000
       350     FEDRAT/MMPR,0.7000
       360     GOTO/152.4700,90.4700,0.0000
       370     FEDRAT/MMPR,0.8000
       380     GOTO/69.2169,90.4700,0.0000
       390     CIRCLE/51.8551,78.3042,0.0000,0.0000,0.0000,-1.0000
,21.2000,0.0508,0.5000,2.4000,0.0000
       400     GOTO/65.5703,94.4700,0.0000
       410     FEDRAT/MMPR,5.0000
       420     GOTO/65.5703,109.4700,0.0000
       430     FEDRAT/200.0000
       440     GOTO/162.4700,109.4700,0.0000
       450     GOTO/162.4700,86.4700,0.0000
       460     FEDRAT/MMPR,0.7000
       470     GOTO/152.4700,86.4700,0.0000
       480     FEDRAT/MMPR,0.8000
       490     GOTO/71.4193,86.4700,0.0000
       500     CIRCLE/51.8551,78.3042,0.0000,0.0000,0.0000,-1.0000
,21.2000,0.0508,0.5000,2.4000,0.0000
       510     GOTO/69.2169,90.4700,0.0000
       520     FEDRAT/MMPR,5.0000
       530     GOTO/69.2169,105.4700,0.0000
       540     FEDRAT/200.0000
       550     GOTO/162.4700,105.4700,0.0000
       560     GOTO/162.4700,82.4700,0.0000
```

```
570     FEDRAT/MMPR,0.7000
580     GOTO/152.4700,82.4700,0.0000
590     FEDRAT/MMPR,0.8000
600     GOTO/72.6418,82.4700,0.0000
610     CIRCLE/51.8551,78.3042,0.0000,0.0000,0.0000,-1.0000
,21.2000,0.0508,0.5000,2.4000,0.0000
620     GOTO/71.4193,86.4700,0.0000
630     FEDRAT/MMPR,5.0000
640     GOTO/71.4193,101.4700,0.0000
650     FEDRAT/200.0000
660     GOTO/162.4700,101.4700,0.0000
670     GOTO/162.4700,78.4700,0.0000
680     FEDRAT/MMPR,0.7000
690     GOTO/152.4700,78.4700,0.0000
700     FEDRAT/MMPR,0.8000
710     GOTO/73.0544,78.4700,0.0000
720     CIRCLE/51.8551,78.3042,0.0000,0.0000,0.0000,-1.0000
,21.2000,0.0508,0.5000,2.4000,0.0000
730     GOTO/72.6418,82.4700,0.0000
740     FEDRAT/MMPR,5.0000
750     GOTO/72.6418,97.4700,0.0000
760     FEDRAT/200.0000
770     GOTO/162.4700,97.4700,0.0000
780     GOTO/162.4700,74.4700,0.0000
790     FEDRAT/MMPR,0.7000
800     GOTO/152.4700,74.4700,0.0000
810     FEDRAT/MMPR,0.8000
820     GOTO/73.0551,74.4700,0.0000
830     GOTO/73.0551,78.3042,0.0000
840     CIRCLE/51.8551,78.3042,0.0000,0.0000,0.0000,-1.0000
,21.2000,0.0508,0.5000,2.4000,0.0000
850     GOTO/73.0544,78.4700,0.0000
860     FEDRAT/MMPR,5.0000
870     GOTO/73.0544,93.4700,0.0000
880     FEDRAT/200.0000
890     GOTO/162.4700,93.4700,0.0000
900     GOTO/162.4700,70.4700,0.0000
910     FEDRAT/MMPR,0.7000
920     GOTO/152.4700,70.4700,0.0000
930     FEDRAT/MMPR,0.8000
940     GOTO/73.0551,70.4700,0.0000
950     GOTO/73.0551,74.4700,0.0000
960     FEDRAT/MMPR,5.0000
970     GOTO/73.0551,89.4700,0.0000
980     FEDRAT/200.0000
990     GOTO/162.4700,89.4700,0.0000
1000    GOTO/162.4700,66.4700,0.0000
1010    FEDRAT/MMPR,0.7000
1020    GOTO/152.4700,66.4700,0.0000
1030    FEDRAT/MMPR,0.8000
1040    GOTO/73.0551,66.4700,0.0000
1050    GOTO/73.0551,70.4700,0.0000
1060    FEDRAT/MMPR,5.0000
1070    GOTO/73.0551,85.4700,0.0000
1080    FEDRAT/200.0000
1090    GOTO/162.4700,85.4700,0.0000
1100    GOTO/162.4700,62.4700,0.0000
1110    FEDRAT/MMPR,0.7000
1120    GOTO/152.4700,62.4700,0.0000
1130    FEDRAT/MMPR,0.8000
1140    GOTO/73.0551,62.4700,0.0000
1150    GOTO/73.0551,66.4700,0.0000
1160    FEDRAT/MMPR,5.0000
1170    GOTO/73.0551,31.4700,0.0000
```

```
1180      FEDRAT/200.0000
1190      GOTO/162.4700,81.4700,0.0000
1200      GOTO/162.4700,58.4700,0.0000
1210      FEDRAT/MMPR,0.7000
1220      GOTO/152.4700,58.4700,0.0000
1230      FEDRAT/MMPR,0.8000
1240      GOTO/73.0551,58.4700,0.0000
1250      GOTO/73.0551,62.4700,0.0000
1260      FEDRAT/MMPR,5.0000
1270      GOTO/73.0551,77.4700,0.0000
1280      FEDRAT/200.0000
1290      GOTO/162.4700,77.4700,0.0000
1300      GOTO/162.4700,54.4700,0.0000
1310      FEDRAT/MMPR,0.7000
1320      GOTO/152.4700,54.4700,0.0000
1330      FEDRAT/MMPR,0.8000
1340      GOTO/73.3713,54.4700,0.0000
1350      CIRCLE/79.8551,56.5196,0.0000,0.0000,0.0000,1.0000,
6.8000,0.0508,0.5000,2.4000,0.0000
1360      GOTO/73.0551,56.5196,0.0000
1370      GOTO/73.0551,58.4700,0.0000
1380      FEDRAT/MMPR,5.0000
1390      GOTO/73.0551,73.4700,0.0000
1400      FEDRAT/200.0000
1410      GOTO/162.4700,73.4700,0.0000
1420      GOTO/162.4700,50.4700,0.0000
1430      FEDRAT/MMPR,0.7000
1440      GOTO/152.4700,50.4700,0.0000
1450      FEDRAT/MMPR,0.8000
1460      GOTO/76.7499,50.4700,0.0000
1470      CIRCLE/79.8551,56.5196,0.0000,0.0000,0.0000,1.0000,
6.8000,0.0508,0.5000,2.4000,0.0000
1480      GOTO/73.3713,54.4700,0.0000
1490      FEDRAT/MMPR,5.0000
1500      GOTO/73.3713,69.4700,0.0000
1510      FEDRAT/200.0000
1520      GOTO/162.4700,69.4700,0.0000
1530      GOTO/162.4700,46.4700,0.0000
1540      FEDRAT/MMPR,0.7000
1550      GOTO/152.4700,46.4700,0.0000
1560      FEDRAT/MMPR,0.8000
1570      GOTO/150.1023,46.4700,0.0000
1580      GOTO/148.3161,49.7196,0.0000
1590      GOTO/79.8551,49.7196,0.0000
1600      CIRCLE/79.8551,56.5196,0.0000,0.0000,0.0000,1.0000,
6.8000,0.0508,0.5000,2.4000,0.0000
1610      GOTO/76.7499,50.4700,0.0000
1620      FEDRAT/MMPR,5.0000
1630      GOTO/76.7499,65.4700,0.0000
1640      FEDRAT/200.0000
1650      GOTO/162.4700,65.4700,0.0000
1660      GOTO/162.4700,42.4700,0.0000
1670      FEDRAT/MMPR,0.7000
1680      GOTO/152.4700,42.4700,0.0000
1690      FEDRAT/MMPR,0.8000
1700      GOTO/152.3009,42.4700,0.0000
1710      GOTO/150.1023,46.4700,0.0000
1720      FEDRAT/MMPR,5.0000
1730      GOTO/150.1023,61.4700,0.0000
1740      FEDRAT/200.0000
1750      GOTO/163.2000,61.4700,0.0000
1760      GOTO/163.2000,-0.0000,0.0000
1770      FEDRAT/MMPR,0.7000
1780      GOTO/153.2000,0.0000,0.0000
```

```
1790      FEDRAT/MMPR,0.8000
1800      GOTO/153.2000,38.4700,0.0000
1810      FEDRAT/200.0000
1820      GOTO/162.4700,38.4700,0.0000
1830      FEDRAT/MMPR,0.7000
1840      GOTO/152.4700,38.4700,0.0000
1850      FEDRAT/200.0000
1860      GOTO/163.2000,61.4700,0.0000
1870      GOTO/163.2000,38.4700,0.0000
1880      FEDRAT/MMPR,0.7000
1890      GOTO/153.2000,38.4700,0.0000
1900      FEDRAT/MMPR,0.8000
1910      GOTO/153.2000,40.8343,0.0000
1920      GOTO/148.3161,49.7196,0.0000
1930      GOTO/79.8551,49.7196,0.0000
1940      CIRCLE/79.8551,56.5196,0.0000,0.0000,0.0000,1.0000,
6.8000,0.0508,0.5000,2.4000,0.0000
1950      GOTO/73.0551,56.5196,0.0000
1960      GOTO/73.0551,78.3042,0.0000
1970      CIRCLE/51.8551,78.3042,0.0000,0.0000,0.0000,-1.0000
,21.2000,0.0508,0.5000,2.4000,0.0000
1980      GOTO/51.8551,99.5042,0.0000
1990      GOTO/-0.0000,99.5042,0.0000
2000      FEDRAT/MMPR,5.0000
2010      GOTO/0.0000,114.5042,0.0000
2020    * END-OF-PATH
2030      GOHOME
2040      FINI
```

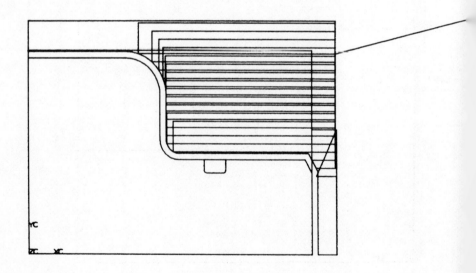

Step 11. Tool path plot also indicating the "FROM location".

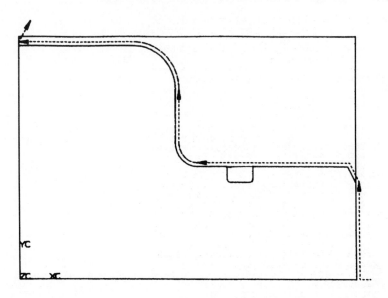

Step 12. Finishing contouring tool path.

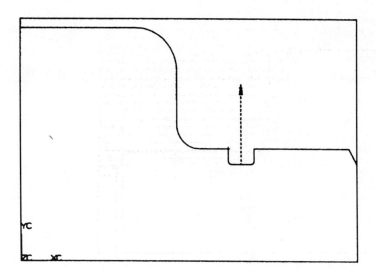

Step 13. The tool path of the grooving operation.

Step 14. The automatically generated finishing and grooving
UNIAPT program.

```
UNIGRAPHICS  CL-SOURCE FILE  3305PAULF      12--72--84


    10   PAINT/OFF
    20   PAINT/OFF
    30 * TOOL PATH/1.5748,0.0000,0.0000,P2
    40     MSYS/0.0000000,0.0000000,0.0000000,1.0000000,0.0000
000,0.0000000,0.0000000,1.0000000,0.0000000
    50     PAINT/TOOL
    60     PAINT/ARROW
    70     PAINT/DASH
    80     FEDRAT/200.0000
    90     PAINT,COLOR,CYAN
   100     GOTO/156.5748,-0.0000,0.0000
   110     FEDRAT/MMPR,5.0000
   120     GOTO/151.5748,0.0000,0.0000
   130     FEDRAT/MMPR,0.1200
   140     GOTO/151.5748,40.4171,0.0000
   150     GOTO/147.3549,48.0944,0.0000
   160     GOTO/79.8551,48.0944,0.0000
   170     CIRCLE/79.8551,56.5196,0.0000,0.0000,0.0000,1.0000,
8.4252,0.0508,0.5000,3.1496,0.0000
   180     GOTO/71.4299,56.5196,0.0000
   190     GOTO/71.4299,78.3042,0.0000
   200     CIRCLE/51.8551,78.3042,0.0000,0.0000,0.0000,-1.0000
,19.5748,0.0508,0.5000,3.1496,0.0000
   210     GOTO/51.8551,97.8790,0.0000
   220     GOTO/-0.0000,97.8790,0.0000
   230     FEDRAT/MMPR,5.0000
   240     GOTO/5.0000,106.5393,0.0000
   250     PAINT/TOOL,NOMORE
   260     PAINT/ARROW,NOMORE
   270     PAINT/DASH,NOMORE
   280 * END-OF-PATH
   290     PAINT/ON
   300     PAINT/ON
   310     PAINT/OFF
   320 * TOOL PATH/0.0000,0.0000,0.0000,P5
   330     MSYS/0.0000000,0.0000000,0.0000000,1.0000000,0.0000
000,0.0000000,0.0000000,1.0000000,0.0000000
   340     PAINT/TOOL
   350     PAINT/ARROW
   360     PAINT/DASH
   370     FEDRAT/200.0000
   380     PAINT/COLOR,CYAN
   390     GOTO/98.6817,72.9575,0.0000
   400     FEDRAT/MMPR,0.1200
   410     GOTO/98.6817,40.0128,0.0000
   420     FEDRAT/MMPR,0.1000
   430     GOTO/98.6817,72.9575,0.0000
   440     PAINT/TOOL,NOMORE
   450     PAINT/ARROW,NOMORE
   460     PAINT/DASH,NOMORE
   470 * END-OF-PATH
```

Step 15. The manually edited program.

```
UNIGRAPHICS  CL-SOURCE FILE  3305PAULF        12--72--84


    10   PARTNO/PAULLFG
    20   MACHIN/40,OPTION,106,40,41
☞   30   FROM/1064,320,0
    40   PAINT/OFF
    50   PAINT/OFF
    60 * TOOL PATH/1.5748,0.0000,0.0000,P2
    70     MSYS/0.0000000,0.0000000,0.0000000,1.0000000,0.0000
000,0.0000000,0.0000000,1.0000000,0.0000000
    80       PAINT/TOOL
    90       PAINT/ARROW
   100       PAINT/DASH
   110       FEDRAT/200.0000
   120       PAINT/COLOR,CYAN
   130       GOTO/156.5748,-0.0000,0.0000
   140       FEDRAT/MMPR,5.0000
   150       GOTO/151.5748,0.0000,0.0000
   160       FEDRAT/MMPR,0.1200
   170       GOTO/151.5748,40.4171,0.0000
   180       GOTO/147.3549,48.0944,0.0000
   190       GOTO/79.8551,48.0944,0.0000
   200       CIRCLE/79.8551,56.5196,0.0000,0.0000,0.0000,1.0000,
8.4252,0.0508,0.5000,3.1496,0.0000
   210       GOTO/71.4299,56.5196,0.0000
   220       GOTO/71.4299,78.3042,0.0000
   230       CIRCLE/51.8551,78.3042,0.0000,0.0000,0.0000,-1.0000
,19.6748,0.0508,0.5000,3.1496,0.0000
   240       GOTO/51.8551,97.8790,0.0000
   250       GOTO/-0.0000,97.8790,0.0000
   260       FEDRAT/MMPR,5.0000
   270       GOTO/5.0000,106.5393,0.0000
   280       PAINT/TOOL,NOMORE
   290       PAINT/ARROW,NOMORE
   300       PAINT/DASH,NOMORE
   310 * END-OF-PATH
   320   PAINT/ON
   330   PAINT/ON
   340   PAINT/OFF
   350 * TOOL PATH/0.0000,0.0000,0.0000,P5
   360     MSYS/0.0000000,0.0000000,0.0000000,1.0000000,0.0000
000,0.0000000,0.0000000,1.0000000,0.0000000
   370       PAINT/TOOL
   380       PAINT/ARROW
   390       PAINT/DASH
   400       FEDRAT/200.0000
   410       PAINT/COLOR,CYAN
   420       GOTO/98.6817,72.9575,0.0000
   430       FEDRAT/MMPR,0.1200
   440       GOTO/98.6817,40.0128,0.0000
   450       FEDRAT/MMPR,0.1000
   460       GOTO/98.6817,72.9575,0.0000
   470       PAINT/TOOL,NOMORE
   480       PAINT/ARROW,NOMORE
   490       PAINT/DASH,NOMORE
   500 * END-OF-PATH
☞  501   GOHOME
   502   FINI
```

Step 16. The listing of the post-processed NC tape file
incorporating the roughing, the finishing and the
grooving operations.

```
              The generated NC program  (turning job)
                                        Page No: 1
+----------------------------------------------------------------+


    FILE = PAULL:LIB1


    000010   GIGAA!FIOII!IIIIF!AAAAO!AAAAO!!!!!!!!!!!!!!!!!!!!!!!
             !!!!!!!!!!!!!!!!!!!!!!!!!!!!
    000020   !!!!!!!!!!!!!!!!!!!!!!!!!!!!!!!!!!!!!!!!!!!!!!!!!!!!!
             !!!!!!!!!!!!!!!!!!!!!!!!!!!!
    000030   !!!!!!!!!!!!!!!!!!!!!!!!!!!!!!!!!!!7!!!!!!!!!!!!!!!!!!
             !!!!!!!!!!!!!!!!!!!!!%<$
    000040   $
    000050   PAULL<$
    000060   :0010G71T100M17<$
    000070   N0020G90<$
    000080   N0030G4X40000T100M53<$
    000090   N0040G94M9<$
    000100   N0050M90<$
    000110   N0060G92S2000<$
    000120   N0070G96R320000S220<$
    000130   N0080G95<$
    000140   N0090G21X-1530Z62470F5397<$
    000150   N0100G1Z52470F700<$
    000160   N0110Z-41604F500<$
    000170   N0120G3X-496Z-48145I20166K6541F800<$
    000180   N0130G1Z-100000<$
    000190   N0140G21X14504F5000<$
```

```
+------------------------------------------------------------+
```

```
000200   N0150Z62470F5397<$

000210   N0160X-5530<$

000220   N0170G1Z52470F700<$

000230   N0180Z-34430F800<$

000240   N0190G3X-1530Z-41604I16166K13715<$

000250   N0200G21X13470F5000<$

000260   N0210Z62470F5397<$

000270   N0220X-9530<$

000280   N0230G1Z52470F700<$

000290   N0240Z-30783F800<$

000300   N0250G3X-5530Z-34430I12166K17362<$

000310   N0260G21X9470F5000<$

000320   N0270Z62470F5397<$

000330   N0280X-13530<$

000340   N0290G1Z52470F700<$

000350   N0300Z-28581F800<$

000360   N0310G3X-9530Z-30783I8166K19564<$

000370   N0320G21X5470F5000<$

000380   N0330Z62470F5397<$

000390   N0340X-17530<$

000400   N0350G1Z52470F700<$

000410   N0360Z-27358F800<$

000420   N0370G3X-13530Z-28581I4166K20787<$

000430   N0380G21X1470F5000<$

000440   N0390Z62470F5397<$

000450   N0400X-21530<$

000460   N0410G1Z52470F700<$

000470   N0420Z-26946F800<$
```

+---+

```
000480   N043063X-17530Z-273581166K21199<$

000490   N044062IX-2530F5000<$

000500   N045026247OF5397<$

000510   N0460X-25530<$

000520   N047061Z52470F700<$

000530   N0480Z-26945F800<$

000540   N0490X-21696<$

000550   N050063X-21530Z-26946I0K21200<$

000560   N0510621X-6530F5000<$

000570   N0520Z62470F5397<$

000580   N0530X-29530<$

000590   N054061Z52470F700<$

000600   N0550Z-26945F800<$

000610   N0560X-25530<$

000620   N0570621X-10530F5000<$

000630   N0580Z62470F5397<$

000640   N0590X-33530<$

000650   N060061Z52470F700<$

000660   N0610Z-26945F800<$

000670   N0620X-29530<$

000680   N0630621X-14530F5000<$

000690   N0640Z62470F5397<$

000700   N0650X-37530<$

000710   N066061Z52470F700<$

000720   N0670Z-26945F800<$

000730   N0680X-33530<$

000740   N0690621X-18530F5000<$
```

The generated NC program   (turning job)
+---+

```
000750   N0700Z62470F5397<$
000760   N0710X-41530<$
000770   N0720G1Z52470F700<$
000780   N0730Z-26945F800<$
000790   N0740X-37530<$
000800   N0750G21X-22530F5000<$
000810   N0760Z62470F5397<$
000820   N0770X-45530<$
000830   N0780G1Z52470F700<$
000840   N0790Z-26629F800<$
000850   N0800G2X-43480Z-26945I2050K6484<$
000860   N0810G1X-41530<$
000870   N0820G21X-26530F5000<$
000880   N0830Z62470F5397<$
000890   N0840X-49530<$
000900   N0850G1Z52470F700<$
000910   N0860Z-23250F800<$
000920   N0870G2X-45530Z-26629I6050K3105<$
000930   N0880G21X-30530F5000<$
000940   N0890Z62470F5397<$
000950   N0900X-53530<$
000960   N0910G1Z52470F700<$
000970   N0920Z50102F800<$
000980   N0930X-50280Z48316<$
000990   N0940Z-20145<$
001000   N0950G2X-49530Z-23250I6800K0<$
001010   N0960G21X-34530F5000<$
001020   N0970Z62470F5397<$
```

The generated NC program (turning job)
                                          Page No: 5
+--+

001030   N0980X-57530<$

001040   N0990G1Z52470F700<$

001050   N1000Z52301F800<$

001060   N1010X-53530Z50102<$

001070   N1020G21X-38530F5000<$

001080   N1030Z63200F5397<$

001090   N1040X-100000<$

001100   N1050G1Z53200F700<$

001110   N1060X-61530F800<$

001120   N1070G21Z62470F5397<$

001130   N1080G1Z52470F700<$

001140   N1090G21X-38530Z63200F5397<$

001150   N1100X-61530<$

001160   N1110G1Z53200F700<$

001170   N1120X-59166F800<$

001180   N1130X-50280Z48316<$

001190   N1140Z-20145<$

001200   N1150G2X-43480Z-26945I6800K0<$

001210   N1160G1X-21696<$

001220   N1170G3X-496Z-48145I0K21200<$

001230   N1180G1Z-100000<$

001240   N1190G21X14504F5000<$

001250   N1200M80<$

001260   N1210G1X320000Z1064000F0<$

FILE = PAULLG:LIB1

The generated NC program   (turning job)
                                      Page No: 6
+--+

```
000010   GIGAA!FIOII!IIIIF!AAAAO!AAAAO!OAGAA!NAMIF!!!!!!!!!!!

         !!!!!!!!!!!!!!!!!!!!!!!!!!!!

000020   !!!!!!!!!!!!!!!!!!!!!!!!!!!!!!!!!!!!!!!!!!!!!!!!!!!!!!

         !!!!!!!!!!!!!!!!!!!!!!!!!!!!

000030   !!!!!!!!!!!!!!!!!!!!!!!!!!!!!!!!!!!!!!!!!!!!!!!!!!!!!!

         !!!!!!!!!!!!!!!!!!!!!!!!!!!!

000040   !!%<$

000050   $

000060   PAULLFG<$

000070   N0010G97M90<$

000080   N0020G4X40000S53M4<$

000090   N0030G95<$

000100   N0040G21X0Z156575F5397<$

000110   N0050Z151575F5000<$

000120   N0060G1X40417F120<$

000130   N0070X48094Z147355<$

000140   N0080Z79855<$

000150   N0090G2X56520Z71430I8425K0<$

000160   N0100G1X78304<$

000170   N0110G3X97879Z51855I0K19575<$

000180   N0120G1Z0<$

000190   N0130G21X106539Z5000F5000<$

000200   N0140X72958Z98682F5397<$

000210   N0150G1X40013F120<$

000220   N0160X72958F100<$

000230   N0170X320000Z1064000F0<$
```

References and further reading

[5.1] W.H.P. Leslie: Numerical Control User's Handbook, McGraw Hill, 1970.

[5.2] P. Bezier: Numerical Control: Mathematics and Applications, Wiley, 1972.

[5.3] H.L. Chambers: Drafting and manual programming for Numerical Control, Prentice-Hall, 1980.

[5.4] J.J. Childs: Principles of Numerical Control: Harper and Row, 1982.

[5.5] Electronic Industries Association: EIA Standards, EIA 1976-79.

[5.6] D. French: Numerical Control of Machine Tools. British Council of Productivity Associates, 1981.

[5.7] D. Gibbs: CNC Machining, Cassel, 1984.

[5.8] Institution of Mechanical Engineers: An annotated bibliography on CNC/DNC machine tools, IME, 1983.

[5.9] P.L. Blake: PROLAMAT '79, Advanced Manufacturing Technology, Conference on programming research and operations logistics in advanced manufacturing technology. North Holland, 1980.

[5.10] IFIP/IFAC Conference: Computer Languages for Numerical Control, Budapest April 10-13, North Holland 1973.

[5.11] B.J. Davies (ed.): 23rd International Machine Tool Design and Research Conference, Manchester, 1982. Macmillan, 1983.

[5.12] Japanese Machine Tool Guide, 1983-84.

[5.13] Morgan Grampian: Fifth survey of machine tools and production equipment in Britain, Metalworking Production 1982.

[5.14] J. Moorhead: Numerical Control, SME, 1980.

[5.15] PERA report No.382: Tool failure monitoring, Development of a system for semi-manned and unmanned NC CNC machining centers, PERA, 1984.

[5.16] J. Pusztai: Computer Numerical Control, Prentice-Hall, 1983.

[5.17] A.D. Roberts: Programming for Numerical Control Machines, McGraw Hill, 1978.

[5.18] R. Shah: Numerical Control Handbook. NCA Verlag, 1983.

[5.19] B. Gross: Electrical feed drives for machine tools, Wiley, 1983.

[5.20] W. Simon: The Numerical Control of Machine Tools, Basic Principles, Systems Analysis and Industrial Applications, E. Arnold 1973.

[5.21] Unimation's VAL Programming manual. (A Unimation/ Westinghouse publication.)

[5.22] Paul G. Ránky: and C. Y. Ho: Robot Modelling, Control and Applications with Software, IFS(Publications) Ltd. and Springer Verlag, 1985.

[5.23] A. Sorge: Microelectronics and manpower in manufacturing, applications of computer numerical control in Great Britain and West Germany, Gower, 1982.

[5.24] T. Toth and D. Vadasz: The TAUPROG System Family, Application Experiences and New Results of Development. Printed by GTI Budapest, Hungary, 1985.

[5.25] M. Horvath and J. Somlo: Optimization and Adaptive Control of Machining Operations (in Hungarian), Muszaki Konyvkiado Budapest, 1979.

[5.26] J.L. Alty: Use of Expert systems, Computer Aided Engineering Journal, Vol.2, No.1, 1985.

[5.27] McAuto Unigraphics: Unigraphics Operational Description and other manuals, MCDONNEL Douglas Corp. Box 516, St. Louis, MO 63166, USA.

CHAPTER SIX

Flexible Manufacturing Systems (FMS)

Flexible Manufacturing Systems offer the most fascinating production method for the computer controlled factory, since by integrating them into CIM (Figure 6.1) one can:

* Increase productivity (often by a factor of 2 - 3.5).

* Decrease production cost (often by 50 %).

* Manufacture (i.e. not only machine, but test, assemble, weld, paint, package, etc.) single parts and/or batches in random order, i.e. on order, rather than on stock.

* Decrease inventory and work in progress (WIP) to a lower level than ever before.

* Provide 100% inspection, thus increasing the quality the product.

* Decrease the amount of often repetitive, or hazardous physical work and increase the need for intelligent, human work.

* Provide a reprogrammable, often almost entirely unmanned manufacturing facility (again underlined not only for machining, but for many other processes and for a wide range of products).

The manufacturing industry and all those involved in its research and development have tried hard to implement cost-performance effective manufacturing facilities in the past.

The most obvious choices for increasing productivity and

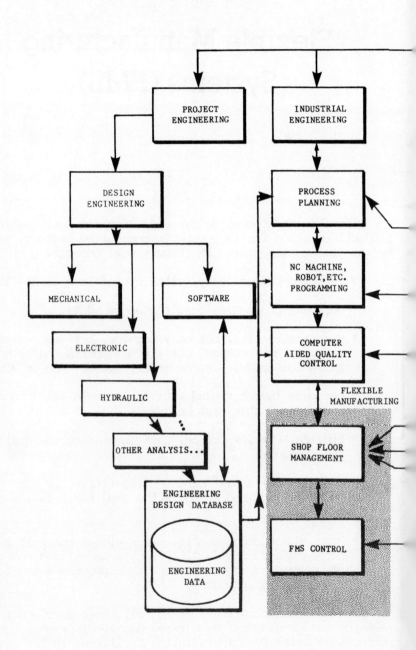

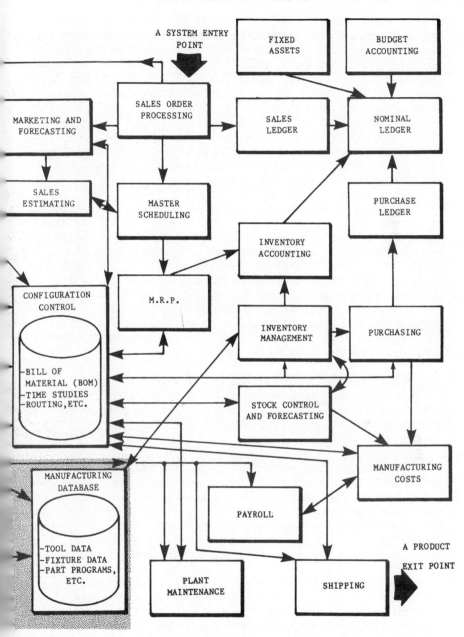

Figure 6.1 An overall Computer Integrated Manufacturing system model and the links to Flexible Manufacturing Systems (FMS.)

CIM-K

decreasing manufacturing cost before the 1960s were the single
and multi-spindle automatic machines and the integrated in-
line type and rotary type transfer lines. As the (mechanical
component) manufacturing industry realized around 1960 that
mass production is only approximately 20 to 30% of the total
output, at the same time the market trend began to shift
towards personalized products manufactured in small batches,
(which as a matter of interest became the major trend in the
1980s, e.g. domestic appliances, Hi-Fi, motor car, furniture,
textile manufacturing, etc.), the automation of small and
medium batch manufacturing methods came to the forefront of
the research and development.

In the 1960s NC machines reached a stage when they became
reasonably reliable and productive. Computers and machine
controllers were developed further, together with production
concepts such as "group technology". By the mid 1960s the
first DNC (Direct Numerical Control) system has emerged in
Japan and the first in 1973 in Europe, in Hungary. Their
control systems have become more powerful along with the
development of tool, material handling and transportation
systems (e.g. automated tool changing systems, industrial
robots, Automated Guided Vehicles, tool magazines, pallet
pools and pallet changing devices).

The first major step towards FMS was made in 1975 when
the first NC machining center successfully operated unmanned
utilizing an automatic tool changing (ATC) system as well as a
10 station pallet pool and automatic pallet changing (APC)
facility.

Since then, but particularly in the 1980s several different highly flexible FMS systems have been installed worldwide. Despite the fact that FMS has emerged from machining and
the traditional machine tool industry, this flexible production method and know-how is invading every other field of
manufacturing, including assembly, test and inspection, welding, packaging and many other production and other fields.

Considering that the number of people employed by the
manufacturing industry is decreasing because most young people
tend to prefer the service industries where the speed of
profit making seems to be faster, and that by the year 2000
the manufacturing industry is going to replace its current
production methods and equipment, FMS technology is of extreme
importance.

Figure 6.2 gives a perfect example showing how the FMS concept is widening in Japan, where even the fashion and textile industry is adopting flexible computer controlled production facilities. We are not too far from the time when not only manufacturing engineers, but also fashion designers will design their new models utilizing three dimensional CAD, solid modeling and animation, and make them using a flexible (and almost unmanned) factory equipped with CNC controlled cutting and sawing machines, computer controlled test equipment using vision, three dimensional sawing and ironing robots, etc.

The aim of this Chapter is to discuss the major components of different FMS systems, concentrating on the computer system architecture, on cells consisting of machine tools, pallet changing and tool changing systems, part washing stations, sheet metal manufacturing cells, welding cells, assembly robots with automated hand changing and part loading facility and other new developments.

Besides material handling and storage systems, full size FMS installations are also discussed in four case studies, these being the Comao FMS, the KTM (Kearny Tracker Marwin) FMS, a Cincinnati-Milacron FMS installation and the Yamazaki FMS factory. The FMS operation software system is also introduced providing an overall picture for further discussion of FMS part loading sequencing and scheduling, FMS capacity planning, lotsize analysis, balancing and secondary optimization (i.e. dynamic) scheduling methods in subsequent Chapters.

6.1 FMS system architecture

To be able to achieve "truly" random production and all the above listed economic benefits, the FMS system architecture must conform to the following basic rules:

1. Utilize highly automated and programmable cells (i.e. CNC machines, robots, etc.) capable of "taking care of themselves" (i.e. incorporating powerful controllers and self diagnostic systems), capable of changing their tools and parts, preferably unmanned, and keeping in touch with a central computer or with a node controller (refer to Chapter 2) from where the production plan, the part programs and further neces-

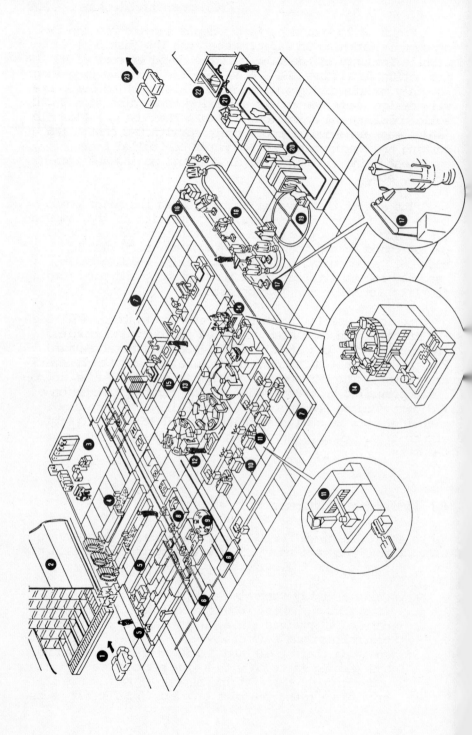

The plant incorporates the following main modules and equipment:

1. Arrival of material
2. Warehouse
3. Computer control room
4. Computer controlled laser cutter
5. Feeder table
6. Transportation truck
7. Transportation system
8. CNC sawing machine
9. Programmable dart jig machine
10. Body processing line
11. Programmable body cutting machine
12. Collar processing line
13. Collar feeder centre
14. Programmable sawing center
15. Sleeve processing line
16. Rejector for defective clothing
17. Three dimensional sawing robot
18. Assembly line
19. Loading/unloading device
20. Finishing stations
21. Final quality control
22. Packaging line
23. Shipment of products to distributors

Figure 6.2 The Japanese textile industry's concept of a flexible, computer controlled garment design and manufacturing plant. (Courtesy of the Ministry of Industrial Science and Technology Ministry of International Trade and Industry, Japan.)

sary data arrive and are fed back.

2. Link these cells into a system by providing preferably direct access, or random material handling
   systems (e.g. AGV), rather than serial access (e.g.
   conveyor line) between them.

3. Create a part, tool and pallet (fixture and clamping
   device) storage facility (i.e. warehouse).

4. Provide high level computer control "inside" and
   "outside" the system, based on a distributed processing system, on databases and on the necessary links
   to other subsystems, such as CAD, CAM and the business
   system as discussed in previous Chapters (see Figure
   6.1).

5. Ensure that if any cells break down the production
   planning and control system can reroute and reschedule
   the production. In other words design in the system not
   only programmable (i.e. flexible) production facilities, but also ensure that the routing of parts can be
   dynamically altered. (Note that this aspect is discussed in more detail in the next Chapter.)

One should realize that FMS is not only applicable to
machining, but is a universal flexible production method which
is relatively easy to control, schedule, operate and maintain
if the above outlined (and to some extent in the previous
Chapters discussed) system design rules are followed.

It must be mentioned that there is some level of confu-
sion in the way different manufacturers and authors use and/or
define terms such as "FMS cell", or "FMS" itself. Most manufacturers define the cell as the smallest building block of the
FMS system consisting of a computer controlled (i.e. reprogrammable) machine (note: not necessarily a machine tool, but
other machines too) capable of executing a single, or a series
of processes on parts automatically loaded and unloaded on the
machine, using tools also automatically changed.

Also note that an FMS cell is preferably unmanned but
does not necessarily exclude some level of human supervision
if the economics of the automation of the involved process, or
processes demand that the designer to employ human operators,
rather than 100% computer controlled equipment.

For a typical FMS cell example taken from the machine
tool industry refer to Figure 6.3 in which the Cincinnati T10
processing center is shown. The most important features of
this cell include:

* Menu driven CNC control system, with built-in CRT
  display and communication ports, program editing
  facility and user-friendly operator interface, resident
  diagnostics monitoring over 200 machine control,
  operator and programmer functions and software for
  multiple program storage for unmanned shifts, probing
  for establishing and/or compensating for pallet and
  fixture offsets, etc.

* Electronic tool length gauging for measuring tool
  dimensions off-line while the machine is producing
  parts.

* Torque controlled machining, consisting of programmable
  Adaptive Control (AC) hardware and software monitoring
  and maintaining optimum metal cutting conditions,
  turning coolant on and off as required, reducing the
  risk of tool or machine damage.

* Automatic tool changing.

* Automatic work changing of palletized components.

* Indexable pallet load station for manual or AGV pallet
  loading, providing the material handling system links
  to the "outside world".

FMS may be defined as a system dealing with high level
distributed data processing and automated material flow using
cells (i.e. computer controlled machines, assembly cells,
industrial robots, inspection machines and so on), together
with computer integrated materials and storage systems. In
this sense FMS is a production know-how offering "on-order"
rather than "on-stock" manufacturing facility for a large
variety of components.

Note that in data processing terms an FMS "cell" is a
node in the distributed control system. Sometimes the terms
"cell" and "unmanned workstation" are mixed too. We prefer the
word "cell" simply because an FMS workstation is often but not
necessarily unmanned. For example consider a manual work mounting station, or a co-ordinate measuring machine with some

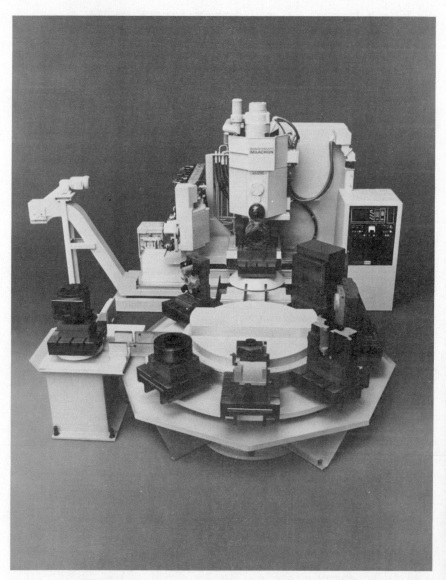

Figure 6.3 The Cincinnati T10 CNC processing center is an FMS
cell in its true sense, capable of receiving workparts mounted on pallets and changing tools automatically, as well as keeping data and material handling system links with the "outside world".

level of human assistance and supervision, etc., which are
"cells" in our general terminology, but are not "unmanned
workstations". (To summarize, we use "cell","FMS workstation",
"computer controlled machine" for roughly the same FMS module.
We believe that any terminology used is acceptable as long as
it is consistent.)

There are large number of photographs attached to this
section showing different FMS cell concepts. Illustrated
applications range from machining to welding, flexible sheet
metal manufacturing and robotized assembly but these by no
means exhaust all possible applications. It has been decided
to keep these illustrations together (Figures 6.4 to 6.20) on
subsequent pages, since many of them show more than one
important aspect discussed above and they represent a kind of
a slide show of different processing cells of FMS.

6.2 Material handling and storage systems in FMS

Generally, requirements set against the FMS materials handling
system include part transportation, raw material and final
product transportation and storage of workpieces, empty
pallets, auxiliary materials, wastes, fixtures and tools.

Materials handling systems in FMS provide palletized
component transfer between cells (Figure 6.4) and include:

* Automated Guided Vehicles (Figures 6.11, 6.12, 6.13 and
  6.21).

* Conveyors, (Figures 6.19/a and b, and 6.25).

* Industrial robots (Figures 6.9, 6.14 and 6.15).

* Special purpose pallet changing and pallet transportation systems (e.g. AGV: Figures 6.11 to 6.13 and 6.21,
  automated pallet carrier on rails: Figures 6.8, 6.28
  and 6.35).

* Forklift trucks (Figure 6.21) and other solutions.

Note that since many of the illustrations show more than
one important aspect discussed in this section they are
grouped at the end of the section or within the case studies.

CIM-K*

Figure 6.4 Photographs showing rigid, modular part fixturing methods and two different pallet designs.

Figure 6.5 The Sandvik-Coromant block tooling system offering
the possibility of storing up to 60 tools per magazine at the machine.

Figure 6.6 The Yamazaki Mazak V10-TS FMS turning cell with
automated tool changing, part loading/unloading
robot and chain type of part magazine.
(Courtesy of Yamazaki Machinery Works Ltd.)

Figure 6.7 Sheet metal processing cell with built-in part
loading/unloading manipulator and CNC control.

Figure 6.8 CNC machining cell with direct access pallet transportation and changing cart and an expandable,
linear buffer store.

Figure 6.9/a Details of the OKUMA CR-30 CNC turning cell.
(Part loading robot prior to pick-up.)

Figure 6.9/b Details of the OKUMA CR-30 CNC turning cell.
(The part loading operation.)

Figure 6.9/c Details of the OKUMA CR-30 CNC turning cell.
(The chain type part magazine.)

Figure 6.10 The vertically indexing, four station pallet magazine and changing system of the "Cubotic" FMS machining cell.

Since computer controlled material handling systems offer
a wide range of solutions in FMS and because the inadequately
designed materials handling system can limit an otherwise
highly flexible manufacturing environment, it is important to
analyze this problem very carefully when designing the FMS.

Probably due to the strong influence of conventional
transfer line developments, which are less flexible than FMS,
between 1974 and 1980 the serial access workpiece transport
systems (i.e. conveyors) dominated the FMS scene. The direct
access, or random materials handling systems are not only more
flexible than the closed loop, or sequential workpiece
transport systems, but can also simplify dynamic scheduling
and flexible operation control of the system.

In systems designed to manufacture prismatic parts, the
workpieces are located on pallets by means of fixtures and
clamping devices. The workpiece must therefore be loaded,
unloaded and transported between FMS cells and workstations
with these reasonably heavy additional elements (see again
Figures 6.3 and 6.4).

In the case of rotational part manufacturing systems,
workpieces are usually carried in small batches or stored in
workpiece magazines, in a similar way to tools (see again
Figures 6.6 and 6.9).

AGVs (Automated Guided Vehicles) are trucks which usually
pick up loads from a work mounting station (where components
are fixtured and clamped on pallets), from the machine table,
a buffer store location or a warehouse input/output location.
Components of AGV systems include the carts, the battery-
powered driverless vehicles capable of selecting their own
path and the cart and traffic control system.

Typical AGV control tasks involve:

* Steering control.

* Sensing of obstacles and control signals transmitted
  via wires or wireless.

* Communication with the traffic control system. AGV
  systems provide the most flexible, direct access
  material handling system in FMS.

Figure 6.11 Sajo FMS cell loaded and unloaded by a BT-Handling
AGV (Automated Guided Vehicle.) The benefit of
this system design is that this combined AGV and
forklift truck can be used not only to transfer
parts between machines, but also to access the
pallet rack, or warehouse. (See also Figure 6.21.)

Figure 6.12 AGV (Automated Guided Vehicle) docking station and
pallet pool of the TI Matrix FMS cell.

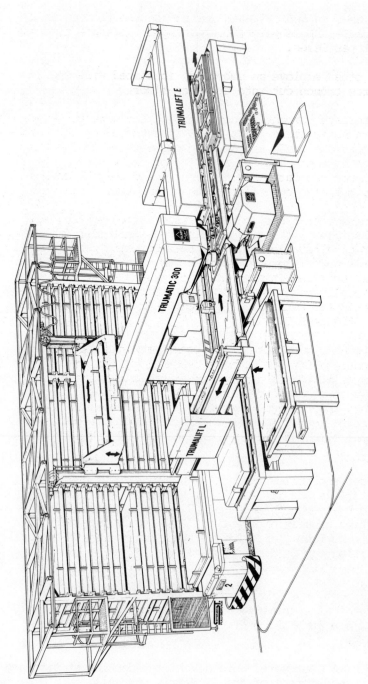

Figure 6.13 The Trumpf TMS Material handling system with the Automated Guided Vehicle is a sheet metal processing cell in its own right. (Drawing by Steiner Full Service, Germany-W.)

There are two different ways of guiding the AGVs.

* One uses an optical path painted on the floor with a special-purpose paint, containing for example fluorescent particles.

* The other employs an embedded, insulated wire in a narrow trench cut into the shop-floor.

An alternate current is sent through the wire to generate an electromagnetic field. This field is detected by the vehicles and electric signals are sent to the servomechanism that controls the movement and the steering of the vehicle. In complex systems the carts are not controlled only by on-board processors, but also by a central host computer.

AGVs can offer a horizontal transportation system, such as towing vehicles with trailers attached, unit load transporters carrying pallet mounted components between the warehouse, buffer stores and different machines and light load transporters, mainly utilized in the automated assembly industry, and the automatically positioned warehouse carriers and trucks featuring computer controlled movement both horizontally and vertically (Figure 6.21).

An automated warehouse is a system providing addressable storage locations for one or more types of pallets on which material maybe moved and handled under computer supervision. Automated warehouses are required when the raw materials, semi-finished and finished products, and the necessary tools, fixtures, spares etc., have to be stored and retrieved using computer controlled stacker cranes, drives and lifts (Figures 6.21/a, 6.22 and 6.23/a-f).

Computer controlled stacker cranes integrate computer controlled drives and lifts, fine positioning and independent lift and fork movements (see again Figures 6.23/e-f). Positioning systems usually employ two different steps, i.e. the primary and the fine positioning steps. Primary positioning is usually carried out by dual pulse generators and a work reader system. Fine positioning needs precise photocell alignment with a reflecting label at each pallet stack.

For material storage purposes automated warehouses are integrated into flexible production systems, covering functions such as:

Figure 6.14 Automated Robot Hand Changing (ARHC) system of an
unmanned assembly cell, designed by the author.The
robot is capable of automatically reconnecting
different robot hands not only mechanically, but
also electronically and pneumatically in under 4
seconds.

* Communication with the FMS material handling system,

* Materials handling system control (e.g. stacker crane control, AGV control, conveyor control, etc.).

* Real-time control of the warehouse input/output stations.

* Recording individual pallet loads and locations.

* Rearrangment of stored items in the warehouse.

* Optimization of crane and truck movements for maximum throughput.

* Stock control, order picking.

* Packaging and shipping of the product.

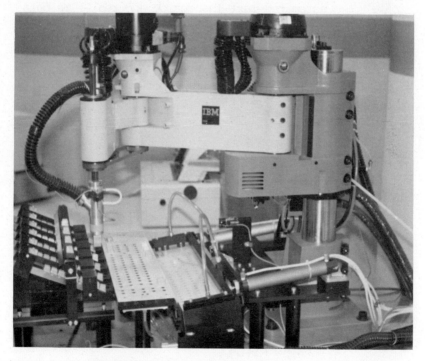

Figure 6.15 IBM 7535 Scara type assembly robot working on a keyboard assembly job.

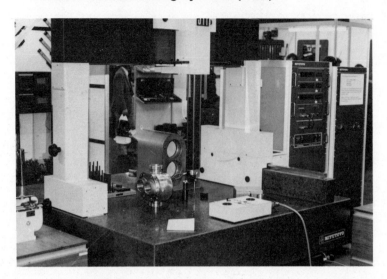

Figure 6.16/a Co-ordinate Measuring Machines represent an
         important part of the FMS because they provide
         automated part inspection and real-time data
         feedback to the quality control module of the
         DNC system.

Figure 6.16/b Gearbox measurement using a Co-ordinate Measuring
         Machine (CMM.)

Figure 6.16/c The photograph shows the PH5 touch-trigger probe, with external adjustable over-travel unit and with a selection of accessories. The TP2 probe is omni-directional and the probe body diameter is 13 mm. (Courtesy of Renishaw Electrical Ltd., England.)

Figure 6.17 The LK-Cincinnati Milacron Co-ordinate Measuring Machine is integrated into the Cincinnati Flexible Manufacturing system. The inspection cell can be loaded and unloaded by an Automated Guided Vehicle. (The complete system is discussed in section 6.3 and further photographs are shown in Figures 6.30 to 6.33.)

Figure 6.18 A flexible welding cell incorporating the Daros PT-300V industrial robot and the HU550 welding manipulator. The manipulator
is capable of carrying 550 Kg. Both the robot and the manipulator
are controled from the robot control unit. (Courtesy of Dainichi
Sykes Ltd., England.)

Figure 6.19 SCAMP FMS cells linked together by means of a conveyor line representing a serial access material handling system. (Courtesy of SCAMP Systems Ltd., Colchester, England.)

Figure 6.20 Automated Guided Vehicles deliver workpieces and fixtures to large, six spindle gantry type milling centres manufactured by FPA Droop and Rein with Bendix control. The machines are equipped with twin tables allowing fixturing on one, during

From the CIM point of view the most important system objectives of automated warehouses in FMS can be summarized as follows:

* Shorter manufacturing throughput times, because the parts are available within the system, as well as being available for access by the FMS to FMS material handling systemsx.

* Reduced inventory because of the better organization and the minimum (i.e. almost nil) work in progress (WIP) level required in FMS.

Figure 6.21/a This BT-Handling Automated Guided Vehicle is capable not only accessing the machining cells in the FMS, but also the warehouse where workparts are stored on pallets. The AGV is equipped with an automatic side shift unit for precise lateral movement. Lifting capacity of the AGV is 3 meters. (Courtesy of BT-Handling, Sweden.)

* Elimination of costly repetitive handling and data management of material found in conventional systems.

* Increased productivity.

To summarize, automated material handling and storage systems are essential in FMS as well as in the overall CIM model because they affect all major processes including receiving raw material, inspection of raw material, part manufacturing, test and inspection, assembly, packaging shipping and linking together different FMS systems, warehouses and different shops in a factory.

6.3 Auxiliary devices in FMS

Auxiliary devices in FMS include buffer stores, swarf removal or part washing stations, work mounting stations and other

Figure 6.21/b The forks of this "high-lift AGV" are equipped with a light source and a photocell for fork positioning when docking at the machining cells, as well as when loading and unloading palettised workparts at the pallet rack. (Courtesy of BT-Handling, Sweden.)

Figure 6.21/c The total positioning error of this solution is
within 4 milimeters in both x and y directions,
which is sufficiently accurate since the machine
docking station is capable of automated self-
alignment. (Courtesy of BT-Handling, Sweden.)

stations such as AGV recharge, tool maintenance room, tool
setup stations, etc.

Even when aiming at a minimum level of WIP buffer stores
cannot be avoided, partly because of production control
constraints, and partly because of technological reasons (for
example the part must cool down before the next operation). In
this case the palletized part must wait in a buffer store,
which can of course be used at the same time as an in-process
pallet rack, or small warehouse in order to save investment
costs. As examples refer to Figures 6.3 and 6.5 again, where
the pallet pool is a buffer store as well as a "little
warehouse", or Figure 6.7 where the machine has a modularly
extendable line of pallet buffers handled by a pallet changing
and transportation cart.

Pallet mounting and fixturing are complex, but not impossible tasks to be resolved automatically. Automated fixturing

(utilizing hydraulically operated devices, or mechanically operated elements moved by robots) and general purpose workholding devices can solve this problem, but because of the weight, required precision and the complexity of movements in most cases it is done manually, especially within a family of prismatic parts.

Special purpose workholding devices can be fully automated, but they place great restrictions on flexibility if one hopes to manufacture a large variety of parts in a random order. The designers of the components can also contribute a great deal to solving this problem, by constructing appropriate surfaces on parts by which they can be clamped and located on the pallet (see again Figures 6.4 and 6.10).

The swarf clearance and retrieval is done at the cells themselves (Figure 6.24) as well as at the washing station of the FMS. Perfect workpiece cleaning is important, especially before inspection. Unremoved swarf can cause several unexpected problems during the automated inspection cycles.

Figure 6.22 Automated warehouse in the Yamazaki Flexible Manufacturing System. (Courtesy of Yamazaki Machinery Co., Japan).

(a)

(b)

Figure 6.23/a-f Photographs taken inside a computer controlled
              warehouse, illustrating the conveyor based part
              transportation system (a, b and d) in various
              parts of the warehouse, the warehouse manipu-
              lator (c and e), and a docking operation using
              sensory controlled positioning techniques (f).
              (Courtesy of SCICON Ltd., England.)

(c)

(d)

(e)

(f)

In some systems the pallet is loaded by an AGV, or conveyor line with the fixtured part on the washing station, where it is docked. Then it is tilted by a mechanism, while being rinsed under high pressured coolant or air. In other

Figure 6.24 Mori Seiki FMS with automated swarf removal and transportation facility.

designs the washing station follows a shape similar to that of a bell. This is then lowered onto the part from above, utilizing vacuum while it is waiting, on a wired cart for example.

In order to improve washing quality recently robots have been programmed to perform this job utilizing vacuum tools and/or washing liquid jet guns. This solution is providing a better result than those previously discussed since the robot can execute a washing cycle which was programmed for the particular part, or batch. (To raise another important aspect of FMS programming one could generate part washing programs automatically using the CAD/CAM system and an expert system to ensure high quality of cleaning.) Once the part is clean it can be taken away by a robot or Automated Guided Vehicle together with its pallet. (The first of its kind in the world, the Comao robotized washing cell is shown inside and outside in Figures 6.27/a and b.)

6.4   Case study: A COMAO FMS installation

The COMAO FMS described in this case study was developed by
Dott. Ing. Sergio Romanini and his collegues at the Italian
firm COMAO in Modena for Borg Warner (USA) in 1982-83. The
system has been in industrial use since then. It consists of
five multi-function machining centers capable of manufacturing
86 different parts (representing seven families of components). (See Figure 6.25 showing the system during its final
assembly phase at the Modena plant prior to it's shipment to
Borg-Warner in the USA).

   According to part availability the DEC PDP11/44 central
DNC computer loads and schedules the components from the
automated warehouse. The calculated order is continuously
reviewed and optimized according to casting availability and
changing priorities, or constraints in the system.

   Besides loading and scheduling the computer sends instructions to the work mounting and tooling stations regarding
which part requires what fixtures, clamping devices and tools.
The MSR type machines in the system can store up to 140
randomly selected tools in a chain type tool magazine (see
Figure 6.26 where the machine is shown during assembly). The
total number of tools in the real-time controlled system can
be as high as 572. (Recall the case study discussed regarding
tool management in section 2.2.3.)

   The system performs turning, milling, drilling, boring
and tapping of components including crankshafts and compressor
housings ranging in size from 200x140x20 mm to 1300x800x700mm
and in weight of 1Kg to several hundred kilogrammes and
washing (see Figures 6.27/a and b).

   Machining cells in the system incorporate infra-red tool
break sensors and adaptive control system altering feedrates
to compensate cutting forces.

   The average system efficiency is 70% representing a
production volume of approximately 160 components per day (in
three shifts) and 39000 components in a year.

Figure 6.25 A COMAO FMS design under installation in Modena, Italy. (Courtesy of COMAO SpA, Modena, Italy.)

Figure 6.26 Twin chain type tool magazine of one of the milling
centers in the COMAO FMS under construction.
(Courtesy of COMAO SpA, Modena, Italy.)

Figure 6.27/a The five axis COMAO Smart robot is programmed
for each different component and manipulates a
pressurized coolant system to clean parts. The
robot is capable of washing out even tapered
holes to tight specifications. (The photograph
was taken outside the washing cell.) (Courtesy of
COMAO SpA, Modena, Italy.)

Figure 6.27/b Inside view of the COMAO robotized washing cell,
which is claimed to be the first of its kind
in the world. (Courtesy of COMAO SpA, Modena,
Italy.)

6.5 Case study: The KTM FMS

Kearney Trecker Marwin represents a very flexible and modular
FMS approach which can be developed and further expanded in
many steps. This is a suitable approach not only from the
system design point of view, but also from that of capital
investment distribution.

The basic machining cell of their system is the Fleximatic FM100 machining center with rotary indexing four station
pallet pool with automated part loading and tool changing
facility. The next stage is the replacement of the four station pallet pool is by a twin pallet changer and the addition
of a rail guided pallet transport vehicle controlled by the
CNC system handling up to 15 pallet stations arranged along a
line (Figures 6.28/a and b and Figures 6.29 and 6.30). If at
least two or more cells are integrated, further cells can be
added to this system architecture, including a control
computer dealing with:

* Off-line scheduling and loading workparts onto the system.

* DNC control, meaning the possibility of editing CNC part programs and downloading them to the controllers, as well as receiving and storing important part program, inspection, tooling, etc. data for further analysis.

* On-line control, which is a real-time scheduling program determining the workflow between all cells in the system.

* Tool management system, containing a library of tools available for use in the system whether assembled or not, together with the tool requirements associated with each part program.

* Management Information Module offering system monitoring data on a dynamic basis of production quantities, break-down of equipment, tool break, quality control information, etc.

The data processing equipment and software modules in the KTM FMS rely on Siemens 8M CNC controllers and the Siemens R30 mini computer with Siemens and KTM software and also on Programmable Logic Controllers (PLC) supervising cells and various ancilliary equipment.

6.6 Case study: A Cincinnati Milacron FMS installation

The purpose of this case study is to illustrate the Cincinnati solution to building up FMS from cells and in particular to introduce their FMCC, FMS controller.

This discussion of a Cincinnati Milacron FMS installation is based on visits payed to the MACH'84 International Machine Tool Exhibition in Birmingham and on several discussions, demonstrations and visits between the chief designer of the system, Mr. Ian Taylor and the author of this text. The FMS solution represented by Cincinnati is currently one of the most modular, expandable and flexible.

Figure 6.28/a KTM/Roevatran railed pallet carrier in the KTM
Flexible Manufacturing System. (Note the mechanically set, floor mounted coding system designed
for positioning the cart accurately in front of
a cell or a buffer store, etc.)

Figure 6.28/b The KTM FM 100 Fleximatic machining cell in the
Kearney and Trecker Marwin FMS.

CIM-L*

Figure 6.29/a KTM/Roevatran pallet carrier in front of machining cells and buffer stores.

Figure 6.29/b Pallet docking operation at a buffer store in the KTM FMS.

In the pilot plant discussed in this text (see Figures 6.30 to 6.33) workpieces are fixtured to machine pallets and loaded by the operator at remotely situated pallet-load (i.e. work mounting) stations (Figure 6.30). The palletized component is transferred automatically between the work mounting station and the machines by an Automated Guided Vehicle (AGV). Before returning to the work-mounting station for dismantling, the part may travel along a route between several machines in the DNC system including milling and boring centers, washing and inspection cells.

The principle controller of the system is the FMCC incorporating two Cincinnati Milacron Programmable Logic Controllers and an optional IBM/XT microcomputer. One of the PLCs is dedicated to solving the real-time control tasks, whereas the other deals with sequencing operations of the load-stations and the pallet transfer between the vehicle and receiving unit. The IBM/XT, programmed in Pascal, is an optional add-on to give more powerful and user friendly FMS control features as well as handling the data links to CAD/CAM systems, or other FMS.

6.6.1 Work mounting (AWC)

The work mounting, or pallet loading station is an Automatic Work Changer (AWC) adapted for vehicle docking. Up to two AWC units may be incorporated in a system representing a total of 24 pallets. Each pallet is uniquely coded with a both man and machine readable number, which corresponds to a fixed position within the AWC, thus the maximum capacity of the AWC units determines the maximum number of pallets within any system.

During system operation the work mounting station is accessed either by the human operator, or by the AGV. If the AWC is free the operator may select a pallet and transfer it to the AWC. When the AGV arrives it unloads its pallet, if it carried one, and waits until a new load arrives from the AWC unit. If no further pallet is available the vehicle will wait at the AWC until either a pallet is ready for loading onto the vehicle, or another vehicle journey is requested by one of the cells of the system.

Before pallet transfer to the vehicle and after pallet transfer from the vehicle, the 5 bit pallet number is read and compared with the pallet code specified by the FMCC controller, thus ensuring that no pallet is "lost" in the system. The AWC has a pallet decoder at the fixed vehicle docking position, so that the presence and absence of a pallet may be correlated with the arrival or removal of the pallet being investigated by the system. (The zero pallet code is taken to be the absence of a pallet.)

After the operator has returned a pallet to the AWC he must update the pallet status within the FMCC to reflect his action. This will show that the pallet is empty, or unloaded. The status of each pallet is shown on the FMCC monitor which is always positioned next to the AWC(s). The AWC(s) themselves are controlled by an Acramatic Programmable Controller, which is additional to the FMCC. This extra CNC controller also synchronizes the vehicle docking operation at both the AWC and other machines in the FMS.

6.6.2 Vehicle transport

The vehicles' function is to transfer one palletized component at a time between the docking stations of the machines in the FMS (Figures 6.31 to 6.33). The routing is controlled by the FMCC, but the docking procedure is controlled by a combination of the vehicle, docking station and docking controller unit. (The actual procedure varies depending on whether AGV, railed shuttle cart, or other part carrier is used in the system.)

Feedback that the pallet has arrived and the pallet transfer has been completed is reported to the FMCC via the docking controller. Collision of vehicles is eliminated by allowing only one vehicle in the FMS at a time. This means that up to eight docking stations can be served, including the ones at the machining cells, washing machine, co-ordinate measuring machine and the AWC(s).

6.6.3 Machining cells in the FMS

The machining cells are controlled by the Accramatic 900 CNC controller (Figure 6.31). Each machining center is equipped with a pallet buffer unit, allowing palletized work to be kept at the machine to permit maximum continuity in machining.

When a new component arrives the 5 bit pallet code is decoded at the outer pallet shuttle position and the result is sent directly to the Acramatic 900 CNC controller.

Figure 6.30 Automated Work Changer and pallet mounting cell in
the Cincinnati Milacron FMS.

Figure 6.31 The machining cells of the Cincinnati Milacron FMS
provide AGV docking facility and incorporate the
Acramatic 900 CNC controller with DNC (Direct
Numerical Control) link to the FMCC system supervisor computer.

The appropriate part program is selected by an 8 bit code, containing the 5 bit pallet code, a 2 bit program identifier and a parity bit, sent to the CNC by the FMCC controller. The 5 bit pallet code of this 8 bit code is compared with the 5 bit code, decoded by the Acramatic 900 controller from data received from the outer shuttle position, to ensure that the correct part program is executed.

The full 8 bit code, received from the FMCC is used for part program selection from either the Acramatic 900 program storage or the data management facility of the system. (It is anticipated that the part program would always be downloaded prior to machining.)

The additional 2 bit code is to allow for cases where the pallet is either re-fixtured for a second operation, or makes more than one journey to the same machine during its route within the system.

When M02 is read at the end of the currently executed part program the Acramatic 900 CNC informs the FMCC that the program is complete. The FMCC may then issue the pallet transfer request to the CNC controller of the machine, assuming that no vehicle journey or docking activity is pending. If the CNC informs the FMCC of a machining cycle error the rest of the program is terminated, the remaining route is aborted and a direct return command is sent to the AGV to take the pallet back to the AWC. If the error at the machine requires operator intervention, the operator can list, analyze and edit any of the part programs at the FMCC controller. Reaching a satisfactory solution the error signal sent previously by the CNC controller can be cancelled and a reinstated route assigned to the component.

6.6.4 The part washing cell

The part washing machine is controlled by the docking controller. For cell mode to be selected the docking station must be aligned. The selection of stand-alone mode does not cause an immediate interruption of the washing process. The washing station is accessed by the AGV in a similar way to the machining cells, but it has no pallet buffer (Figure 6.32).

6.6.5 The inspection cell

The inspection cell is based on a CNC Co-ordinate Measuring Machine capable of receiving palletized components and measuring programs from the FMCC controller in a similar way to the machining cells,even though all functions are not present. The inspection cell does not have a pallet buffer (Figure 6.33).

6.6.6 The FMCC controller

The FMCC synchronizes at a system level the pallet trans-port and cell activity inside the DNC controlled area. It also provides an interface for the FMS operator to establish the schedule of work and to perform other system level activities via the keyboard of the IBM/XT and the three available VDU screens and several "soft" function keys (i.e. function keys configured by means of software), in an interactive mode accessing:

* The initial power-on format screen.

* The process status screen.

* The pallet status screen.

* The routing page.

* The assign page.

The system is initialized by turning a key selector switch to cold start simultaneously with pressing start on the console. Before starting automated operation the pallets must

Figure 6.32 Robotized part washing cell in the Cincinnati Milacron FMS.

Figure 6.33 Pallet loading and unloading operation with an AGV
at the Cincinnati Milacron/LK tools Co-ordinate
Measuring Machine.

be physically available in the AWC of the pallet loading
station. System initialization clears all previous status
information and reinitialization is used normally only after a
complete shut-down, or unrecoverable system error.

Each cell (e.g. machining, inspection, washing and the
vehicle system) may be individually included or removed from
the FMS on a logical basis. (The AWC is considered to be a
part of the operator station, thus it cannot be removed
logically from the system). If a cell is removed from the FMS
no more pallets are despatched to it and pallets are removed
from it on reaching the end of the process.

Each process can be both physically and logically separated from the FMCC controller and external logic by selecting
stand-alone mode at the FMCC at a press of a function key. If
this is the case all signals are inhibited from passing
between the selected processing cell(s) and the FMCC control-

ler in both directions, with the exeption of the AGV "pallet complete" and "docking complete" commands. At the time of switching to stand-alone mode there may be vehicle journeys in progress to the machine, which are completed before switching off the cell from the system.

6.7 Case study: Yamazaki FMS installation

The Yamazaki FMS and FMF (Flexible Manufacturing Factory) represent large scale integration of a variety of different computer controlled equipment and processes. Within this case study we introduce one of their first systems at the Oguchi plant, incorporating two parallel flexible machining lines and we underline the economics of FMS by listing savings published by Yamazaki.

To illustrate the savings Yamazaki claims that by utilising FMS technology, they have reduced the building time of a medium sized CNC machining center from scratch from 18 weeks to just 4 weeks, because flexible manufacture has eliminated downtime for setting and greatly reduced the necessary machine setups, typically from 10 to just 3. Besides the increased productivity the inventory and work in progress levels were reduced to 1/20th of its previous (i.e. "pre - FMS") value.

To underline the major benefits FMS can bring if properly designed and utilized, let us list some more economic data (published by Yamazaki) for 543 different types and 11,120 items of workpieces manufactured in their Minokamo FMS:

* Compared to the conventional system where they have utilized 90 machines, FMS needs only 43.

* Instead of 170 operators they need only a total of 36 in the computer room, tool room and on the shop floor, and only 3 dealing with workpiece transfer and production control, compared with the 25 before. (This represents a total of 39 operators when in FMS and 195 when conventional.)

* In-process times have been reduced dramatically too. In FMS they spend only 3 days on machining, 7 days on unit assembly and 20 days on total assembly, compared to 35, 14 and 42 days in the conventional system.

(Representing a total of 30 days in FMS compared to 91 in the conventional system.)

* Finally the floor space was reduced too, from 16,500 square meters to 6,600 square meters.

By analyzing the layout of their FMS and FMF system, as shown in Figure 6.34, one can identify the following cells and equipment:

* Six YMS H-25Q machining centres equipped with ATC and APC.

* Three sets of pallet loading and carrier carts with a load capacity of 3 tons (6600lbs) each, (Figure 6.35).

* Two sets of pallet loading and carrier carts with a load capacity of 3, or 6 tons (6600/35200 lbs).

* Two workpiece storage racks and loading stations.

* Storage rack for complete parts (59 locations) and stations for work reversing on the pallet.

* Two carriage grinding cells.

* Twelve power centers (H-22).

* Two tool drum loading/unloading robots (Figure 6.36).

* One grinding and hardening machine.

* One central computer room (Mazak).

* The tool room.

Standard features in the system include DNC control of all computer controlled equipment, automatic part and tool transport, automatic tool life monitoring, adaptive control system (operated by altering feedrate), tool break sensing, automatic pallet centreing and alignment, self diagnostics.

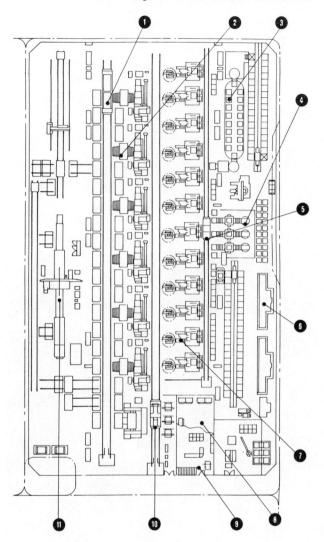

Figure 6.34 The Yamazaki Flexible Manufacturing System layout
and FMF (Flexible Manufacturing Factory) concept.

Legend: (1) Pallet loading cart, (2) Six YMS H-25Q machining
centres, (3) Workpiece storage and part loading station, (4)
Workpiece reversing station and storage rack, (5) Pallet
loading cart, (6) Carriage grinding machine, (7) Twelve
machining centers H-22, (8) Computer control room, (9) Tool
room, (10) Tool drum changing device, (11) Bed grinding
machine. (Overall length=95 m, width=50 m, approximately.)

Figure 6.35 Pallet loading and carrier carts in the Yamazaki Flexible Manufacturing System. (Courtesy of Yamazaki Machinery Co., Japan.)

Figure 6.36 Machining centres and tool drum loading/unloading
manipulators in the Yamazaki FMS. (Courtesy of
Yamazaki Machinery Co.)

6.8 FMS operation control

So far we have discussed the major components of CIM including
those used by the FMS subsystem of the computer integrated
factory, i.e. data base management, computer communication
networks, CNC control and programming of computer controlled
machines and robots.

   We have also seen case studies demonstrating different
features of FMS systems, underlining the flexibility offered
by cells. In the case of the Cincinnati FMS also some details
of its operation control, using the FMCC controller were
discussed.

Let us now briefly extend the already discussed principles with some new rules regarding FMS operation control and
summarize them in order to understand more details of the ways
FMS systems can be operated and controlled, and to introduce
further Chapters where FMS scheduling (Chapter 7), manufacturing system capacity planning (Chapter 8), lotsize analysis
(Chapter 9), and operation balancing (Chapter 10) are discussed in detail.

6.8.1 Summary of the FMS operation control activities

A summary of the most important operation control activi ties
is shown in Figure 6.37, identifying three levels at which
simulation and optimization is required prior to, or during
FMS part manufacturing, the three levels being:

* The factory level, or business level handled by the
  business system of CIM.

* The FMS off-line level, representing simulation and
  optimization activities prior to loading a batch or
  a single component on the FMS, handled sometimes by
  the CAM system, sometimes by the FMS part programming
  computer.

* The real-time controlled level, handled by the FMS
  operation control system, representing a situation
  where the parts are already physically as well as logically in the DNC controlled environment.

One can also see from this Figure that to satisfy the
demand generated by the business system modules of CIM (i.e.
the MRP and the MPS programs) the first major task is the
selection of the appropriate part mix for a defined period
of time, which can be as short as a few hours, a shift
(normally eight hours), a day, a few days, or a few weeks, but
usually not longer. (Note that this period of time very much
depends on how random is the part arrival at the FMS, what
the batch sizes are, whether there are batches at all, etc.)

If there are a sufficient number of different batches of
components and if the time span allows, (i.e. due dates are
not too close) it is worth while doing a batchsize, or lotsize
analysis to group the components into economic batches and to
establish economic cycle times.

Batchsize analysis is less important in FMS compared to conventional methods, because the aim is a single component, rather than batch production, because the setup and down times are low anyway, and because the entire production is better organized, there is practically no WIP and the inventory holding times are much shorter too. However it can prove to be useful and can save production cost and time, depending on the components, on the setup costs, on inventory holding costs and on other less important factors. (Note that lotsize analysis is explained in Chapter 9.)

Balancing can also be useful in particular if there are many operations to be done on the part, and if the resources allow different operations to be performed in a random order at no additional cost (e.g. if machines have an adequate selection of tools in their magazines, or if robots can automatically change their hands, if machines and assembly robots can change parts automatically, etc.). Balancing has a potential in the machining oriented FMS too, when one cell is more "popular" than the others, and thus becomes overloaded.

Balancing in robotized assembly systems (or FAS-Flexible Assembly Systems) is generally very beneficial because of the large amount of different operations assembly systems usually perform and because it helps to decide how the operations should be distributed between the assembly robots. Finally in the case of conventional mechanized assembly line development balancing is an essential design tool to evaluate the number and arrangement of different assembly heads and the overall cycle time of the line. (Balancing for assembly type operations is discussed in detail in Chapter 10).

To avoid unrealistic planning, capacity planning is important, but capacity checking is even more important shortly after the above discussed processes in order to ensure that the established part mix can be processed on a given system. This process is indicated by feedback loops at the three major modules of the off-line process (Figure 6.37).

Loading sequencing, or FMS off-line scheduling is a simulation, or optimization method to establish the best order in which parts should be processed by the system in a short period of time, e.g. few hours, or upto the length of a shift.

If parts arrive at the FMS in a random order, in very

small batches, or as single components and if they must be executed as soon as they arrive, there is not much point in selecting the appropriate part mix, nor in establishing an optimum loading sequence, because there is simply no time for that. However in the case of most FMS, parts arrive in a fairly well planned order, there is time to load sequence and then it is essential not only for utilizing the system at a level close to 100%, but also for processing the parts without delay.

The input to the loading sequencing program can be the MRP, or directly the MPS output, or alternatively if higher level optimization is possible, the result of the part mix evaluation procedure using lotsize analysis and operation balancing with capacity checks. Loading sequencing should also take care of some real-time disturbances, such as which cell is not operational, etc. but is not a real-time program, although it is valid for a very short time span only. (Note that FMS loading sequencing is described in more detail in Chapter 7).

Once the loading sequence is known parts can be sent in the calculated order to the FMS for processing. At this point there is a need for a DNC program which deals with data communications tasks, part program editing and downloading and others as we have seen in the Cincinnati FMS case study in section 6.6 and as described in [6.12].

FMS must be able to react to real-time changes; to mention a few, what should happen to its operation control:

* If the processing priorities of a part or a batch must be changed?

* If a part, or a batch must be deleted from the schedule and removed from the system?

* If one or more cell(s) break(s) down?

This is an area when fast "secondary optimization" is required using a dynamic FMS scheduling program. (Note that further aspects of this and the software requirements are described in detail in Chapter 7.)

6.8.2 FMS versus job shop operation control

To summarize what has been said in previous Chapters as well as in this section, Flexible Manufacturing Systems deal with high level distributed data processing and automated material flow using computer controlled machines with the aim of combining the benefits of a highly productive, but inflexible transfer line and a highly flexible, but inefficient job shop.

The operation control of the job shop is guided by a schedule which is more often a desired, or theoretical plan only, rather than a real one because of the continuous disruptions and unplanned discontinuities in production. Eventually when the "panic" is high enough the job shop is partly controlled by (usually wrong) decisions taken under heavy pressure, or as an equally bad alternative, a new schedule is prepared in order to re-establish the theoretically correct plans.

The scheduling algorithm used in the job shop environment is off-line, because it is applied at the beginning of the scheduling period and the results are prepared for the entire shift, or for an even longer period of time, for example a week or a month. If an unexpected event happens, such as a tool break, machine tool or robot break down, a part not setup on planned time, etc., then because of the deterministic scheduling methods used, the production is disrupted.

A properly designed and implemented FMS need to and can perform operations under the control of a dynamic scheduling system.

This means that decisions concerning which workpiece is manufactured next on which cell are made close to the end of the operation currently being performed by the particular cell. In other words, in FMS part scheduling need not necessarily be carried out in full detail and has not necessarily be made in advance, because it is capable of responding to real-time decisions.

Cells in FMS can be accessed by part and tool carrier robots, or AGVs in random order, or in other words in any order (as long as the actual order is programmed).

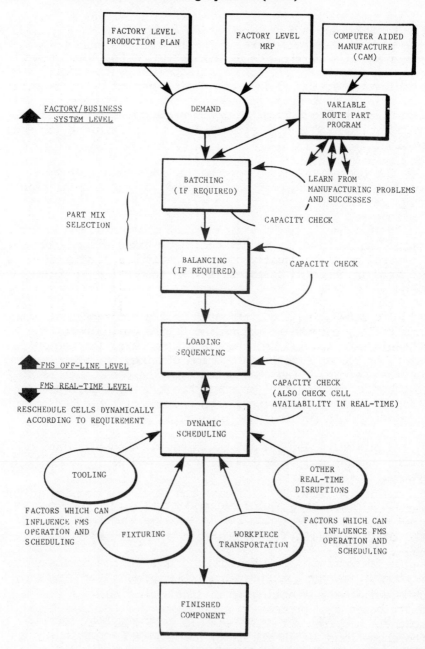

Figure 6.37 Summary of FMS data processing links to CIM modules. (FMS part mix selection, off-line operation planning and real-time operation control tasks.)

To demonstrate this very important principle refer to Figures 6.38/a and 6.38/b. Figure 6.38/a shows an FMS layout where cells have pallet pools and to which the workpieces are taken from the warehouse and loaded onto pallets and then

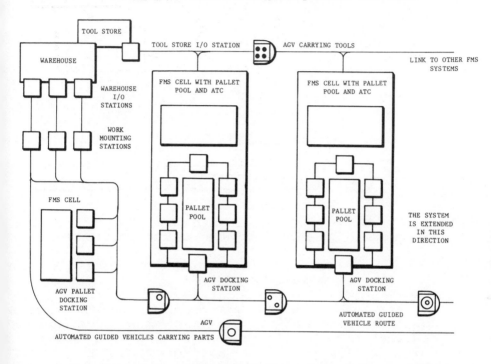

Figure 6.38/a Flexible Manufacturing System design with AGV part transport between the cells and the warehouse.

carried into the DNC system. Figure 6.38/b represents a different approach. In this case the parts are mounted at the warehouse and a combination of fork-lift trucks and AGV(s) are used to take the components directly from the warehouse to the cells and back. (For a possible solution see again Figure 6.21.)

Cells in FMS are also capable of sending and receiving data and generally acting as an intelligent node of the distributed data processing system. When following a fixed schedule prepared off-line regardless of real-time changes on the shop-floor, it is virtually impossible to fully load all processing stations. This kind of imbalance results in higher operating costs due to under utilization of the system.

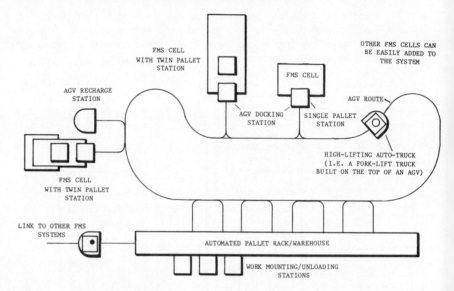

Figure 6.38/b FMS design utilysing a high-lifting auto-truck
capable of accessing not only the cells in the
system but also the pallet rack, where parts are
mounted and stored.

When applying the more advanced dynamic scheduling system
and the "variable route" FMS part programming method (as
described in [6.1]) offering alternatives for both loading
scheduling as well as for secondary optimization as mentioned
above, a much higher level of flexibility and equipment
utilization can be achieved even in cases when part processing
priorities change, or cells break down, etc.

Figure 6.39/a shows a sample structure of an FMS part
program providing "variable routing". Figure 6.39/b and 6.39/c
describe this part program using a language, in structure
similar to Pascal, developed by the author. This part program
description is very important because it contains all
necessary information in one single structure. It describes
alternatives, their order, when can they be executed, what
fixtures and finally what tools are required. In other words,
it represents the "production rules" relating to a single
component, or maybe a batch and must be therefore accessed by
the dynamic scheduling program before selecting current (alternative) production routes. (Obviously an FMS part program
can contain more than one pallet program with a single operation description, as well as can be more complex than the one
shown in Figure 6.39).

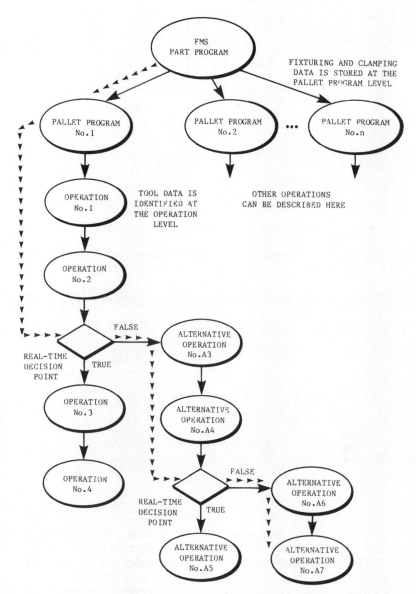

Figure 6.39/a The "variable route" FMS part programming concept and program structure, allowing alternative
operations to be taken (and production routes to
be assigned) if necessary. (Note that an operation can mean machining, washing, inspection,
assembly, welding, grinding and others.)

```
BEGIN FMS_Part_program/Code:ABCO8
(* NOTE THAT THIS IS THE TOP LEVEL OF THE PART PROGRAM,
   INCORPORATING ALL PALLET LEVEL AND OPERATION LEVEL
   PROGRAMS FOLLOWING A USER DEFINED STRUCTURE AND A PASCAL
   LIKE SYNTAX *)

   BEGIN Pallet_program/Code:No.1,Pallet_code:P1;

          (* NOTE THAT PALLET AND FIXTURE DATA ARE STORED AT
             THIS LEVEL FOR THE ENTIRE PALLET PROGRAM AND THAT
             ALL CODES USED ARE FREE FORMAT,USER DEFINED CODES *)

          Execute_operation/Code:No.1, Tool_file:P1No.1T01;

          (* NOTE THAT TOOL DATA FOR EACH OPERATION IS STORED
             AT THIS LEVEL IN A TOOL FILE OBTAINED FROM THE
             CNC PART PROGRAM, OR A HIGH LEVEL APT MACRO
             PROGRAM, ETC. *)

          Execute_operation/Code:No.2, Tool_file:P1No.2T02;

          IF <Condition true> THEN

             (* NOTE THAT A CONDITION ANALYSIS CAN MEAN A CELL
                AVAILABILITY CHECK, A TOOL BREAK CHECK, RESULT
                OF A PROBING CYCLE, RESULT OF AN INSPECTION,
                ETC. IF MORE THAN ONE CONDITION MUST BE
                EVALUATED THE CASE-OF... INSTRUCTION CAN BE
                USED TOO *)
                             BEGIN
          Execute_operation/Code:No.3, Tool_file:P1No.3T03;
          Execute_operation/Code:No.4, Tool_file:P1No.4T04;
                             END (* PALLET PROGRAM END *)
                          ELSE
                             BEGIN
          Execute_operation/Code:No.A3, Tool_file:P1No.A3T13;
          Execute_operation/Code:No.A4, Tool_file:P1No.A4T14;

             IF <Condition true> THEN
             Execute_operation/Code:No.A5, Tool_file:P1No.A6T26
                             ELSE
                             BEGIN
                Execute_operation/Code:No.A6, Tool_file:P1No.A6T16;
                Execute_operation/Code:No.A7, Tool_file:P1No.A7T17;
                             END; (* 2nd ELSE *)
                          END; (* 1st ELSE *)
   END; (* PALLET PROGRAM No.1 *)

      BEGIN Pallet_program/Code:No.2,Pallet_code:P12;
          (* PALLET PROGRAM No.2 DATA STRUCTURE DESCRIPTION *)
      END: (* PALLET PROGRAM No.2 *)

      BEGIN Pallet_program/Code:No.3,Pallet_code:P33;
          (* PALLET PROGRAM No.3 DATA STRUCTURE DESCRIPTION *)
      END: (* PALLET PROGRAM No.3 *)

          (* FURTHER PALLET PROGRAMS CAN BE ADDED HERE... *)

   END. (* FMS PART PROGRAM ABCO8 *)
```

Figure 6.39/b Description of the FMS part program structure
(or "production rules relating to a workpart")
given in Figure 6.39/a, using a free format,
Pascal-like language designed by the author.

Note that in this Figure condition analysis can
mean the interpretation of a message sent by the
DNC host, or by an other cell via the DNC host,
or direct (depending on the type of network
being used), a tool break check, dimension out of
tolerance message from the CNC inspection cell,
capacity overload or underload message from cell
cell breakdown message, etc.

```
FMS PART PROGRAM (FIG.6.39/b without comments)

+--------------------------------------------------------------------+


BEGIN FMS_Part_program/Code:ABC08

  BEGIN Pallet_program/Code:No.1,Pallet_code:P1;
            Execute_operation/Code:No.1, Tool_file:P1No.1T01;
            Execute_operation/Code:No.2, Tool_file:P1No.2T02;

        IF <Condition true> THEN
                     BEGIN
            Execute_operation/Code:No.3, Tool_file:P1No.3T03;
            Execute_operation/Code:No.4, Tool_file:P1No.4T04;
                     END (* PALLET PROGRAM END *)
                     ELSE
                     BEGIN
        Execute_operation/Code:No.A3, Tool_file:P1No.A3T13;
        Execute_operation/Code:No.A4, Tool_file:P1No.A4T14;

            IF <Condition true> THEN
            Execute_operation/Code:No.A5, Tool_file:P1No.A6T26
                     ELSE
                     BEGIN
            Execute_operation/Code:No.A6, Tool_file:P1No.A6T16;
            Execute_operation/Code:No.A7, Tool_file:P1No.A7T17;
                     END; (* 2nd ELSE *)
                     END; (* 1st ELSE *)
  END; (* PALLET PROGRAM No.1 *)

  BEGIN Pallet_program/Code:No.2,Pallet_code:P12;
        (* PALLET PROGRAM No.2 DATA STRUCTURE DESCRIPTION *)
  END; (* PALLET PROGRAM No.2 *)

  BEGIN Pallet_program/Code:No.3,Pallet_code:P33;
        (* PALLET PROGRAM No.3 DATA STRUCTURE DESCRIPTION *)
  END; (* PALLET PROGRAM No.3 *)

        (* FURTHER PALLET PROGRAMS CAN BE ADDED HERE... *)

END. (* FMS PART PROGRAM ABC08 *)
```

Figure 6.39/c FMS program structure description. (Note, that this source listing is the same as in Figure 6.39/b, but the comments have been deleted to be able to follow the structure better.)

6.9 The development concept of the FMS Software Library

The FMS Software Library has been created by the author with the aim of providing modular and portable software development tools and turnkey programs for designing, implementing, controlling and maintaining FMS.

The Library, capable of running on over thirty different micro-, and mini-computers, contains user friendly programs for FMS project planning, DNC control, robotized assembly line and manufacturing system balancing, off-line and dynamic scheduling, FMS batch size analysis, FMS capacity planning, pallet alignment and positioning error calculation for machining centers, robot positioning and orientation error analysis and other programs. (For further information about the programs please refer to [6.12] to [6.18]).

The FMS Software Library follows a modular design philosophy, similar to that of building up an FMS from cells, with the aim of providing system development tools and turnkey packages for development engineers in industry and researchers and students dealing with Computer Integrated Manufacture at universities and polytechnics.

Each program starts with a brief introduction, explaining to the user the function of the program, the terminology used in the screen-by-screen, menu-driven input and the computed output. Safe, run-time checked input is assured by having an interactive screen management system incorporated into each program. This means that each input data item is checked run-time, thus eliminating fatal data input combinations.

Calculated output is first listed on the screen, but can be printed on a printer attached to the particular system or stored in a user specified file on the disk. (Allowing further processing of data and also chaining of programs together via this file.)

There is a command to generate formatted output with user-defined page heading, page numbering, etc. Each time the output is printed, the relevant user input data is echoed (i.e. printed), thus allowing easy to understand listings. The output from the programs can be downloaded into a communications network for further processing. The operating system of the library allows different software-controlled serial and parallel interfaces as well as communication speeds to be identified.

Where feasible, programs offer "What if?" questions, allowing experimentation, simulation and the further optimization of results with different input data. This feature is useful, since the user can learn a lot by re-running the program with modified input data.

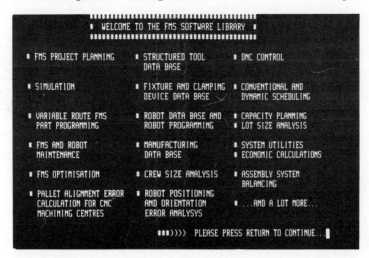

Figure 6.40 The logo of the FMS Software Library created by
the author, indicating the areas where software
support is essential in FMS design, simulation
and operation.

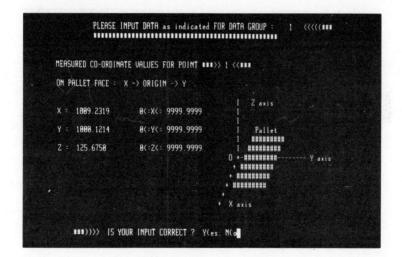

Figure 6.41 The photograph taken from the screen shows one of
the input screens of the pallet alignment error
analysis program. (In each package of the Library
·input data is checked not only for syntax, but for
semantics too, using pre-programmed, or run-time
calculated range error limits.)

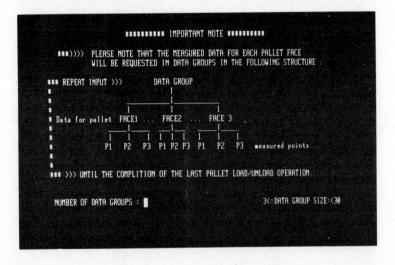

```
              INPUT DATA FOR PRODUCT TYPE: 1
              ******************************

PLEASE IDENTIFY THE BATCH WITH A CODE = BATCH 1          MAX. 25 CHARS
**) NOTE: a DAY should always mean a user specified LENGTH OF TIME

TOTAL PRODUCTION PERIOD IN DAYS      =   60.00      1<=PERIOD<=365

DAILY PRODUCTION IN UNITS            =    5.00      1<=UNITS<=3000

SETUP COST PER BATCH                 =  453.60      1<=COST<=1000

DAILY DEMAND RATE IN UNITS           =    2.50      0.1<=UNITS<=  4.50

THE INVENTORY HOLDING COST PER
COMPONENT PER DAY                    =    0.89      0.05 <=COST<=1000


              ***))))))  IS YOUR INPUT CORRECT ?  Y(es. N(o
```

Figure 6.42 One of the input screens of the FMS lotsize
           analysis program. Note that each data is checked
           during the interactive input procedure and each
           data and screen can be edited.

```
              ********** IMPORTANT NOTE **********

   ***))))  PLEASE NOTE THAT THE MEASURED DATA FOR EACH PALLET FACE
            WILL BE REQUESTED IN DATA GROUPS IN THE FOLLOWING STRUCTURE :

*** REPEAT INPUT >>>        DATA GROUP
 *                              I
 *                              I
 *                   _____
 *                   I          I         I
 * Data for pallet  FACE1 ...  FACE2 ...  FACE 3
 *                  __I__      __I__      __I__
 *                  I  I  I    I I I      I   I   I
 *                  P1 P2 P3   P1 P2 P3   P1  P2  P3   measured points..
 *
 *** >>> UNTIL THE COMPLITION OF THE LAST PALLET LOAD/UNLOAD OPERATION..

   NUMBER OF DATA GROUPS =                      3<=DATA GROUP SIZE<(30
```

Figure 6.43 Screen indicating the data structure of the pallet
           alignment error analysis program of the Library.
           (This program is capable of calculating three
           dimensional pallet positioning and alignment
           errors and providing the results in vector
           format.)

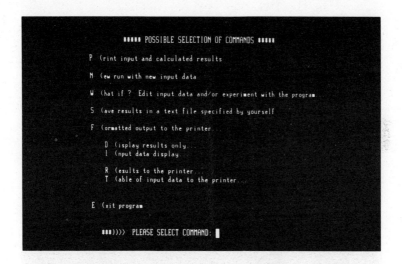

```
POSSIBLE SELECTION OF COMMANDS
*********************************

P (rint input and calculated results

R (un program again with new input data

W (hat if ? ... Experiment with different input data

S (ave results in a text file specified by yourself

F (ormatted output to the printer

T (able of results to the printer


E (xit program

***))))>> YOUR CHOICE ? █
```

Figure 6.44/a Programs are menu driven in this Library. The
photograph shows one of the command screens.

```
***** POSSIBLE SELECTION OF COMMANDS *****

P (rint input and calculated results

N (ew run with new input data

W (hat if ? Edit input data and/or experiment with the program

S (ave results in a text file specified by yourself

F (ormatted output to the printer ...

        D (isplay results only...
        I (nput data display...

        R (esults to the printer...
        T (able of input data to the printer....


E (xit program

    ***))>> PLEASE SELECT COMMAND: █
```

Figure 6.44/b The photograph shows a single character
command driven screen of the Library.

```
EQUIPMENT REQUIREMENTS FOR THE SPECIFIED ROUTE IN THE FMS
**********************************************************
FINISHED PARTS required per production period =   34.00 <<<(**

PRODUCTION    RAW        CELL     DEFECT    UNADJUSTED   ADJUSTED
  CELL    PARTS REQUIRED  EFF.[%]  RATE [%]  REQUIREMENT  REQUIREMENT

   1        70.00        99.50     5.50        4.98        5.00
   2        66.00        76.00     5.00       10.49       11.00
   3        63.00        51.00    13.00        1.54        2.00
   4        55.00        34.50     8.00       14.61       15.00
   5        51.00        67.00    14.00        3.69        4.00
   6        44.00        65.00     6.00        0.00        0.00
   7        41.00        47.00    10.30       24.44       25.00
   8        37.00        56.00     8.00        4.82        5.00


***))>> END OF OUTPUT... PLEASE PRESS RETURN TO CONTINUE...
```

Figure 6.45 Output screen of the FMS capacity planning and checking program, indicating the calculated loads of different cells involved in the selected route in the FMS.

```
*** TABLE OF RESULTS  >>> NOTE: values are not rounded

**)> The OVERALL ECONOMIC BATCH SIZE is based on the optimal production
     cycle time for for all components of = 24.81 DAYS

PRODUCT   MIN. COST    TOTAL PERIOD   OVERALL ECON.   ECON. PROD.
 TYPE     BATCH SIZE   BATCH SIZE     BATCH SIZE      CYCLE TIME
          [UNIT]       [UNIT]        **)> [UNIT]      [DAY]

   1       71.39        300.00         62.02          28.56
   2      156.77        420.00        198.45          19.60
   3      136.00         60.00        272.88          12.36
   4       34.00        150.00         24.81          34.00
   5       99.83        120.00        124.03          19.97


****)))>> PRESS RETURN TO CONTINUE...
```

Figure 6.46 Output screen of the FMS lotsize analysis program indicating calculated economic batch sizes and their optimum manufacturing cycle times for each individual batch (or component) and the part mix. (Note that a "day" means a user defined period of time, thus could be hours, days, weeks, etc.)

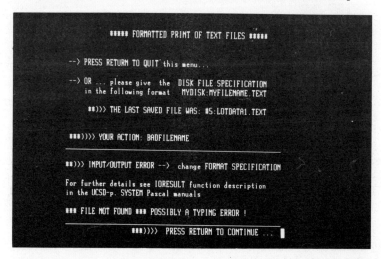

Figure 6.47 User friendliness and robustness are very important features of the Library. The photograph shows
that input/output and disk errors are handled
from the program.

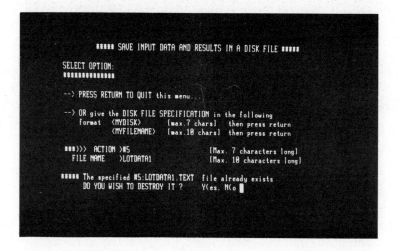

Figure 6.48 Disk directories are also accessed by the programs
ensuring that existing files are not deleted
accidently.

The packages can also be integrated into the software system of sophisticated FMS cell controllers, CNC machine tool controllers, robot and co-ordinate inspection machine controllers, etc., allowing real-time problem solving on the shop floor.

The Library is in industrial as well as in educational use both in Europe and overseas.

References and further reading

[6.1] Paul G. Ránky: The Design and Operation of Flexible Manufacturing Systems, IFS(Publications) Ltd. and North-Holland, 1983.

[6.2] Robert J. Thierauf: Distributed Processing Systems, Prentice-Hall, 19878.

[6.3] C. Gaude,J. Langet, S. Palssin, C. Kaiser: Distributed Processing as a key to reliable end evolving software for real-time applications. Information Processing 1980. (ed. S. H. Lavington), North Holland, IFIP 1980.

[6.4] Paul G. Ránky: Increasing Productivity with robots in Flexible Manufacturing Systems, The Industrial Robot, December 1981, p. 234-237.

[6.5] K. Hitomi: Manufacturing Systems Engineering, Taylor and Francis, 1979.

[6.6] V.P. Valeri: The case of gantry robots in machine loading, Robotics Today (USA) Vol.6, No.4, August 1984, p. 23-26.

[6.7] R.R. Schreiber: Machine loading applications at P and W's factory in the future, Robotics Today (USA) Vol.6, No.4, August 1984, p. 26-27.

[6.8] H.J. Bullinger, K.P. Fahnrich, M. Sprenger: User-oriented andtask-consistent programming interfaces for CNC machines, Proc. of the 1st International Conference on Human Factors in Manufacturing, London, England, 3-5 April 1984. IFS(Publications) Ltd. 1984.

[6.9] N. Percival: Safety aspects of robots and Flexible

Manufacturing Systems, Proc. of the 1st International
Conference on Human Factors in Manufacturing, London,
England, 3-5 April 1984. IFS(Publications) Ltd. 1984.

[6.10] J. Barbic, F. Dacar, M. Spegel: Oriented geometric
objects in computer graphics and numerical control,
Computer Graphics Forum (Netherlands) Vol.3, No.1,
March 1984.

[6.11] J. Schmidt, W. Westerteicher: Machine tool organization
in a flexible manufacturing system, Ind. Anz (W-Germany)
Vol.106, No.66, 17 Aug. 1984.

[6.12] Paul G. Ránky: The FMS software Library, The FMS
Magazine, IFS(Publications) Ltd., Vol.2, No.2, January
1984, p. 19-22.

[6.13] Paul G. Ránky: A Software Library for Designing and
Controlling Flexible Manufacturing Systems, IFAC'84,
9th World Congress of the International Federation of
Automatic Control, Vol. VI, Budapest, Hungary, July 2-6
1984. p. 147-153.

[6.14] Paul G. Ránky: Pallet Alignment Error Correction in FMS
by Means of Software, International Conference on the
Development of Flexible Automation Systems, organised
by the Institution of Electrical Engineers and MITRA,
London, July 1984.

[6.15] Paul G. Ranky: Programming Industrial Robots in FMS,
ROBOTICA (1984) Vol.2, Cambridge University Press,
Cambridge, UK, p. 87-92.

[6.16] Paul G. Ránky: A Software Library for the Design and
Optimization of Computer Integrated Systems, AUTOFACT
Europe Conference, Basel, Switzerland, September 1984.

[6.17] Paul G. Ránky: Libreria Software FMS, PIXEL (Computer
Graphics, CAD/CAM, Image Processing), Vol.5, No.1,
1984. p. 49-54.

[6.18] Paul G. Ránky: The FMS Software Library, User and
System Manuals. Available from the Author, from MALVA,
Nottingham, or COMPORGAN, Budapest.

CHAPTER SEVEN

Introduction to scheduling models, computation methods and their application in FMS

As we have already suggested scheduling at CIM level is considered to be a process that relates specific events to specific times or to a specific span of time (refer to section 3.4 in Chapter 3). In general scheduling involves the order and timing for assignment of resources to specific orders.

Scheduling problems can be categorized in a variety of different ways. For our purpose we have adopted the classification method of Graves ([7.1]) with some minor changes, in order to explain the differences between conventional manufacturing system scheduling and FMS loading sequencing and dynamic part scheduling.

Within this Chapter scheduling means the allocation of jobs to be processed on specified machines, or FMS cells, or FMS systems in a given time span.

As already defined, a job is a workpart to be manufactured through one or more different stages, including machining, washing, inspection, assembly, test, packaging, etc. At each stage (i.e. FMS cell) an operation is (or a series of operations are) executed on the component. The sum of operations, or processes performed in the proper order produces the job, or component.

In the case of FMS, the answers (or the "knowledge") given to the question "How to manufacture the part?", and "In what order should the operations be executed?" is given for each job by its "variable route" FMS part program, as discussed in the previous Chapter and illustrated in Figures 6.39/a to 6.39/c. This structure can be considered as a kind of "production rule base" for storing and updating different alternative production routes and manufacturing conditions under which the part can be produced by the system. In other words it is an input data structure attached to each job, describing "what can and should be done with it".

If this structure is generated automatically, or is updated and extended in real-time as the system "learns" more and more about its own production limits and rules, such a rule base can be a great asset to the FMS, because it solves the most difficult part of dynamic FMS scheduling in terms of production control engineering.

The purpose of this Chapter is to give an overview of different scheduling models and to discuss with sample runs examples of part loading sequencing and dynamic operation scheduling in FMS based on the "variable route" part programming method and on the cellular structure of "truly" flexible systems, as introduced in the previous Chapter.

7.1 An overview of production scheduling with particular emphasis towards FMS

In order to understand the problem of FMS loading sequencing and dynamic part scheduling, let us first introduce production scheduling methods by giving a brief review, based on [7.1].

Production scheduling can be classified in the following way, regarding:

1. Requirements generation

   * Open shop
   * Closed shop

2. Processing complexity

   * "n" jobs single resource problem
   * "n" jobs parallel resources problem
   * The multistage flow shop problem
   * The multistage job-shop problem
   * FMS, or random manufacturing problem in CIM

3. Sceduling criteria

  * Scheduling and rescheduling cost
  * Performance

4. Nature of the requirement specification

  * Deterministic
  * Stochastic

5. Scheduling environment

  * Static
  * Dynamic

Let us discuss the above list with particular regard to flexible manufacturing.

Requirements may be generated either by direct orders, considering an "open shop" scheduling model, or by inventory replenishment decisions, relying on a "closed shop" scheduling model.

In an open shop no inventory is stocked, whereas in the case of the closed shop the orders are fulfilled from an inventory, consequently the manufacturing system produces parts for inventory, rather than on order. This conventional manufacturing method is an expensive solution from the manu- facturing point of view, because often a high level of inventory must be kept, but it is also convenient, because the inventory acts as a safety buffer so that various shortages, disruptions and organizational faults can be balanced within certain limitations.

Since one of the major sources of savings in CIM is achieved by keeping the inventory at an extremely low level, the low inventory and work in progress are the rules to be followed at each level of the computer integrated factory, including the FMS.

Is FMS an open shop, or a closed shop? The answer depends on how the FMS was designed, as well as how it is operated. FMS as a "standalone island" in an otherwise fairly conventional shop is more an open shop than a closed shop, but is not an entirely open shop. In a CIM environment the FMS can be a truly open shop because part orders and part arrivals and resources in general are better organized.

This result is important, because open shop scheduling

means sequencing only, whereas closed shop scheduling means not only sequencing, but economic lotsizing and sometimes balancing too. Of course a "well designed FMS" can be operated as an open shop or as a closed shop, but it must be operated according to the part mix it receives, in other words in a flexible way according to the orders. (This is where the real strength of FMS is compared to any other manufacturing systems.)

Processing complexity of parts is mainly concerned with the number of operations associated with each part.

In the case of the "n" jobs one machine problem all jobs can be manufactured on a single processor. This model applies for example to an FMS cell, as shown in Figures 6.3 or 6.5, offering a pallet pool and sufficient amount and variety of tools to manufacture several often different parts, usually unmanned.

The "n" jobs parallel processor problem is similar to the one machine problem, except that each job requires an operation which can be performed on any of the parallel production facilities. This model is important in the case of transfer lines.

The flow shop model assumes that parts of the same kind are moving in one direction only, in other words that all "n" jobs are to be processed on the same set of resources ("m" machines) with an identical precedence ordering of the operations. This production method, mainly applied in the case of transfer lines, is very efficient and productive, but rigid too, particularly compared to FMS, and requires large batch sizes to offer an economic solution.

In the job shop there are "n" jobs waiting to be processed on "m" machines. This is a flexible situation since the jobs can be different and there are no restrictions on their routings.

The major disadvantage of the job-shop scheduling method that it is off-line, since it applies for a fixed period of time, throughout which it is meant to be valid in its unchanged form, and job arrivals and real-time changes cannot be accurately planned because of the lack of an overall material handling and real-time operated computer control system.

To summarize the most important features of the job-shop

the following points must be emphasized:

* It can handle a variety of jobs at the same time. (Note, that in the case of the flow-shop the workparts are of the same type.)

* The resources are shared by different workparts.

* Different jobs, or batches can have different priorities.

7.2   What solutions are available for FMS scheduling?

The answer is that there are many solutions available, most of them representing an "n" job, "m" machine job shop scheduling method extended for FMS. Unfortunately above two machines mathematically these scheduling models are only approximations, and thus are inaccurate. Most of them cannot consider real-time changes in the FMS at the required speed, and/or the new schedule frequently generated for the entire FMS system cannot be executed by the FMS control system, the tool management system and the material handling system.

The other problem is the processing time it takes to schedule or reschedule even a relatively low number of jobs on a few machines only, (because scheduling "n" jobs on "m" machines is a combinatorical problem: for example in the case of 5 jobs and 5 machines the number of alternatives represent a value of 25,000,000).

Even if heuristic calculations are used, they can take a relatively long time, e.g. 3 to 20 minutes in the case of an "average FMS problem", or longer depending on the number of jobs and machines. As soon as the new, updated schedule is available, the environment, or FMS status is different, and so the schedule becomes obsolete.

To give an example, Lageweg in [7.12] reports solving a 6 job 6 machine scheduling problem in a few seconds (CPU time), but was unable to solve the 10 job, 10 processor task in under 5 minutes (CPU time) using a Control Data Cyber 73-78 computer, which was one of the fastest machines available at the time. Since most FMS systems are controlled by minis, or by smaller mainframes in a multi-user environment, the duration time (i.e. the CPU time plus the time the job (note in computing terms) is waiting for input/output and to be processed in different parts of the computer) of such calculations can easily double, or triple in the case of an FMS. This is unfortunately inadequate in the case of any real-time controlled system.

When sequencing jobs for the job-shop and when scheduling jobs at the overall FMS planning level, or when writing computer programs for this purpose, the following data must be specified and considerations taken into account for the analysis and/or the program design:

* The number of machines (m), or FMS cells assigned to the number of operations (see Figure 6.39/a) in the selected route.

* The number of jobs (n) to be processed during the time span (usually a few hours, or a shift) of the analysis.

* The job arrival pattern, which is usually dynamic, except when starting to work with the FMS or the single cell, when it is static.

The arrival pattern is dynamic if jobs are arriving in a random order while there are jobs in the system waiting to be processed, for example in the pallet pool or in the buffer store of an FMS.

Static arrival means, that all scheduled jobs to be processed are available at the beginning of the analyzed period (e.g. shift). This arrival pattern is more typical for the job shop, although a machining cell with a twelve station pallet magazine working unmanned in the third shift could be well scheduled using this assumption.

* The routing, or job-flow pattern, specifying the sequence of manufacturing.

For the FMS this should be altered according to the constraints (i.e. changing production requests, urgent jobs, deleted jobs, system faults, tool break, etc.). Note, that following the "variable route" FMS part programming method, the part programming system should contribute equally to solving this problem, often in real-time.

The job-shop is not flexible enough to be able to cope
with real-time decisions, thus the routing must be fixed
for the analysed period of time (e.g. shift).

* The scheduling rule, these being:

 - First-come, first-served rule, if there is no precedence constraint, or if the precedence is the same
   for all jobs, (note that this rule is used by several
   operating FMS, because it is very simple and also
   because it is adaptive to real-time changes in the
   scheduling enviromnment).

 - SPT (Shortest Processing Time) rule.

 - WSPT (Weighted Shortest Processing Time) rule.

 - EDD (Earliest Due Date) rule.

 - SST (Shortest Slack Time) rule, etc.

* The schedule evaluation criteria, considering:

 - The average waiting time.

 - The average processing time.

 - The percentage of late jobs.

 - Resource utilization levels, etc.

Most scheduling, or sequencing models are unfortunately
deterministic and static, in other words were developed as if
the manufacturing environment was static and its behavior
"fully known" for at least a finite length of time, whereas in
reality, manufacturing systems are stochastic and dynamic,
(with the exception of FMS which is dynamic, but not stochastic, or at least less stochastic than any other manufacturing
system was in the past).

Unfortunately, scheduling theory and practice are often
far apart and many mathematically perfect models do not work
in practice. This is mainly because of the problems listed
above, and because up-to-date machining, assembly, etc. cells
are no longer restricted either in terms of available tools
(since they can use block tooling systems capable of storing
60 to 240 tool heads per magazine at each cell), or in terms

of receiving parts when required (because of the availability
of random, or direct access part loading and inter-cell transportation systems by AGVs).

7.3 FMS scheduling rules

The scheduling method to be used in FMS should be a multi-
level, dynamic scheduling method. This means that loading
sequencing in FMS should rely on information sources provided
from different levels of the organization, thus there should
be an overall planning level and a dynamic, or real-time
level.

1. The overall planning level should be applied

   * preferably to a relatively longer period of time (e.g.
     usually a month, but often only a week) and to

   * a shorter period of time (often some days, or one or
     two shifts only) allowing appropriate part mix selec-
     tion and optimization, lotsizing and balancing, as
     required, and part loading sequencing (see again
     Figure 6.37).

2. A dynamic, real-time level

   * at which the scheduling program causes decisions
     concerning which component is manufactured on which
     cell to be made when the operation currently being
     performed by the particular FMS cell is almost
     finished.

   * Realistically, if the FMS operates reliably and there
     is no change in part priorities, this real-time
     schedule can also last an hour or even longer, for
     example a shift. Obviously this concept relies on a
     "truly" flexible FMS, where jobs can be processed
     along alternative production routes in a scheduled
     order and where production alternatives are described
     by a rule base, formerly introduced in Figure 6.39/b.

The major benefit of this FMS scheduling concept is that

   * The overall planning level can utilize the scheduling

system of the CIM business data processing system, thus the input and the output of the FMS will be in accordance with the overall CIM planning levels.

* The FMS loading sequencing program can be a relatively simple and fast (meaning a few seconds CPU time for over 30 jobs) "n" job, one processor scheduler (the single processor being the whole FMS, as a system).

* The FMS dynamic scheduler can be a single processor "n" job scheduler too, but applied for each of those cells on which the component is going to be processed every time and immediately after the disrup tion occurs in the system.

Note that the possible order of operations and the alternatives are given in the "variable route" FMS part program. (Disruptions mean changes in the part manufacturing priorities, deleted jobs in the plan, cell breakdowns, AGV breakdowns, etc.)

To summarize FMS scheduling is much simpler than the general "n" job, "m" machine job shop scheduling problem:

1. If the design of the FMS is based on the cellular principle utilising reprogrammable machines with automated tool, or hand changing (in the case of assembly cells), and automated part, or pallet changing.

2. If a direct access (rather than serial access) material handling system is used.

3. If there are buffer stores in the system which can also be part of a pallet rack (see again Figures 6.38/a and 6.38/b).

4. If the cells contain a sufficient number and selection of tools for the schedule horizon to be a few hours, or a shift (refer to the case study given in section 2.2.3 in Chapter 2).

5. If the production rules for every job are described in a "variable route" part program, similar in structure and contents to those shown in Figures 6.39/a and 6.39/b.

These rules, extended by those discussed in previous
Chapters regarding database management, computer networks,
macro programming, etc., and one more new rule, regarding FMS
capacity planning (listed below and discussed also in Chapter
8), will ensure that the FMS will operate in a "truly"
flexible way, meaning that:

* Mixed batches or single parts can be manufactured,
  without major limitations in terms of productivity,
  or FMS utilization.

* Parts can arrive in random order, in other words the
  FMS can react to urgent orders, without disruptions
  within a limited time span (limitations mainly due to
  the availability of tools in the tool magazines of the
  machines, although this can be foreseen and automatically recognized by a system software if the tool files
  in the "variable route" FMS part program are compared to
  the tool magazines of the cells in the system, as
  discussed in the case study in section 2.2.3 in Chapter
  2).

* The system can be loaded and utilized at practically
  100% capacity, (due to the capacity planning rule saying
  that parts can wait in a buffer store, or in a pallet
  pool, but there should be at least one part available
  for every cell before the part currently being processed
  is complete, and that no cell is permitted to be idle)
  which is only an " unachievable dream" in the case of
  any previous manufacturing method.

  Note that even if following these rules machine utilization may not be close enough to 100%, then this can be
  due to the fact that either the reliability of the cells
  in the system is low, or the system is not loaded with
  the appropriate parts mix, or tool management does not
  work properly, but scheduling should not restrict system
  utilization in FMS. (Note that it usually does in any
  other manufacturing system.)

* Job priorities can be altered, jobs can be deleted,
  new jobs can be inserted, cells can be deleted and
  maintained whilst the others are working, the system can
  be reconfigured, cells can be modified and tested, etc.
  without any major limitations within a time span,
  usually in the region of a few hours, maybe a whole

shift (e.g. 8 hours).

Before demonstrating some of these rules by running simulation programs for FMS loading sequencing and dynamic scheduling, let us introduce our key model, the mathematically well defined "n" job single machine scheduling problem and the available different optimization rules.

7.4 Scheduling "n" jobs on one machine

The "n" jobs, single machine scheduling problem is very important in the case of loading sequencing the FMS, because the entire FMS can be considered as one single resource, (i.e. as one processor), and also in the case of FMS cells incorporating automated tool and work changing, working partly, or entirely unmanned. (Note again, that following the previously introduced terminology, an FMS cell can be other than machining, for example assembly, in which case automatic tool changing means automated robot hand changing, see again Figure 6.14, or inspection with automatic probe changing, etc.)

The "n" jobs single processor problem can be analysed by using the following assumptions:

* There are "n" independent jobs available for processing at time zero (t=0), meaning that there are jobs arriving from another shop, or an FMS, or warehouse and that they are available when work mounting must begin, prior to loading at the FMS.

* Operations are not divided or interrupted while processed. In other words, if processing has already started, (e.g. the part is being machined on a cell, or being assembled by a robot, or being tested by a CMM, etc.) the operation must be finished on the previously selected cell.

* Setup times are independent of the job sequence and can be included in processing times. (Note that in the case of automated pallet changing systems palletized components are loaded and unloaded. Thus the time spent for this operation is in the region of 5 to 8 seconds for each pallet load/unload operation, which is usually only a fraction of the operation time, thus can be eliminated in practice, particularly if there is some

level of buffering built into the system.)

* All CNC part programs, or robot programs are available
  in the controller, or via a DNC link before the processing of the component in the form of a "variable route"
  part program, containing the production rules by means
  of a structure of macro programs, NC/CNC machine
  instructions, robot programs, CMM macros, tool requirement and fixture description files, etc. as discussed
  in the previous Chapter.

* The FMS cell is continuously available for processing
  during the analyzed period of time (i.e. it is in
  operation for the horizon of the loading sequencing, or
  schedule) although it can break down in which case
  parts scheduled for it must be dispatched to other
  cells and a new schedule, relative to t=0, must begin.

* Parts can wait in a buffer store, or in a pallet
  magazine, but the processing station (i.e. the machine)
  is not permitted to be idle. (In other words there is a
  component always waiting, or rather available when
  required for processing for every oerational cell in
  the system, thereby ensuring continuous operation, or
  in other words 100% loading of each cell.)

Before discussing the different mathematical models and
optimization rules, let us summarize the most important data
and terminology used in this Chapter, relating to the scheduling problem:

* Processing time (t_j) is defined as the calculated time
  or the forecasted estimate of how long it will take to
  complete job i.

  Note, that processing times can be very accurately
  calculated from the CNC part program, or robot program
  in particular if macro programming is used. If the
  process includes some stochastic variables, expert
  systems can be of great use in providing a good estimate. It must be emphasized that without accurate
  processing times no accurate schedule can be provided.

* Waiting time (w_i) is the time that job i must wait
  before being processed.

* Flow time (F_i), or manufacturing interval is defined as the time span between the point at which job i is available for processing and the point at which it is completed. In other words, F_i is the total time job i spends in the system, or cell, and it is equal to the sum of the processing time (t_i) and the waiting time (w_i) of job i ($F_i = t_i + w_i$).

* Due date (d_i), or delivery date for job i, is the deadline at which the processing of job i is due to be completed, and beyond which it would be considered as tardy.

* Lateness (L_i) is the deviation between the actual completion time (C_i) and the due date (d_i) of operation i ($L_i = C_i - d_i$)

  Lateness can be positive or negative. Positive lateness means that the operation has been completed after its due date (and it is called tardiness). Negative lateness means that the operation has been completed earlier than its due date.

* Tardiness (T_i) is the measure of positive lateness of the completion of operation i.

* Completion time (C_i) is the span between the beginning of the first operation to be performed on the machine and the time when job i is finished plus the waiting time (w_i) ($C_i = L_i + d_i$).

* Slack (SL_i) is defined as the difference between the remaining time to an operations due date (d_i) and its processing time (t_i), $SL_i = d_i - t_i$.

Scheduling "n" jobs in its most elementary situation means the decision of the sequence of different operations on the machine, or FMS cell.

Assuming that all "i" jobs are available when the schedule is started (i.e. at t=0) then for schedule "s" the flow time ($F_{i,s}$) equals its completion time ($C_{i,s}$) for each job (following the notation of Bedworth and Bailey in [7.24])

$$F_{i,s} = C_{i,s}$$

392 Scheduling models, computation methods and application

The choice of sequence of the schedule will affect the completion of each job, but the total time spent on processing all jobs will for each different sequence be the same, i.e. is always constant. This value is called the makespan (M), and it is the sum of the processing times of all jobs ("n") considered in the schedule.

For schedule "s" the makespan is

$$Ms = \sum_{i=1}^{n} t_i$$

The mean flow-time for schedule "s" is given by

$$Fs = \frac{1}{n} \sum_{i=1}^{n} F_{i,s}$$

Assuming that all due dates are measured from t=0 the lateness and tardiness of each job is given by

Lateness: $L_{i,s} = C_{i,s} - d_i$

Tardiness: $T_{i,s} = \max(0, C_{i,s} - d_i)$,

The mean for both lateness and tardiness are given by dividing them through the number of tasks in the schedule

Mean lateness: $$Ls = \frac{1}{n} \sum_{i=1}^{n} L_{i,s}$$

Mean tardiness: $$Ts = \frac{1}{n} \sum_{i=1}^{n} T_{i,s}$$

The number of tardy jobs is given by

$$Ntardy = \sum_{i=1}^{n} k_i$$

where $k_i = 1$ if $T_i > 0$, else $k_i = 0$,

and finally the maximum lateness or tardiness is given by

Maximum tardiness: $T_{max} = max(0, L_{max})$

Maximum lateness: $L_{max} = max(L_{i,s})$,

where 1<=i<=n for all "i" in "n".

The optimum sequence affects the above shown values, except the makespan, which is always constant (for a single FMS part program route) in the case of the single processor scheduling problem.

7.4.1 The Shortest Processing Time (SPT) rule

Following the Shortest Processing Time rule for loading sequencing "n" jobs on a single processor (i.e. on the FMS as a whole), or when scheduling "n" jobs on individual cells, the flow time is minimized by scheduling the shorter tasks before the longer ones.

In other words it means that following the SPT rule

$$t_1 <= t_2 <= t_3 <= ... <= t_n$$

(Note that a sample run for this case can be found in the case studies, and that the proof is described in [7.8]).

The SPT rule in addition to minimizing mean flow time also minimizes mean-lateness and mean-waiting time. The SPT rule in general also performs well when measured in mean flow time and work in progress inventory.

Its major disadvantage is that by using it some jobs might be delayed for a long time, because it suffers from high variance with respect to flow time. To avoid this a "time out", or "cut-off time" limit should be introduced in FMS, meaning that if any job has waited the specified "cut-off time" it is automatically assigned to the front of the queue. Another solution is to change automatically the weighting factor (see discussion below) of those jobs which had to wait longer than a specified time limit.

7.4.2 The Weighted Shortest Processing Time (WSPT) rule

The Weighted Shortest Processing Time rule is a variation of the SPT rule and is used when the completion order of different jobs carry different importance, or weight.

When scheduling "n" jobs on a single resource such as the whole FMS, or an FMS cell then the weighted mean flow-time is minimised by sequencing jobs in order of:

$$\frac{t_1}{w_1} <= \frac{t_2}{w_2} <= ... <= \frac{t_i}{w_i} <= ... <= \frac{t_n}{w_n}$$

where each job has an importance weight "wi".

(Note that a sample run for this case can be found in the case studies below, and that the proof is described in [7.24]).

7.4.3 The Earliest Due Date (EDD) rule

The Earliest Due Date rule minimizes the maximum job lateness, or job tardiness, but unfortunately it tends to make more tasks tardy and increases the mean tardiness too.

When scheduling "n" jobs on a single machine, maximum job lateness and maximum job tardiness are minimized by sequencing the jobs in an order for which the due dates are as follows:

$$d_1 <= d_2 <= d_3 <= ... <= d_n$$

(Note that a sample run for this case can be found in the case studies, and that the proof is described in [7.8]).

7.5 Case study: FMS loading sequencing

The purpose of this case study is to demonstrate one possible way in which FMS loading sequencing can be performed. It is not claimed either to be the only, or the best method, but is definitely the simplest way of FMS loading sequencing which is applied in practice by several companies running FMS.

Loading sequencing here means finding an order in which jobs should be mounted on pallets and loaded for execution into the system. If there are no other constraints, such as the availability of fixtures, or empty pallets, etc., which should not be the case if an appropriate modular fixturing system is utilized, then one of the most important purposes of loading sequencing could be to minimize mean flow time in terms of getting onto the system.

As we have discussed earlier, prior to loading sequencing one might need to perform economic lotsize analysis and operation balancing, both with capacity checks. These procedures are described in subsequent Chapters. The reason for discussing loading sequencing first, and then dynamic scheduling in the next case study is that this is the simplest way of solving the problem, and that these programs are the most important for "driving" the FMS. (As already explained, lotsizing and balancing assumes the availability of data for at least a small selection of parts prior to the analysis, thus it is more complicated and is discussed in Chapters 9 and 10.)

The key source of information for loading sequencing is the release order of components for the FMS, arriving from the CIM business management system (MRP and MSP) and the "variable route" FMS part program (Figure 6.39), giving the production rules for each job (i.e. work mounting requirements, tool files for each operation, operation times and alternative sequences of operations for each job).

The major steps of the FMS work loading sequencing program, without lotsizing and balancing, are as follows:

1. Access the production rules (i.e. the "variable route" FMS part program) of each job in the database.

2. Run the loading sequencing program using the sum of the operation times of each operation involved as the total operation time for each job in the primarily selected production route.

3. In any case check whether any of the cells in the FMS are overloaded or under-utilized by means of the capacity check program (see detailed discussion in the next Chapter). If any of the cells are overloaded, (this is the more important case) run the loading sequencing program again for the second alternative route combination and try to avoid over-, or under-loading any of the cells. (Note, that alternative production route combinations between different jobs

can be generated mathematically, by means of balancing programs, by using human expertise, or by means of expert systems.)

The simplest way of avoiding overloading is by deleting jobs from the input queue for the current loading horizon. Most FMS systems are limited by the number of pallets anyway, thus this is not a major restriction.

The other point to note is that the system can be run at full capacity as long as there is at least one job prepared for each cell before it finishes the previous job. This is again not a difficult condition to ensure as long as there is a buffer store in the system, or cells are equipped with twin pallet changing systems, making this point not as complex as it might look at first...

4. Repeat loading sequencing as described above until a satisfactory solution is achieved. Note that the solution does not necessarily need to be perfect. The important condition to ensure is that each cell has at least a job to work on, and that another part always follows, so that the system is continuously utilized. When running the system in real-time adjustments (in this text also called "secondary optimization", or "dynamic scheduling") can be made enabling job priority changes in real-time.

Note again the difference between a transfer line and an FMS. Transfer lines are very sensitive for loading sequencing, FMSs are not.

Having discussed the most important principles, let us introduce a loading sequencing program and let us also demonstrate the capacity check program for one part and one single route only, despite the fact that the mathematical model of the capacity check program is given in the next Chapter.

In our first run of the FMS loading sequencing program, shown in Figure 7.1, we set up 24 jobs and we have also defined their processing times. The indicated processing times mean the sum of operation times following a production route, or the first production route of the part, whichever is feasible. Note that since each job can have alternative routes, and each of them can provide a possibly different sum

Figure 7.1 FMS loading sequencing using mean flow time minimization.

```
FMS LOADING SEQUENCING USING MEAN FLOW TIME MINIMIZATION
                                                    Page 1
+------------------------------------------------------------------+


         ****************************************************
         *         FMS PART LOADING SEQUENCING PROGRAM       *
         *--------------------------------------------------*
         *            THE FMS SOFTWARE LIBRARY               *
         ****************************************************


            *** MEAN FLOW TIME MINIMIZED ***

         ***************************************
         *        CALCULATED JOB SEQUENCE       *
         ***************************************
            SCHEDULED POSITION    JOB NUMBER
         ----------------------+--------------
                    1           JOB   3
                    2           JOB   1
                    3           JOB  16
                    4           JOB   2
                    5           JOB  18
                    6           JOB  13
                    7           JOB  15
                    8           JOB   4
                    9           JOB  14
                   10           JOB  17
                   11           JOB   6
                   12           JOB   9
                   13           JOB  11
                   14           JOB   7
                   15           JOB  12
                   16           JOB  10
                   17           JOB  19
                   18           JOB   5
                   19           JOB  22
                   20           JOB  20
                   21           JOB  23
                   22           JOB  21
                   23           JOB  24
                   24           JOB   8
         ----------------------+--------------

TOTAL NUMBER OF JOBS SCHEDULED FOR THE FMS IN THIS RUN = 24


            EVALUATION OF THE CALCULATED SEQUENCE
            *************************************


MAKESPAN                 =    262.00

NUMBER OF TARDY JOBS     =     0

      LATENESS           =     0.00
MEAN LATENESS            =     0.00
```

FMS LOADING SEQUENCING USING MEAN FLOW TIME MINIMIZATION

Page 2

+---+

```
*******************************************************************
*                        TABLE OF RESULTS                         *
*******************************************************************
```

JOB NUMBER	PROCESSING TIME	STARTING...FINISHING TIME		DUE TIME	LATENE
1	4	2.00	6.00		
2	5	10.00	15.00		
3	2	0.00	2.00		
4	8	34.00	42.00		
5	14	141.00	155.00		
6	10	59.00	69.00		
7	12	91.00	103.00		
8	20	242.00	262.00		
9	11	69.00	80.00		
10	13	115.00	128.00		
11	11	80.00	91.00		
12	12	103.00	115.00		
13	7	20.00	27.00		
14	8	42.00	50.00		
15	7	27.00	34.00		
16	4	6.00	10.00		
17	9	50.00	59.00		
18	5	15.00	20.00		
19	13	128.00	141.00		
20	17	171.00	188.00		
21	18	205.00	223.00		
22	16	155.00	171.00		
23	17	188.00	205.00		
24	19	223.00	242.00		

```
**********************************************
*     INPUT DATA ECHO AND JOB RECORD          *
**********************************************
```

JOB No. 1
****

       PROCESSING TIME   =   4

JOB No. 2
****

       PROCESSING TIME   =   5

JOB No. 3
****

       PROCESSING TIME   =   2

JOB No. 4
****

       PROCESSING TIME   =   8

JOB No. 5
****

       PROCESSING TIME   =   14

FMS LOADING SEQUENCING USING MEAN FLOW TIME MINIMIZATION
Page 3
+--+

```
        JOB No.  6
        *********

            PROCESSING TIME      =    10

        JOB No.  7
        *********

            PROCESSING TIME      =    12

        JOB No.  8
        *********

            PROCESSING TIME      =    20

        JOB No.  9
        *********

            PROCESSING TIME      =    11

        JOB No. 10
        *********

            PROCESSING TIME      =    13

        JOB No. 11
        *********

            PROCESSING TIME      =    11

        JOB No. 12
        *********

            PROCESSING TIME      =    12

        JOB No. 13
        *********

            PROCESSING TIME      =     7

        JOB No. 14
        *********

            PROCESSING TIME      =     8

        JOB No. 15
        *********

            PROCESSING TIME      =     7

        JOB No. 16
        *********

            PROCESSING TIME      =     4

        JOB No. 17
        *********

            PROCESSING TIME      =     9
```

```
JOB No. 18
**********

        PROCESSING TIME      =      5

JOB No. 19
**********

        PROCESSING TIME      =      13

JOB No. 20
**********

        PROCESSING TIME      =      17

JOB No. 21
**********

        PROCESSING TIME      =      18

JOB No. 22
**********

        PROCESSING TIME      =      16

JOB No. 23
**********

        PROCESSING TIME      =      17

JOB No. 24
**********

        PROCESSING TIME      =      19
```

of processing time at the end of the route, this process
should definitely be performed by a computer program utilising
the "variable route" FMS part program structure as the input.

The second run (Figure 7.2) shows what can be done if one
requires a job, or more jobs to be pushed ahead in the part
loading queue. As one can see from the results the order has
been changed due to the fact that jobs 20 and 24 have received
a fairly high importance weight (i.e. weight=50) compared to
the remaining jobs, which were left as before (i.e. weight=1).

Note that this run also demonstrates the Weighted
Shortest Processing Time - WSPT rule applied in FMS loading
sequencing.

The next two runs demonstrate the possible effects of
introducing due dates. (Note that these due dates represent a
deadline by whch time jobs must be loaded onto the FMS). The
first run (Figure 7.3) shows mean flow time minimization with
due dates. The second one (Figure 7.4) generates a part
loading order in which the maximum lateness values for each
part are minimized. (Note the way the maximum tardiness of 67
minutes of job 22 in Figure 7.3 has shrunk to 38 (at job 19)
in Figure 7.4, at the price of increasing the number of tardy
jobs from 8 to 10 and the total lateness from 16 minutes to 41
minutes.)

Demonstration of the way cell capacity and loading level
can be calculated is shown in a sample run for a single job in
the FMS for a selected production route. This program is
basically a planning tool helping managers to evaluate how
many jobs can the FMS be loaded with during a time period (in
this run 4.36 hours, because of the previously calculated 262
minutes loading horizon, or FMS level makespan time, see
Figure 7.4 again represents 4.36 hours).

Note in the listings given in Figure 7.5, (calculations
discussed in more detail in the next Chapter), that:

* Downtime means a period whilst the cell is not
  producing parts and it includes cell maintenance
  times too.

* Setup times are very short because of automated part
  changing devices applied in FMS.

* For cell efficiency calculations 24 hours has been

Figure 7.2 This run demonstrates the way important jobs (in
this case jobs marked 20, 24 and 9) can be pushed
ahead in the loading queue by increasing their
weighting factor and using the WSPT rule.

```
FMS LOADING SEQUENCING WITH IMPORTANT WEIGHTS (MEAN FLOW TIME MIN.)
                                                            Page 1
+------------------------------------------------------------------+



          ****************************************************
          *        FMS PART LOADING SEQUENCING PROGRAM      *
          *------------------------------------------------*
          *           THE FMS SOFTWARE LIBRARY             *
          ****************************************************


             *** MEAN FLOW TIME MINIMIZED ***

          ****************************************
          *        CALCULATED JOB SEQUENCE       *
          ****************************************
             SCHEDULED POSITION     JOB NUMBER
          ---------------------------+----------------
                          1          JOB  20
                          2          JOB  24
                          3          JOB   9
                          4          JOB   3
                          5          JOB   1
                          6          JOB  16
                          7          JOB   2
                          8          JOB  18
                          9          JOB  13
                         10          JOB  15
                         11          JOB   4
                         12          JOB  14
                         13          JOB  17
                         14          JOB   6
                         15          JOB  11
                         16          JOB   7
                         17          JOB  12
                         18          JOB  10
                         19          JOB  19
                         20          JOB   5
                         21          JOB  22
                         22          JOB  23
                         23          JOB  21
                         24          JOB   8
          ---------------------------+----------------


TOTAL NUMBER OF JOBS SCHEDULED FOR THE FMS IN THIS RUN = 24


             EVALUATION OF THE CALCULATED SEQUENCE
             **************************************


MAKESPAN                =     262.00

NUMBER OF TARDY JOBS    =       0

      LATENESS          =       0.00
MEAN LATENESS           =       0.00
```

```
FMS LOADING SEQUENCING WITH IMPORTANT WEIGHTS (MEAN FLOW TIME MIN.)
                                                          Page 2
+-----------------------------------------------------------------------+
```

```
***********************************************************************
*                         TABLE OF RESULTS                            *
***********************************************************************
JOB NUMBER   PROCESSING TIME   STARTING...FINISHING TIME   DUE TIME   LATENESS
----------+-----------------+----------+-----------------+----------+----------
```

JOB NUMBER	PROCESSING TIME	STARTING...FINISHING TIME		DUE TIME	LATENESS
1	4	49.00	53.00		
2	5	57.00	62.00		
3	2	47.00	49.00		
4	8	81.00	89.00		
5	14	177.00	191.00		
6	10	106.00	116.00		
7	12	127.00	139.00		
8	20	242.00	262.00		
9	11	36.00	47.00		
10	13	151.00	164.00		
11	11	116.00	127.00		
12	12	139.00	151.00		
13	7	67.00	74.00		
14	8	89.00	97.00		
15	7	74.00	81.00		
16	4	53.00	57.00		
17	9	97.00	106.00		
18	5	62.00	67.00		
19	13	164.00	177.00		
20	17	0.00	17.00		
21	18	224.00	242.00		
22	16	191.00	207.00		
23	17	207.00	224.00		
24	19	17.00	36.00		

```
----------+-----------------+----------+-----------------+----------+----------
```

```
***********************************************
*      INPUT DATA ECHO AND JOB RECORD         *
***********************************************

    JOB No.  1
    **********

        PROCESSING TIME        =     4
        IMPORTANCE WEIGHTING   =     1

    JOB No.  2
    **********

        PROCESSING TIME        =     5
        IMPORTANCE WEIGHTING   =     1

    JOB No.  3
    **********

        PROCESSING TIME        =     2
        IMPORTANCE WEIGHTING   =     1

    JOB No.  4
    **********

        PROCESSING TIME        = '   8
        IMPORTANCE WEIGHTING   =     1
```

CIM-N

FMS LOADING SEQUENCING WITH IMPORTANT WEIGHTS (MEAN FLOW TIME MIN.)
Page 3

+--+

```
        JOB No.  5
        *********
                PROCESSING TIME        =      14
                IMPORTANCE WEIGHTING   =       1

        JOB No.  6
        *********
                PROCESSING TIME        =      10
                IMPORTANCE WEIGHTING   =       1

        JOB No.  7
        *********
                PROCESSING TIME        =      12
                IMPORTANCE WEIGHTING   =       1

        JOB No.  8
        *********
                PROCESSING TIME        =      20
                IMPORTANCE WEIGHTING   =       1

        JOB No.  9
        *********
                PROCESSING TIME        =      11
                IMPORTANCE WEIGHTING   =      10  ◀

        JOB No. 10
        *********
                PROCESSING TIME        =      13
                IMPORTANCE WEIGHTING   =       1

        JOB No. 11
        *********
                PROCESSING TIME        =      11
                IMPORTANCE WEIGHTING   =       1

        JOB No. 12
        *********
                PROCESSING TIME        =      12
                IMPORTANCE WEIGHTING   =       1

        JOB No. 13
        *********
                PROCESSING TIME        =       7
                IMPORTANCE WEIGHTING   =       1

        JOB No. 14
        *********
                PROCESSING TIME        =       8
                IMPORTANCE WEIGHTING   =       1

        JOB No. 15
        *********
                PROCESSING TIME        =       7
                IMPORTANCE WEIGHTING   =       1

        JOB No. 16
        *********
                PROCESSING TIME        =       4
                IMPORTANCE WEIGHTING   =       1
```

+--+

```
JOB No. 17
*********

    PROCESSING TIME        =      9
    IMPORTANCE WEIGHTING   =      1

JOB No. 18
**********

    PROCESSING TIME        =      5
    IMPORTANCE WEIGHTING   =      1

JOB No. 19
**********

    PROCESSING TIME        =     13
    IMPORTANCE WEIGHTING   =      1

JOB No. 20
**********

    PROCESSING TIME        =     17
    IMPORTANCE WEIGHTING   =     50  ◀

JOB No. 21
**********

    PROCESSING TIME        =     18
    IMPORTANCE WEIGHTING   =      1

JOB No. 22
**********

    PROCESSING TIME        =     16
    IMPORTANCE WEIGHTING   =      1

JOB No. 23
**********

    PROCESSING TIME        =     17
    IMPORTANCE WEIGHTING   =      1

JOB No. 24
**********

    PROCESSING TIME        =     19
    IMPORTANCE WEIGHTING   =     50  ◀
```

Figure 7.3 FMS loading sequencing with mean flow time minimization and due dates. (Note that the due dates represent the deadline by which the job must be loaded
and that in this run the maximum tardiness is 67
minutes.)

```
FMS LOADING SEQ. WITH IMP. WEIGHTING AND MEAN FLOW TIME MINIMIZATION
                                                            Page 1
+-------------------------------------------------------------------+
```

```
         ****************************************************
         *        FMS PART LOADING SEQUENCING PROGRAM       *
         *--------------------------------------------------*
         *              THE FMS SOFTWARE LIBRARY            *
         ****************************************************


             *** MEAN FLOW TIME MINIMIZED ***

             ****************************************
             *        CALCULATED JOB SEQUENCE       *
             ****************************************
             SCHEDULED POSITION    JOB NUMBER
             ----------------------+--------------
                          1         JOB   20
                          2         JOB   24
                          3         JOB    9
                          4         JOB    3
                          5         JOB    1
                          6         JOB   16
                          7         JOB    2
                          8         JOB   18
                          9         JOB   13
                         10         JOB   15
                         11         JOB    4
                         12         JOB   14
                         13         JOB   17
                         14         JOB    6
                         15         JOB   11
                         16         JOB    7
                         17         JOB   12
                         18         JOB   10
                         19         JOB   19
                         20         JOB    5
                         21         JOB   22
                         22         JOB   23
                         23         JOB   21
                         24         JOB    8
             ----------------------+--------------

TOTAL NUMBER OF JOBS SCHEDULED FOR THE FMS IN THIS RUN = 24
```

```
1S LOADING SEQ. WITH IMP. WEIGHTING AND MEAN FLOW TIME MINIMIZATION
                                                         Page 2
+-------------------------------------------------------------------+
```

```
            EVALUATION OF THE CALCULATED SEQUENCE
            ************************************
```

```
AKESPAN                 =      262.00

UMBER OF TARDY JOBS     =        8

      LATENESS          =       16.00
EAN LATENESS            =        0.67
```

```
***********************************************************************
                       TABLE OF RESULTS                              *
***********************************************************************
JB NUMBER   PROCESSING TIME   STARTING...FINISHING TIME   DUE TIME  LATENESS
---------+-----------------+----------+----------------+---------+---------
    1             4            49.00        53.00           55      -2.00
    2             5            57.00        62.00           60       2.00 *
    3             2            47.00        49.00           41       8.00 *
    4             8            81.00        89.00           90      -1.00
    5            14           177.00       191.00          200      -9.00
    6            10           106.00       116.00          150     -34.00
    7            12           127.00       139.00          145      -6.00
    8            20           242.00       262.00          280     -18.00
    9            11            36.00        47.00           35      12.00 *
   10            13           151.00       164.00          155       9.00 *
   11            11           116.00       127.00          100      27.00 *
   12            12           139.00       151.00          142       9.00 *
   13             7            67.00        74.00           78      -4.00
   14             8            89.00        97.00           99      -2.00
   15             7            74.00        81.00           90      -9.00
   16             4            53.00        57.00           70     -13.00
   17             9            97.00       106.00          110      -4.00
   18             5            62.00        67.00           80     -13.00
   19            13           164.00       177.00          155      22.00 *
   20            17             0.00        17.00           25      -8.00
   21            18           224.00       242.00          250      -8.00
   22            16           191.00       207.00          140      67.00 *
   23            17           207.00       224.00          230      -6.00
   24            19            17.00        36.00           39      -3.00
---------+-----------------+----------+----------------+---------+---------
*)>> PLEASE NOTE that jobs marked with  *  are TARDY
```

Figure 7.4 FMS loading sequencing with maximum lateness minimization and due dates. Note that in this run the maximum tardiness has been reduced from 67 minutes to 38 minutes for job 13. (Also note the indicated dynamic changes and disturbances.)

FMS LOADING SEQ. WITH IMP.WEIGHTING AND MAX.LATENESS MINIMIZATION

+--+
```
Page 1
```

```
**************************************************
*        FMS PART LOADING SEQUENCING PROGRAM      *
*------------------------------------------------*
*            THE FMS SOFTWARE LIBRARY             *
**************************************************


        *** MAXIMUM LATENESS MINIMIZED ***

        ****************************************
        *      CALCULATED JOB SEQUENCE         *
        ****************************************
        SCHEDULED POSITION     JOB NUMBER
        ------------------+----------------
                       1      JOB   20 ◄───
                       2      JOB    9 ◄───
                       3      JOB   24 ◄───
                       4      JOB    3
                       5      JOB    1
                       6      JOB    2
                       7      JOB   16 ◄───
                       8      JOB   13        ┌──────────────┐
                       9      JOB   18        │ JOBS SELECTED│
                      10      JOB    4        │ FOR CELL No.k│
                      11      JOB   15        │(SEE FIGURE 7.6)│
                      12      JOB   14        └──────────────┘
                      13      JOB   11
                      14      JOB   17
                      15      JOB   22        ┌──────────────┐
                      16      JOB   12 ◄──── │ NEWLY ARRIVED JO│
                      17      JOB    7        │ REPLACING JOB 3 │
                      18      JOB    6        │ IN FIGURE 7.7   │
                      19      JOB   10 ◄──── ┌──────────────┐
                      20      JOB   19        │ NEWLY ARRIVED JO│
                      21      JOB    5 ◄───  │ REPLACING OLD JO│
                      22      JOB   23        │ IN FIGURE 7.7   │
                      23      JOB   21 ◄───  └──────────────┘
                      24      JOB    8 ◄───
        ------------------+----------------
```

TOTAL NUMBER OF JOBS SCHEDULED FOR THE FMS IN THIS RUN = 24

FMS LOADING SEQ. WITH IMP.WEIGHTING AND MAX.LATENESS MINIMIZATION
                                                  Page 2
+--+

                EVALUATION OF THE CALCULATED SEQUENCE
                ****

MAKESPAN   =   262.00

NUMBER OF TARDY JOBS   =   10

        LATENESS   =   41.00
MEAN LATENESS   =   1.71

****
*   TABLE OF RESULTS   *
****

JOB NUMBER	PROCESSING TIME	STARTING...FINISHING TIME		DUE TIME	LATENESS
1	4	49.00	53.00	55	-2.00
2	5	53.00	58.00	60	-2.00
3	2	47.00	49.00	41	8.00 *
4	8	74.00	82.00	90	-8.00
5	14	193.00	207.00	200	7.00 *
6	10	157.00	167.00	150	17.00 *
7	12	145.00	157.00	145	12.00 *
8	20	242.00	262.00	280	-18.00
9	11	17.00	28.00	35	-7.00
10	13	167.00	180.00	155	25.00 *
11	11	97.00	108.00	100	8.00 *
12	12	133.00	145.00	142	3.00 *
13	7	62.00	69.00	78	-9.00
14	8	89.00	97.00	99	-2.00
15	7	82.00	89.00	90	-1.00
16	4	58.00	62.00	70	-8.00
17	9	108.00	117.00	110	7.00 *
18	5	69.00	74.00	80	-6.00
19	13	180.00	193.00	155	38.00 *
20	17	0.00	17.00	25	-8.00
21	18	224.00	242.00	250	-8.00
22	16	117.00	133.00	140	-7.00
23	17	207.00	224.00	230	-6.00
24	19	28.00	47.00	39	8.00 *

**)>> PLEASE NOTE that jobs marked with * are TARDY

selected as a reasonably length of time for which
statistics on failures, maintenence, etc. might be
available. (Note the important benefit of FMS, that it
can automatically record cell failure distributions,
helping capacity planning, and other programs, in
similar circumstances).

Regarding loading levels one should realise that because
of mixed part, or batch production, each cell can be accessed
by many different parts during the analyzed period of time
(e.g. shift), thus the total load for each cell is the sum of
loads (see "unadjusted cell requirement" values in Figure 7.5)
calculated from single production routes. (Note that the production cell numbers in Figure 7.5 represent logical cell
codes, rather than a physical order). This is a check against
major mistakes of overloading, or under-utilizing the system
as a whole. For continuous system utilization the most
important condition remains to ensure that each cell has at
least a job to work on, and that another part always follows.

To summarize, it must be underlined again that the
achieved loading sequence more than likely will not be the
operation sequence of jobs on the system, because the operation sequence can, and should be altered dynamically according
to different real-time changes. These sample runs demonstrate
different solutions and rules for loading sequencing. The
decision on which one to choose is not only mathematical, but
involves engineering and management aspects too.

7.6 Case study: FMS dynamic scheduling, or "secondary optimization"

This case study takes further the case of simulating FMS
scheduling rules because it attempts to illustrate different
dynamic changes in the system at cell level.

It is assumed that as a result of loading sequencing,
shown in the previous case study, there are 6 jobs waiting to
be processed by cell No.k in its pallet pool, or in the common
buffer store allocated for it in the system. (See again the
layouts we have in mind in Figures 6.38/a and 6.38/b.)

It must also be noted that at the beginning of our analy-
zis the FMS is operating normally, i.e. cells are processing
parts and there is no shortage in parts, tools, etc. for the

Figure 7.5 FMS capacity check along a selected production route.
(Note that the cell numbers represent logical sequence numbers, rather than a physical order of
machines.)

```
FMS CELL CAPACITY CHECK FOR JOBS 20 AND 23 FOLLOWING THE FIRST ROUTE
                                                              Page 1
+-------------------------------------------------------------------+



       *********************************************
       *         THE FMS SOFTWARE LIBRARY          *
       *-------------------------------------------*
       *    FMS EQUIPMENT REQUIREMENTS AND CELL    *
       *          EFFICIENCY CALCULATION           *
       *********************************************



FMS CELL REQUIREMENTS FOR THE SPECIFIED ROUTE IN THE FMS
***********************************************************


FINISHED PARTS required per production period =     2.00 <<((**
```

PRODUCTION CELL	RAW PARTS REQUIRED	CELL EFF. [%]	DEFECT RATE [%]	UNADJUSTED REQUIREMENT	ADJUSTED REQUIREMENT
1	2.20	95.83	2.50	0.04	1.00
2	2.15	97.29	2.00	0.06	1.00
3	2.10	97.92	2.00	0.02	1.00
4	2.06	92.50	3.00	0.02	1.00

```
          ****************
            FMS CELL 1
          ****************

          INPUT DATA ECHO
          ---------------

          LENGTH OF PRODUCTION PERIOD    =     4.36 HOURS

          AVERAGE DOWNTIME               =    15.00 MINS
          AVERAGE SETUP TIME             =     5.00 MINS
          ANALYSED PERIOD OF TIME        =     8.00 HOURS

          PROCESSING TIME PER COMPONENT  =     5.00 MINS
          DEFECT RATE OF THE CELL        =     2.50 %
          (Assuming that bad components cannot be reworked)
```

CIM-N*

FMS CELL CAPACITY CHECK FOR JOBS 20 AND 23 FOLLOWING THE FIRST ROUTE

+--+

CALCULATED RESULTS

FMS CELL EFFICIENCY	=	95.83 %
RAW PARTS REQUIRED PER PERIOD	=	2.20
UNADJUSTED EQUIIMENT REQUIREMENT	=	0.04 FMS CELLS
ADJUSTED EQUIPMENT REQUIREMENT	=	1.00 FMS CELLS

****
    FMS CELL 2
****

INPUT DATA ECHO

LENGTH OF PRODUCTION PERIOD   =   4.36 HOURS

AVERAGE DOWNTIME	=	10.00 MINS
AVERAGE SETUP TIME	=	3.00 MINS
ANALYZED PERIOD OF TIME	=	8.00 HOURS
PROCESSING TIME PER COMPONENT	=	7.00 MINS
DEFECT RATE OF THE CELL	=	2.00 %

(Assuming that bad components cannot be reworked)

CALCULATED RESULTS

FMS CELL EFFICIENCY	=	97.29 %
RAW PARTS REQUIRED PER PERIOD	=	2.15
UNADJUSTED EQUIPMENT REQUIREMENT	=	0.06 FMS CELLS
ADJUSTED EQUIPMENT REQUIREMENT	=	1.00 FMS CELLS

****
    FMS CELL 3
****

INPUT DATA ECHO

LENGTH OF PRODUCTION PERIOD   =   4.36 HOURS

FMS CELL CAPACITY CHECK FOR JOBS 20 AND 23 FOLLOWING THE FIRST ROUTE
+--+

```
        AVERAGE DOWNTIME              =      5.00 MINS
        AVERAGE SETUP TIME           =      5.00 MINS
        ANALYSED PERIOD OF TIME      =      8.00 HOURS

        PROCESSING TIME PER COMPONENT =     3.00 MINS
        DEFECT RATE OF THE CELL      =      2.00 %
        (Assuming that bad components cannot be reworked)

        CALCULATED RESULTS
        ------------------

        FMS CELL EFFICIENCY              =      97.92 %
        RAW PARTS REQUIRED PER PERIOD    =      2.10
        UNADJUSTED EQUIPMENT REQUIREMENT =      0.02 FMS CELLS
        ADJUSTED EQUIPMENT REQUIREMENT   =      1.00 FMS CELLS



        ****************
           FMS CELL 4
        ****************

        INPUT DATA ECHO
        ---------------

        LENGTH OF PRODUCTION PERIOD   =      4.36 HOURS

        AVERAGE DOWNTIME              =     20.00 MINS
        AVERAGE SETUP TIME           =     16.00 MINS
        ANALYZED PERIOD OF TIME      =      8.00 HOURS

        PROCESSING TIME PER COMPONENT =     2.00 MINS
        DEFECT RATE OF THE CELL      =      3.00 %
        (Assuming that bad components cannot be reworked)

        CALCULATED RESULTS
        ------------------

        FMS CELL EFFICIENCY              =      92.50 %
        RAW PARTS REQUIRED PER PERIOD    =      2.06
        UNADJUSTED EQUIPMENT REQUIREMENT =      0.02 FMS CELLS
        ADJUSTED EQUIPMENT REQUIREMENT   =      1.00 FMS CELLS
```

planning horizon. (Note that the best way of avoiding shortages in any FMS is to check requirements before loading the
part, using the production rules documented in the "variable
route" FMS part program and comparing them with the current
status and setup conditions of the FMS. (The simplest as well
as most economic policy to follow for changing system setup
conditions, for example changing tool magazines or their
contents, is to set fixed intervals at the end of which
conditions are changed, rather then changing everything all
the time).

Note that at this stage operations are assigned to different cells. This must be done in a flexible way too, meaning
that not only the operations can have alternatives, but also
the cells assigned to them. The final assignment of operations
to cells must be done in real-time by a computer program, or
by an expert system, by looking into the production rules and
requirements listed in the "variable route FMS part program"
and the current status of the FMS (i.e current tool mix in the
magazines, current load levels and current queues in front of
each cell, just to mention the most important condition
checks).

The first sample run offered here represents a case where
there are six parts at cell No.k and one would like to produce
them in an order which minimizes mean flow time.

Note that we have assumed that jobs with loading codes of
JOB 20, JOB 9, JOB 24, JOB 16, JOB 5 and JOB 8, from the
loading list shown in Figure 7.4 and respectively numbered
with job codes at cell No.k from 1 to 6, in Figure 7.6, are in
the pallet pool, or in the buffer store assigned to Cell No.k.
at the beginning of this scheduling run.

Also note that the above listed jobs travel further in
the system leaving cell No.k, thus their operation times are
shorter than in the list given in Figure 7.4, with the exception of JOB 16, which has only one operation to be performed
on it at cell No.k and after that is considered to be
complete.

Now that we have explained the conditions, let us run the
program (Figure 7.6). The results show an order which is what
we have expected. Note that the starting and finishing times
given in the schedule represent a relative time, where t=0 is
the beginning of the scheduling horizon. Besides relative
timing an absolute FMS system time must be maintained too, to

be able to predict likely part completion times.

Let us start to disturb the system by changing the environment dynamically. (To follow the changes please look at Figures 7.4 and 7.6.)

Assuming that at the time of the disturbance

* JOB 4 (note this is the job code assigned to cell No.k in Figure 7.6) has already been completed, and is waiting to be carried away by an AGV.

* Meanwhile the cell is working on JOB 2 (cell job code in Figure 7.6).

* Then let us replace JOB 3 (cell job code in Figure 7.6) for a newly arrived job, JOB 12 (loading job code in Figure 7.4) so that it becomes the new JOB 3 (cell job code in Figure 7.6).

* Let us also allow another newcomer, JOB 10 (loading list job code in Figure 7.4) to become the new JOB 4 (cell job code in Figure 7.6).

Now having set these conditions let's run the program again for cell No.k. The results (Figure 7.6) show again the expected changes.

The next trouble we receive is that cell job code 1 (JOB 1 in Figure 7.7) becomes very urgent, thus must be "pushed ahead" in the queue. This is illustrated for cell No.k in a run shown in Figure 7.8.

At this point let us reflect on the fact that by changing priorities at this cell, other cells involved in manufacturing these parts will likely be affected too, unless they have enough work waiting for them in their pallet pools, or in the warehouse and no rescheduling is done at subsequent cells.

Certainly if one job is urgent at a cell where it is currently processed, or waiting for processing, it is likely that it needs to be "pushed through" the entire system as fast as possible. This can be done by the "n" job, single processor scheduler too, in such a way that when the urgent job arrives at the subsequent cell in its route then that cell is rescheduled automatically so that the urgent job gets the highest

Figure 7.6 Simulation of dynamic changes at cell No.k of the system. (See again Figure 7.4.)

JOB SEQUENCING AT CELL No.k

Page 1

```
**************************************************
*  GENERAL PURPOSE FMS CELL SCHEDULING PROGRAM  *
*------------------------------------------------*
*           THE FMS SOFTWARE LIBRARY             *
**************************************************

      *** MEAN FLOW TIME MINIMIZED ***

**************************************
*      CALCULATED JOB SEQUENCE       *
**************************************
```

SCHEDULED POSITION	JOB NUMBER
1	JOB 4
2	JOB 2
3	JOB 3
4	JOB 5
5	JOB 1
6	JOB 6

TOTAL NUMBER OF JOBS SCHEDULED FOR THE CELL IN THIS RUN = 6

EVALUATION OF THE CALCULATED SEQUENCE
****

MAKESPAN	=	53.00
NUMBER OF TARDY JOBS	=	0
LATENESS	=	0.00
MEAN LATENESS	=	0.00

TABLE OF RESULTS

JOB NUMBER	PROCESSING TIME	STARTING...	...FINISHING TIME	DUE TIME	LATEN
1	11	28.00	39.00		
2	6	4.00	10.00		
3	8	10.00	18.00		
4	4	0.00	4.00		
5	10	18.00	28.00		
6	14	39.00	53.00		

+---+

```
************************************************
*      INPUT DATA ECHO AND JOB RECORD          *
************************************************

JOB No.  1
**********

     PROCESSING TIME        =    11

JOB No.  2
**********

     PROCESSING TIME        =     6

JOB No.  3
**********

     PROCESSING TIME        =     8

JOB No.  4
**********

     PROCESSING TIME        =     4

JOB No.  5
**********

     PROCESSING TIME        =    10

JOB No.  6
**********

     PROCESSING TIME        =    14
```

Figure 7.7 Simulation of dynamic changes at cell No.k of the system. (See again Figure 7.4.)

```
            SIMULATION OF DYNAMIC CHANGES AT CELL No.k
                                                    Page 1
+---------------------------------------------------------------+

            ***************************************************
            *   GENERAL PURPOSE FMS CELL SCHEDULING PROGRAM   *
            *-----------------------------------------------*
            *            THE FMS SOFTWARE LIBRARY            *
            ***************************************************

            *** MEAN FLOW TIME MINIMIZED ***

            ***************************************
            *      CALCULATED JOB SEQUENCE        *
            ***************************************
            SCHEDULED POSITION     JOB NUMBER
            ----------------------+-------------
                        1         |  JOB   2
                        2         |  JOB   4
                        3         |  JOB   3
                        4         |  JOB   5
                        5         |  JOB   1
                        6         |  JOB   6
            ----------------------+-------------
```

TOTAL NUMBER OF JOBS SCHEDULED FOR THE CELL IN THIS RUN = 6

```
            EVALUATION OF THE CALCULATED SEQUENCE
            *************************************
```

MAKESPAN   =   57.00

NUMBER OF TARDY JOBS   =   0

        LATENESS   =   0.00
MEAN LATENESS   =   0.00

```
*******************************************************************
*                      TABLE OF RESULTS
*******************************************************************
```

JOB NUMBER	PROCESSING TIME	STARTING...FINISHING TIME		DUE TIME	LATEN
1	11	32.00	43.00		
2	6	0.00	6.00		
3	9	13.00	22.00		
4	7	6.00	13.00		
5	10	22.00	32.00		
6	14	43.00	57.00		

+--+

```
***********************************************
*       INPUT DATA ECHO AND JOB RECORD        *
***********************************************

JOB No.  1
**********

     PROCESSING TIME        =    11

JOB No.  2
**********

     PROCESSING TIME        =    6

JOB No.  3
**********

     PROCESSING TIME        =    9

JOB No.  4
**********

     PROCESSING TIME        =    7

JOB No.  5
**********

     PROCESSING TIME        =    10

JOB No.  6
**********

     PROCESSING TIME        =    14
```

Figure 7.8 Simulation of changing job priorities at cell level.

```
     SIMULATION OF DYNAMIC CHANGES AT CELL No.k  (URGENT JOB)
                                                      Page 1
+----------------------------------------------------------------------+



                 ************************************************
                 *  GENERAL PURPOSE FMS CELL SCHEDULING PROGRAM  *
                 *----------------------------------------------*
                 *             THE FMS SOFTWARE LIBRARY          *
                 ************************************************

                    *** MEAN FLOW TIME MINIMIZED ***

                 ****************************************
                 *      CALCULATED JOB SEQUENCE         *
                 ****************************************
                    SCHEDULED POSITION    JOB NUMBER
                 ---------------------+---------------
                            1              JOB   1
                            2              JOB   2
                            3              JOB   4
                            4              JOB   3
                            5              JOB   5
                            6              JOB   6
                 ---------------------+---------------

TOTAL NUMBER OF JOBS SCHEDULED FOR THE CELL IN THIS RUN = 6


              EVALUATION OF THE CALCULATED SEQUENCE
              **************************************


MAKESPAN                =     57.00

NUMBER OF TARDY JOBS    =      0

     LATENESS           =      0.00
MEAN LATENESS           =      0.00

**************************************************************************
*                          TABLE OF RESULTS                              *
**************************************************************************
JOB NUMBER  PROCESSING TIME  STARTING...FINISHING TIME   DUE TIME   LATE
----------+----------------+---------+----------------+---------+-----
     1            11            0.00        11.00
     2             6           11.00        17.00
     3             9           24.00        33.00
     4             7           17.00        24.00
     5            10           33.00        43.00
     6            14           43.00        57.00
----------+----------------+---------+----------------+---------+-----
```

SIMULATION OF DYNAMIC CHANGES AT CELL No.k (URGENT JOB)
```
                                                    Page 2
+-----------------------------------------------------------------+
```

```
**********************************************
*      INPUT DATA ECHO AND JOB RECORD        *
**********************************************
```

JOB No. 1
****

        PROCESSING TIME   =   11
        IMPORTANCE WEIGHTING =   10

JOB No. 2
****

        PROCESSING TIME   =   6
        IMPORTANCE WEIGHTING =   1

JOB No. 3
****

        PROCESSING TIME   =   9
        IMPORTANCE WEIGHTING =   1

JOB No. 4
****

        PROCESSING TIME   =   7
        IMPORTANCE WEIGHTING =   1

JOB No. 5
****

        PROCESSING TIME   =   10
        IMPORTANCE WEIGHTING =   1

JOB No. 6
****

        PROCESSING TIME   =   14
        IMPORTANCE WEIGHTING =   1

Figure 7.9 Simulation of the effect of new due due dates and mean flow time optimization rules at cell level.

```
SIMULATION OF DYNAMIC CHANGES AT CELL No.k  (NEW DUE DATES ...1)
                                                          Page 1
+------------------------------------------------------------------+


              ****************************************************
              *   GENERAL PURPOSE FMS CELL SCHEDULING PROGRAM   *
              *------------------------------------------------*
              *          THE FMS SOFTWARE LIBRARY               *
              ****************************************************

                 *** MEAN FLOW TIME MINIMIZED ***

              ****************************************
              *      CALCULATED JOB SEQUENCE         *
              ****************************************
                 SCHEDULED POSITION    JOB NUMBER
              ------------------------+---------------
                        1                JOB    1
                        2                JOB    2
                        3                JOB    4
                        4                JOB    3
                        5                JOB    5
                        6                JOB    6
              ------------------------+---------------

TOTAL NUMBER OF JOBS SCHEDULED FOR THE CELL IN THIS RUN = 6


              EVALUATION OF THE CALCULATED SEQUENCE
              *************************************


MAKESPAN               =      57.00

NUMBER OF TARDY JOBS   =      3

     LATENESS          =      3.00
MEAN LATENESS          =      0.50

*******************************************************************
*                      TABLE OF RESULTS
*******************************************************************
JOB NUMBER  PROCESSING TIME  STARTING...FINISHING TIME  DUE TIME  LATE
----------+----------------+----------+----------------+--------+-----
    1            11            0.00          11.00          12      -1.
    2            6            11.00          17.00          20      -3.
    3            9            24.00          33.00          31       2.
    4            7            17.00          24.00          19       5.
    5            10           33.00          43.00          40       3.
    6            14           43.00          57.00          60      -3.
----------+----------------+----------+----------------+--------+-----

**)>> PLEASE NOTE that jobs marked with   *   are TARDY
```

+---+

```
*************************************************
*      INPUT DATA ECHO AND JOB RECORD       *
*************************************************

JOB No.  1
*********

      PROCESSING TIME       =       11
      IMPORTANCE WEIGHTING  =       10
      DUE TIME              =       12

JOB No.  2
*********

      PROCESSING TIME       =        6
      IMPORTANCE WEIGHTING  =        1
      DUE TIME              =       20

JOB No.  3
*********

      PROCESSING TIME       =        9
      IMPORTANCE WEIGHTING  =        1
      DUE TIME              =       31

JOB No.  4
*********

      PROCESSING TIME       =        7
      IMPORTANCE WEIGHTING  =        1
      DUE TIME              =       19

JOB No.  5
*********

      PROCESSING TIME       =       10
      IMPORTANCE WEIGHTING  =        1
      DUE TIME              =       40

JOB No.  6
*********

      PROCESSING TIME       =       14
      IMPORTANCE WEIGHTING  =        1
      DUE TIME              =       60
```

Figure 7.10 Simulation of the effect of due dates and maximum
lateness optimization rules.

```
SIMULATION OF DYNAMIC CHANGES AT CELL No.k (NEW DUE DATES...2)
                                                        Page 1
+-------------------------------------------------------------------+


            ***************************************************
            *   GENERAL PURPOSE FMS CELL SCHEDULING PROGRAM   *
            *-------------------------------------------------*
            *           THE FMS SOFTWARE LIBRARY              *
            ***************************************************

               *** MAXIMUM LATENESS MINIMIZED ***

            ***********************************
            *     CALCULATED JOB SEQUENCE     *
            ***********************************
               SCHEDULED POSITION   JOB NUMBER
            -----------------------+--------------
                        1           JOB   1
                        2           JOB   4
                        3           JOB   2
                        4           JOB   3
                        5           JOB   5
                        6           JOB   6
            -----------------------+--------------

TOTAL NUMBER OF JOBS SCHEDULED FOR THE CELL IN THIS RUN = 6


            EVALUATION OF THE CALCULATED SEQUENCE
            ***********************************

MAKESPAN               =     57.00

NUMBER OF TARDY JOBS   =      3

       LATENESS        =      4.00
MEAN LATENESS          =      0.67

*******************************************************************
*                       TABLE OF RESULTS                          *
*******************************************************************
JOB NUMBER  PROCESSING TIME  STARTING...FINISHING TIME  DUE TIME  LATE
----------+----------------+---------+-----------------+---------+----
    1            11            0.00         11.00          12      -1.
    2             6           18.00         24.00          20       4.
    3             9           24.00         33.00          31       2.
    4             7           11.00         18.00          19      -1.
    5            10           33.00         43.00          40       3.
    6            14           43.00         57.00          60      -3.
----------+----------------+---------+-----------------+---------+----
 **)>> PLEASE NOTE that jobs marked with  *  are TARDY
```

+---+

```
*********************************************
*      INPUT DATA ECHO AND JOB RECORD       *
*********************************************

JOB No.  1
*********

       PROCESSING TIME        =     11
       IMPORTANCE WEIGHTING    =     10
       DUE TIME               =     12

JOB No.  2
*********

       PROCESSING TIME        =      6
       IMPORTANCE WEIGHTING    =      1
       DUE TIME               =     20

JOB No.  3
*********

       PROCESSING TIME        =      9
       IMPORTANCE WEIGHTING    =      1
       DUE TIME               =     31

JOB No.  4
*********

       PROCESSING TIME        =      7
       IMPORTANCE WEIGHTING    =      1
       DUE TIME               =     19

JOB No.  5
*********

       PROCESSING TIME        =     10
       IMPORTANCE WEIGHTING    =      1
       DUE TIME               =     40

JOB No.  6
*********

       PROCESSING TIME        =     14
       IMPORTANCE WEIGHTING    =      1
       DUE TIME               =     60
```

426 Scheduling models, computation methods and application
priority.

In this case the urgent job will suffer only the delay it
takes to complete the currently processed job on this cell.
This process should be followed throughout the entire system.
(Note that the key point is that a cell is only rescheduled
when the urgent job has already arrived, because this is the
only way the scheduling system can take care of other real-
time changes which might occur in the system. Because a good
cell scheduling program is fast, taking on average under 3
seconds but generally nearer 1 second to schedule 6 to 12
jobs at a cell, this represents no restriction in operation at
all).

One must also notice that this urgent job (or other
changes) will disturb most, if not all other job finish times,
but the urgent job will be finished as soon as possible too,
which was the purpose of marking it urgent.

Finally the last two runs in this case study illustrate
what happens if due dates are introduced at cell level (see
Figures 7.9 and 7.10).

One must note that it is impossible to produce due dates
for any jobs which are valid for the entire production period, or scheduling horizon if one allows dynamic changes in
the FMS, which are essential because of the nature of the
manufacturing method. Due date values can however be
calculated dynamically, always using the current production
routing information on each part, but changing every time a
cell rescheduling process occurs in the production route.

To summarize, in this case study we have demonstrated
dynamic changes (such as job is deleted, job is replaced, new
job arrives, job priority changed, due dates changed, optimization rules are changed) at cell level. We have also
explained the way such changes can affect and can be simulated
for other cells involved in making the component(s) along
alternative production routes.

References and further reading

[7.1] Stephan C. Graves: A Review of Production Scheduling,
Operations Research, Vol.29, No.4, July-August 1981.

[7.2] K.R. Baker: Introduction to Sequencing and Scheduling,

John Wiley and Sons, New York, 1974.

[7.3] S.M. Johnson: Optimal Two - and Three Stage Production
Schedules with setup times included, Neval Research
Logistics Quarterly, Vol.1. No.1. March 1954.

[7.4] K.R. Baker: Scheduling with parallel processors and
linear delay costs, Neval Research Logistics Quarterly,
Vol.20, p. 193-204.

[7.5] J.M. Charlton and C.C. Death: A generalised Machine
Scheduling Algorithm, Operational Research Quarterly,
1970. Vol.21, p. 127-134.

[7.6] M. Girnt and E. Szelke: Some Experiences in the Field of
scheduling medium lot size production, COMPCONTROL'74
Conference, Szeged, Hungary, 1974.

[7.7] M. Horvath and J. Somlo: Adaptive Control Systems for
Machining Operations, (In Hungarian), Muszaki Konyvkiado, Budapest, 1979.

[7.8] Frederick S. Hillier and Gerald J. Lieberman:
Introduction to Operations Research, Holden Day Inc.,
1981.

[7.9] P. Langley and J.G. Carbonell: Approaches to Machine
Learning, J. Am. Soc. Inf. Sci (USA), Vol.35, No. 5,
September 1984.

[7.10] W. Fromm: Utilization of flexibility through efficient
control systems, Ind. Anz. (W-Germany), Vol. 106,
No.66, 17 Aug. 1984. (In German).

[7.11] H.G. Campbell at all: A heuristic algorithm for the
"n" job "m" machine sequencing problem, Management
Science, Vol.16, No. 10, June 1970.

[7.12] B.J. Lageweg at all: A general bounding scheme for the
permutation flow-shop problem, Opns. Res. 1978. Vol.26,
p. 53-67.

[7.13] H.J.J. Kals and F.J.A.M. Van Houten: Flexible Manufacture based on a Production Information Management
System (PIMS), Manufacturing Systems, Vol.12, No.3,
p. 187-196.

[7.14] J.R. King: The theory-practice gap in job shop scheduduling, The Production Engineer, March 1976., p.138-146.

[7.15] M. Girnt and E. Szelke: Adaptive Control in Production Scheduling, IFAC, 1982.

[7.16] T. Watanabe and S. Iwai: Real time programming of computer numerical control of machine tool in CAM, IFAC Real-time programming, Kyoto, Japan, 1981.

[7.17] S.L. Hwang: Integration of Humans and Computers in the Operation and Control of Flexible Manufacturing Systems, Int. J. Prod. Res (GB), Vol.22, No.5, September-October 1984.

[7.18] Elwood S. Buffa and William H. Taubert: Production Inventory Systems, Planning and Control. Richard D. Irwin Inc., 1972.

[7.19] R. Granow, R. Schmiedeskamp and J. Balbach: Tool management. First step to a computer-aided workshop communication, VDI. Z. (Germany), Vol. 126, No.17, September 1984.

[7.20] Paul G. Ránky: The FMS Software Library: General Purpose FMS Cell Scheduling Program. Computer program with User and System Manuals, 1983-84.

[7.21] Paul G. Ránky: The FMS Software Library: General Purpose FMS Capacity Planning Program. Computer program with User and System Manuals, 1983-84.

[7.22] K. Hitomi: Manufacturing Systems Engineering, Taylor and Francis, London, 1979.

[7.23] Q. Semeraro: Heuristic Scheduling in a flow shop type manufacturing cell, Manufacturing Systems, Vol. 12, No.3, p. 197-208.

[7.24] David. D. Bedworth at all: Integrated Production Control Systems, Management, analysis, design. John Wiley and Sons, Inc., 1982.

CHAPTER EIGHT

FMS capacity planning and control

As discussed in Chapter 3, capacity planning is a crucial decision because it affects both short and long term planning and control in the entire CIM system.

From the manufacturing system's point of view, capacity planning and control provides a load summary of the manufacturing system, or any other equipment used in the company for designing, fabricating, testing, packaging, etc. the products.

Productive capacity is usually measured in units, and it refers either to the maximum output rate for products or services, or to the resources available in an analyzed period of time for a defined number of products and processes.

In a conventional shop where "crisis management methods" rule, the load summary is usually never level nor matches exact targets in any given week unless by coincidence, thus managers and shop floor supervisors are always interested in the production rate required to finish their overdues.

Capacity control in CIM is a matter of moving from a reactive attitude to an active mode, in which upcoming workload is carefully analyzed and planned by means of computers.

As we have seen in our earlier discussion in Chapter 3 and Chapter 7, capacity planning and control in general can be utilized at many different levels of the computer integrated factory. In this Chapter let us concentrate on its effect on the manufacturing side, including Flexible Manufacturing Systems.

Capacity planning and control in manufacturing can be

applied:

* To the design process of manufacturing systems,

* To the evaluation and optimization of the process of manufacturing systems.

* To the planning process to decide whether existing capacity is large enough to take on a newly arrived order.

* For controlling capacity in existing manufacturing systems.

To summarize, the purpose of this Chapter is to explain the role of capacity planning and control in FMS and to give examples both for fully automated and for mixed manufacturing systems (e.g. manual machines mixed with automated cells, or with CNC machining centers, manual assembly stations mixed with robotised assembly cells, etc.) [8.1].

8.1 Capacity planning and control in FMS

Capacity planning and control in FMS is somewhat similar in principle to the way such techniques are applied in a conventional job shop, with two major differences:

1. In FMS cell capacity, cell and system utilization can be maximized by preparing and loading at least one component onto the cell's buffer store, or pallet pool before the process on the previous component is complete. This method ensures full loading of the cell and continuous production of the system.

   (The only downtime which must be encountered will occur from pallet changing times, representing 5 to 8 seconds in general per pallet changes, from machine maintenance, from real-time disruptions, tool breaks and other type of mostly mechanical, or electro-mechanical faults.)

2. The other important difference is that in flexible production systems jobs can take alternative routes much more easily than in conventional job shop systems thus capacity requirement fluctuates, and because of

this should be checked in real-time before an alternative route is instructed by the control computer.

The other consequence of alternative routes is that
dynamic scheduling and real-time capacity analysis must
co-operate, as demonstrated in the previous Chapter,
otherwise one cell could be overloaded, and another
under-utilized.

As we have already illustrated in Figure 6.37, in a "well
designed" flexible production environment major decisions
regarding loading sequencing, lotsize analysis, manufacturing
system balancing and scheduling are checked using up-to-date
data provided by real-time data logging facilities from the
shop floor, as well as from shared databases of distributed
data processing networks, which link capacity planning and
control to the rest of the CIM software environment.

8.2   The mathematical model of the capacity planning program

The purpose of the capacity program is to calculate a measure
of capacity to enable FMS cell and system requirements to be
identified for each component, or batch, or batches for all
selected production routes in the FMS.

The program can be used as a simulation tool for FMS cell
requirement and efficiency planning. This is useful if the
system is manufacturing a series of batches of components, or
mixed types of components in any order during a user-specified
production period. When analysing the results the bottleneck
cells for each different production route can be identified as
can the number of components the FMS can produce be determined for each production period [8.2].

By utilizing the program as a system architecture design
tool, the capacity of each cell as well as the total system
can be optimized.

The program follows the concept of the "variable route"
FMS part programming method as already discussed. The determination of FMS cell and system requirements is done by taking
care of losses of defective components and various differences
in the efficiency of cells and the operation of the system.

The mathematical model considers several production cells,

used in any user-defined order for the batches or single components. The determination of FMS cell (i.e. machining, part washing, assembly, inspection, etc.) requirements is done by the successive use of the following formula:

$$\text{Requirement} = \frac{\text{Time} * \text{Output rate}}{60 * \text{Shift} * \text{Efficiency}}$$

Where   Requirement = the FMS cell requirement, (i.e. the number of machines required)

Time = the processing time per component per hour per manufacturing cell (i.e. machine)

Output rate = production rate per cell (i.e. manufactured (good) components per production period (i.e. shift)

Shift = user defined duration of an operating period (i.e. shift) in hours

Efficiency = FMS cell efficiency in percentage, as defined later

so that at each cell (i.e. machine) we account for the various operating conditions of the particular machine and the previous cells involved in the given route.

FMS cell efficiency is calculated for each cell as follows

$$\text{Efficiency} = 1 - \frac{\text{Setup time} + \text{Downtime}}{\text{Analyzed period}} \quad [\%]$$

Where   Setup time = is given in minutes per cell per analyzed period of time, meaning work mounting time in the case of manually controlled machines, and time spent on automated pallet changing in the case of FMS cells during the manufacturing of the specified batch, or number of components.

Downtime = is given in minutes per cell per analyzed period of time, meaning the sum of time the machine is not doing productive work because of some fault, because of maintenance, or other reasons.

Analyzed period = time utilized for the evaluation of the average setup and downtime values

(i.e. Note that its value is not necessarily equal to the length of production period considered for an analysis).

To estimate the output rate, we must be aware of the fact that often some defective components cannot be entirely eliminated from the process, unless each cell performs 100% quality control on all those operations it has performed on it.

Although FMS should in general provide a higher level of quality, and defect rates are usually very low (depending on the process, e.g. around 1-2% in the case of machining, around 2-5% in the case of assembly, around 5-7%, or even higher in the case of part washing unless performed by a robot, etc.), because of economic reasons in many systems components are often transferred from one cell to an other in the route until the fault is detected, thus to be on the safe side the total number of components produced on the FMS are calculated for each cell in the specified route as follows:

Total part requirement = Good + Defective components

This calculation must also consider the fact that because of the defective components, there are gradually less and less good components transferred to the next cell (i.e. next as programmed in the current production route), although the system must cope with the part handling tasks of all (i.e. good and defective) components at the same time.

The defect rate is calculated for each successive cell in the selected route as follows:

$$\text{Defect rate} = \frac{\text{Number of defective components}}{\text{Total number of components}}$$

(These two values must be measured during the same time period, i.e. the analyzed time period)

Having introduced the defect rate variable and the total number of products (i.e. the output rate) we must make during the specified time period (e.g. shift), to be able to satisfy the demand for the "good" components the initial load size can be calculated as follows:

$$\text{Output rate} = \frac{\text{Good components}}{1 - \text{Defect rate}}$$

These values must be calculated by the program successively for each cell, following the programmed route in the FMS. These values could be different for each batch and/or process, thus their average value could be used if there is no other reliable data available.

The algorithm does not consider the fact that some defective components can be reworked, thus the given solutions are overestimated by the same percentage as for the reworkable components. This calculation (or rather estimation) has been avoided since there is no reliable mathematical model available to judge this practical decision, which often must be taken on the shop floor. However if the number of reworkable components is known, a run can be made only for the reworkable components and the requirement needs can be then added to the results of the previous calculation.

8.3   Case study: Capacity planning for a highly automated and a mixed production system

In our case study we have a batch of 32 components, which we would like to manufacture. Each component involves several different processes, such as milling, drilling, inspection, washing, etc. and some of these processes could be done in different orders, (i.e. along "alternative routes") depending on the actual load of the system. We shall follow one selected route of the part, keeping in mind the "variable route" FMS part programming concept (see again Figure 6.39 in Chapter 6).

What we would like to know is how many FMS cells (or manufacturing stations) we need to make these components during the given period of time, which is 8 hours, in our example.

The input data of the program will be determined by whether one uses a highly automated flexible system, a dedicated transfer line or a conventional machine tool shop.

In our first example (Figure 8.1) let us assume that we work on an FMS and that each part is mounted onto a pallet and that work mounting is done outside the machining, or inspection, etc. area.

The input data also depends on the type of the cell itself (e.g. if it is a machining cell it could be typically a roughing, or finishing cell; it could depend on the actual operations involved, e.g. assembly, inspection, etc.) on the maintenance policy used and the way the given cells are operated (e.g. programmed at very high feedrates and speeds or properly optimized for efficient but reliable use, etc.)

In an FMS setup times are small if palletized part transfer, or robotised part loading/ unloading is employed. In conventional machine shops setup time and downtime are usually several times greater than in FMS.

Having selected one of the production routes let us see what the CAPACITY program can do for us to analyze our capacity needs and to show cell loading levels. (For the full listing of the sample run refer to Figure 8.1). The results show that one should load more components into the system than the volume of the "good components", (in this example 47, instead of 32) because of non-zero cell defect rate. It is also important to realize that the "unadjusted requirement" column shows the cell utilization levels for this run in the selected production route.

If more batches of components are manufactured on the same system during the same time period in a random or any sche-duled order, then the maximum FMS cell requirement for the given period of time can be obtained of the sum of the

436 FMS capacity planning and control

Figure 8.1 Capacity planning taking a production route with the highest load in the FMS.

```
              FMS CAPACITY PLANNING DEMONSTRATION
                                                        Page 1
+-----------------------------------------------------------------------+


        **********************************************
        *          THE FMS SOFTWARE LIBRARY          *
        *--------------------------------------------*
        *      FMS EQUIPMENT REQUIREMENTS AND CELL    *
        *             EFFICIENCY CALCULATION          *
        **********************************************


FMS CELL REQUIREMENTS FOR THE SPECIFIED ROUTE IN THE FMS
*********************************************************


FINISHED PARTS required per production period =    32.00 <<<<(**
```

PRODUCTION CELL	RAW PARTS REQUIRED	CELL EFF. [%]	DEFECT RATE [%]	UNADJUSTED REQUIREMENT	ADJUSTED REQUIREMENT
1	47.00	98.06	3.00	0.95	1.00
2	46.00	91.00	3.50	0.32	1.00
3	44.00	96.11	6.00	0.57	1.00
4	41.00	89.00	2.00	0.38	1.00
5	40.00	97.00	1.00	0.60	1.00
6	40.00	98.13	2.00	0.76	1.00
7	39.00	88.00	2.00	1.02	1.00
8	38.00	93.00	2.00	0.43	1.00
9	37.00	98.00	1.00	0.20	1.00
10	37.00	96.00	2.00	0.64	1.00
11	36.00	98.61	4.00	0.91	1.00
12	35.00	93.00	2.00	0.64	1.00
13	34.00	87.00	3.00	0.24	1.00
14	33.00	94.00	2.00	0.63	1.00

```
        *****************
          FMS CELL 1
        *****************

        INPUT DATA ECHO
        ---------------

        LENGTH OF PRODUCTION PERIOD    =     8.00 HOURS
        AVERAGE DOWNTIME               =    22.00 MINS
        AVERAGE SETUP TIME             =     6.00 MINS
        ANALYZED PERIOD OF TIME        =    24.00 HOURS

        PROCESSING TIME PER COMPONENT  =     9.50 MINS
        DEFECT RATE OF THE CELL        =     3.00 %
        (Assuming that bad components cannot be reworked)

        CALCULATED RESULTS
        ------------------

        FMS CELL EFFICIENCY               =      98.06 %
        RAW PARTS REQUIRED PER PERIOD     =      47.00
        UNADJUSTED EQUIPMENT REQUIREMENT  =       0.95 FMS CELLS
        ADJUSTED EQUIPMENT REQUIREMENT    =       1.00 FMS CELLS
```

```
                    FMS CAPACITY PLANNING DEMONSTRATION
                                                            Page 2
+------------------------------------------------------------------+


      ****************
        FMS CELL 2
      ****************

      INPUT DATA ECHO
      ---------------

      LENGTH OF PRODUCTION PERIOD   =      8.00 HOURS
      FMS CELL EFFICIENCY           =     91.00 %
      PROCESSING TIME PER COMPONENT =      3.00 MINS
      DEFECT RATE OF THE CELL       =      3.50 %
      (Assuming that bad components cannot be reworked)

      CALCULATED RESULTS
      ------------------

      RAW PARTS REQUIRED PER PERIOD     =     46.00
      UNADJUSTED EQUIPMENT REQUIREMENT  =      0.32 FMS CELLS
      ADJUSTED EQUIPMENT REQUIREMENT    =      1.00 FMS CELLS


      ****************
        FMS CELL 3
      ****************

      INPUT DATA ECHO
      ---------------

      LENGTH OF PRODUCTION PERIOD   =      8.00 HOURS
      AVERAGE DOWNTIME              =     45.00 MINS
      AVERAGE SETUP TIME            =     11.00 MINS
      ANALYZED PERIOD OF TIME       =     24.00 HOURS

      PROCESSING TIME PER COMPONENT =      6.00 MINS
      DEFECT RATE OF THE CELL       =      6.00 %
      (Assuming that bad components cannot be reworked)

      CALCULATED RESULTS
      ------------------

      FMS CELL EFFICIENCY               =     96.11 %
      RAW PARTS REQUIRED PER PERIOD     =     44.00
      UNADJUSTED EQUIPMENT REQUIREMENT  =      0.57 FMS CELLS
      ADJUSTED EQUIPMENT REQUIREMENT    =      1.00 FMS CELLS


      ****************
        FMS CELL 4
      ****************

      INPUT DATA ECHO
      ---------------

      LENGTH OF PRODUCTION PERIOD   =      8.00 HOURS
      FMS CELL EFFICIENCY           =     89.00 %
      PROCESSING TIME PER COMPONENT =      4.00 MINS
      DEFECT RATE OF THE CELL       =      2.00 %
      (Assuming that bad components cannot be reworked)

      CALCULATED RESULTS
      ------------------

      RAW PARTS REQUIRED PER PERIOD     =     41.00
      UNADJUSTED EQUIPMENT REQUIREMENT ,=      0.38 FMS CELLS
      ADJUSTED EQUIPMENT REQUIREMENT    =      1.00 FMS CELLS
```

438 FMS capacity planning and control

```
+------------------------------------------------------------------+

        ****************
          FMS CELL 5
        ****************

        INPUT DATA ECHO
        ----------------

        LENGTH OF PRODUCTION PERIOD    =      8.00 HOURS
        FMS CELL EFFICIENCY            =     97.00 %
        PROCESSING TIME PER COMPONENT  =      7.00 MINS
        DEFECT RATE OF THE CELL        =      1.00 %
        (Assuming that bad components cannot be reworked)

        CALCULATED RESULTS
        ------------------

        RAW PARTS REQUIRED PER PERIOD      =      40.00
        UNADJUSTED EQUIPMENT REQUIREMENT   =       0.60 FMS CELLS
        ADJUSTED EQUIPMENT REQUIREMENT     =       1.00 FMS CELLS



        ****************
          FMS CELL 6
        ****************

        INPUT DATA ECHO
        ----------------

        LENGTH OF PRODUCTION PERIOD    =      8.00 HOURS

        AVERAGE DOWNTIME               =     15.00 MINS
        AVERAGE SETUP TIME             =     12.00 MINS
        ANALYZED PERIOD OF TIME        =     24.00 HOURS

        PROCESSING TIME PER COMPONENT  =      9.00 MINS
        DEFECT RATE OF THE CELL        =      2.00 %
        (Assuming that bad components cannot be reworked)

        CALCULATED RESULTS
        ------------------

        FMS CELL EFFICIENCY                =      98.13 %
        RAW PARTS REQUIRED PER PERIOD      =      40.00
        UNADJUSTED EQUIPMENT REQUIREMENT   =       0.76 FMS CELLS
        ADJUSTED EQUIPMENT REQUIREMENT     =       1.00 FMS CELLS



        ****************
          FMS CELL 7
        ****************

        INPUT DATA ECHO
        ----------------

        LENGTH OF PRODUCTION PERIOD    =      8.00 HOURS
        FMS CELL EFFICIENCY            =     88.00 %
        PROCESSING TIME PER COMPONENT  =     11.00 MINS
        DEFECT RATE OF THE CELL        =      2.00 %
        (Assuming that bad components cannot be reworked)

        CALCULATED RESULTS
        ------------------

        RAW PARTS REQUIRED PER PERIOD      =      39.00
        UNADJUSTED EQUIPMENT REQUIREMENT   =       1.02 FMS CELLS
        ADJUSTED EQUIPMENT REQUIREMENT     =       1.00 FMS CELLS
```

```
****************
    FMS CELL 8
****************

INPUT DATA ECHO
---------------

LENGTH OF PRODUCTION PERIOD    =     8.00 HOURS
FMS CELL EFFICIENCY            =    93.00 %
PROCESSING TIME PER COMPONENT  =     5.00 MINS
DEFECT RATE OF THE CELL        =     2.00 %
(Assuming that bad components cannot be reworked)

CALCULATED RESULTS
------------------

RAW PARTS REQUIRED PER PERIOD       =     38.00
UNADJUSTED EQUIPMENT REQUIREMENT    =      0.43 FMS CELLS
ADJUSTED EQUIPMENT REQUIREMENT      =      1.00 FMS CELLS


****************
    FMS CELL 9
****************

INPUT DATA ECHO
---------------

LENGTH OF PRODUCTION PERIOD    =     8.00 HOURS
FMS CELL EFFICIENCY            =    98.00 %
PROCESSING TIME PER COMPONENT  =     2.50 MINS
DEFECT RATE OF THE CELL        =     1.00 %
(Assuming that bad components cannot be reworked)

CALCULATED RESULTS
------------------

RAW PARTS REQUIRED PER PERIOD       =     37.00
UNADJUSTED EQUIPMENT REQUIREMENT    =      0.20 FMS CELLS
ADJUSTED EQUIPMENT REQUIREMENT      =      1.00 FMS CELLS


****************
    FMS CELL 10
****************

INPUT DATA ECHO
---------------

LENGTH OF PRODUCTION PERIOD    =     8.00 HOURS
FMS CELL EFFICIENCY            =    96.00 %
PROCESSING TIME PER COMPONENT  =     8.00 MINS
DEFECT RATE OF THE CELL        =     2.00 %
(Assuming that bad components cannot be reworked)

CALCULATED RESULTS
------------------

RAW PARTS REQUIRED PER PERIOD       =     37.00
UNADJUSTED EQUIPMENT REQUIREMENT    =      0.64 FMS CELLS
ADJUSTED EQUIPMENT REQUIREMENT      =      1.00 FMS CELLS
```

+---+

```
****************
   FMS CELL 11
****************

INPUT DATA ECHO
-----------------

LENGTH OF PRODUCTION PERIOD    =      8.00 HOURS

AVERAGE DOWNTIME               =     15.00 MINS
AVERAGE SETUP TIME             =      5.00 MINS
ANALYZED PERIOD OF TIME        =     24.00 HOURS

PROCESSING TIME PER COMPONENT  =     12.00 MINS
DEFECT RATE OF THE CELL        =      4.00 %
(Assuming that bad components cannot be reworked)

CALCULATED RESULTS
--------------------

FMS CELL EFFICIENCY                =      98.61 %
RAW PARTS REQUIRED PER PERIOD      =      36.00
UNADJUSTED EQUIPMENT REQUIREMENT   =       0.91 FMS CELLS
ADJUSTED EQUIPMENT REQUIREMENT     =       1.00 FMS CELLS



****************
   FMS CELL 12
****************

INPUT DATA ECHO
-----------------

LENGTH OF PRODUCTION PERIOD    =      8.00 HOURS
FMS CELL EFFICIENCY            =     93.00 %
PROCESSING TIME PER COMPONENT  =      8.10 MINS
DEFECT RATE OF THE CELL        =      2.00 %
(Assuming that bad components cannot be reworked)

CALCULATED RESULTS
--------------------

RAW PARTS REQUIRED PER PERIOD      =      35.00
UNADJUSTED EQUIPMENT REQUIREMENT   =       0.64 FMS CELLS
ADJUSTED EQUIPMENT REQUIREMENT     =       1.00 FMS CELLS



****************
   FMS CELL 13
****************

INPUT DATA ECHO
-----------------

LENGTH OF PRODUCTION PERIOD    =      8.00 HOURS
FMS CELL EFFICIENCY            =     87.00 %
PROCESSING TIME PER COMPONENT  =      3.00 MINS
DEFECT RATE OF THE CELL        =      3.00 %
(Assuming that bad components cannot be reworked)
```

```
                    FMS CAPACITY PLANNING DEMONSTRATION
                                                              Page 6
+---------------------------------------------------------------------+


    CALCULATED RESULTS
    ------------------

    RAW PARTS REQUIRED PER PERIOD      =      34.00
    UNADJUSTED EQUIPMENT REQUIREMENT   =       0.24 FMS CELLS
    ADJUSTED EQUIPMENT REQUIREMENT     =       1.00 FMS CELLS


    *****************
       FMS CELL 14
    *****************

    INPUT DATA ECHO
    ---------------

    LENGTH OF PRODUCTION PERIOD     =      8.00 HOURS
    FMS CELL EFFICIENCY             =     94.00 %
    PROCESSING TIME PER COMPONENT   =      8.60 MINS
    DEFECT RATE OF THE CELL         =      2.00 %
    (Assuming that bad components cannot be reworked)

    CALCULATED RESULTS
    ------------------

    RAW PARTS REQUIRED PER PERIOD      =      33.00
    UNADJUSTED EQUIPMENT REQUIREMENT   =       0.63 FMS CELLS
    ADJUSTED EQUIPMENT REQUIREMENT     =       1.00 FMS CELLS



- - - - - - - - - end of output - - - - - - end of output - - - - - - -
```

unadjusted cell requirement values together for each batch
(or individual component if the batch size equals 1) in the
selected route.

Logically, the sum of this data will provide the total
FMS requirement for the given time period.

To be able to make a perfect analysis for selected
intervals of the production period, one should also use the
scheduling programs in order to simulate the time distribution
of the peak loads with different scheduling methods.

The second run, given in Figure 8.2 is an extension of
our first example, (shown in Figure 8.1) and it considers a
mixed system, in which cells marked 1, 6, 11, and 14 are
manually controlled. One can imagine for example an
assembly system in which there are manual workstations mixed
with automated assembly robots. The results show that because
of the increased setup and downtimes, the cell efficiency of
the indicated cells are reasonably lower, thus more cells, or
machines are required to be able to cope with the capacity
requirements.

Figure 8.2 Mixed manufacturing system capacity planning.

```
        CAPACITY PLANNING FOR MIXED PRODUCTION SYSTEMS
                                                    Page 1
+-----------------------------------------------------------------+


      ***********************************************
      *           THE FMS SOFTWARE LIBRARY          *
      *---------------------------------------------*
      *    FMS EQUIPMENT REQUIREMENTS AND CELL      *
      *           EFFICIENCY CALCULATION            *
      ***********************************************


FMS CELL REQUIREMENTS FOR THE SPECIFIED ROUTE IN THE FMS
************************************************************


FINISHED PARTS required per production period =    32.00 <<((**


PRODUCTION     RAW        CELL       DEFECT    UNADJUSTED     ADJUSTED
   CELL    PARTS REQUIRED EFF. [%]  RATE [%]  REQUIREMENT   REQUIREMENT
_____I_____I_____I_____I_____I_____

    1        47.00       64.58      3.00       1.44         2.00
    2        46.00       91.00      3.50       0.32         1.00
    3        44.00       96.11      6.00       0.57         1.00
    4        41.00       89.00      2.00       0.38         1.00
    5        40.00       54.17      1.00       1.08         1.00
    6        40.00       62.85      2.00       1.19         2.00
    7        39.00       88.00      2.00       1.02         1.00
    8        38.00       93.00      2.00       0.43         1.00
    9        37.00       98.00      1.00       0.20         1.00
   10        37.00       96.00      2.00       0.64         1.00
   11        36.00       58.47      4.00       1.54         2.00
   12        35.00       93.00      2.00       0.64         1.00
   13        34.00       87.00      3.00       0.24         1.00
   14        33.00       44.10      2.00       1.34         2.00


      *****************
         FMS CELL 1
      *****************

      INPUT DATA ECHO
      ---------------

      LENGTH OF PRODUCTION PERIOD     =      8.00 HOURS
      AVERAGE DOWNTIME                =    243.00 MINS
      AVERAGE SETUP TIME              =    267.00 MINS
      ANALYZED PERIOD OF TIME         =     24.00 HOURS

      PROCESSING TIME PER COMPONENT   =      9.50 MINS
      DEFECT RATE OF THE CELL         =      3.00 %
      (Assuming that bad components cannot be reworked)

      CALCULATED RESULTS
      ------------------

      FMS CELL EFFICIENCY              =      64.58 %
      RAW PARTS REQUIRED PER PERIOD    =      47.00
      UNADJUSTED EQUIPMENT REQUIREMENT =       1.44 FMS CELLS
      ADJUSTED EQUIPMENT REQUIREMENT   =       2.00 FMS CELLS
```

```
+-----------------------------------------------------------------------+
```

```
****************
   FMS CELL 2
****************

INPUT DATA ECHO
---------------

LENGTH OF PRODUCTION PERIOD    =      8.00 HOURS
FMS CELL EFFICIENCY            =     91.00 %
PROCESSING TIME PER COMPONENT  =      3.00 MINS
DEFECT RATE OF THE CELL        =      3.50 %
(Assuming that bad components cannot be reworked)

CALCULATED RESULTS
------------------

RAW PARTS REQUIRED PER PERIOD       =      46.00
UNADJUSTED EQUIPMENT REQUIREMENT    =       0.32 FMS CELLS
ADJUSTED EQUIPMENT REQUIREMENT      =       1.00 FMS CELLS



****************
   FMS CELL 3
****************

INPUT DATA ECHO
---------------

LENGTH OF PRODUCTION PERIOD    =      8.00 HOURS
AVERAGE DOWNTIME               =     45.00 MINS
AVERAGE SETUP TIME             =     11.00 MINS
ANALYZED PERIOD OF TIME        =     24.00 HOURS

PROCESSING TIME PER COMPONENT  =      6.00 MINS
DEFECT RATE OF THE CELL        =      6.00 %
(Assuming that bad components cannot be reworked)

CALCULATED RESULTS
------------------

FMS CELL EFFICIENCY                 =      96.11 %
RAW PARTS REQUIRED PER PERIOD       =      44.00
UNADJUSTED EQUIPMENT REQUIREMENT    =       0.57 FMS CELLS
ADJUSTED EQUIPMENT REQUIREMENT      =       1.00 FMS CELLS



****************
   FMS CELL 4
****************

INPUT DATA ECHO
---------------

LENGTH OF PRODUCTION PERIOD    =      8.00 HOURS
FMS CELL EFFICIENCY            =     89.00 %
PROCESSING TIME PER COMPONENT  =      4.00 MINS
DEFECT RATE OF THE CELL        =      2.00 %
(Assuming that bad components cannot be reworked)

CALCULATED RESULTS
------------------

RAW PARTS REQUIRED PER PERIOD       =      41.00
UNADJUSTED EQUIPMENT REQUIREMENT    =       0.38 FMS CELLS
ADJUSTED EQUIPMENT REQUIREMENT      =       1.00 FMS CELLS
```

CIM-O*

```
                CAPACITY PLANNING FOR MIXED PRODUCTION SYSTEMS
                                                              Page 3
+------------------------------------------------------------------+


     *****************
       FMS CELL 5
     *****************

     INPUT DATA ECHO
     ---------------

     LENGTH OF PRODUCTION PERIOD   =     8.00 HOURS

     AVERAGE DOWNTIME              =   324.00 MINS
     AVERAGE SETUP TIME            =   336.00 MINS
     ANALYZED PERIOD OF TIME       =    24.00 HOURS

     PROCESSING TIME PER COMPONENT =     7.00 MINS
     DEFECT RATE OF THE CELL       =     1.00 %
     (Assuming that bad components cannot be reworked)

     CALCULATED RESULTS
     ------------------

     FMS CELL EFFICIENCY              =     54.17 %
     RAW PARTS REQUIRED PER PERIOD    =     40.00
     UNADJUSTED EQUIPMENT REQUIREMENT =      1.08 FMS CELLS
     ADJUSTED EQUIPMENT REQUIREMENT   =      1.00 FMS CELLS



     *****************
       FMS CELL 6
     *****************

     INPUT DATA ECHO
     ---------------

     LENGTH OF PRODUCTION PERIOD   =     8.00 HOURS

     AVERAGE DOWNTIME              =   180.00 MINS
     AVERAGE SETUP TIME            =   355.00 MINS
     ANALYSED PERIOD OF TIME       =    24.00 HOURS

     PROCESSING TIME PER COMPONENT =     9.00 MINS
     DEFECT RATE OF THE CELL       =     2.00 %
     (Assuming that bad components cannot be reworked)

     CALCULATED RESULTS
     ------------------

     FMS CELL EFFICIENCY              =     62.85 %
     RAW PARTS REQUIRED PER PERIOD    =     40.00
     UNADJUSTED EQUIPMENT REQUIREMENT =      1.19 FMS CELLS
     ADJUSTED EQUIPMENT REQUIREMENT   =      2.00 FMS CELLS



     *****************
       FMS CELL 7
     *****************

     INPUT DATA ECHO
     ---------------

     LENGTH OF PRODUCTION PERIOD   =     8.00 HOURS
     FMS CELL EFFICIENCY           =    88.00 %
     PROCESSING TIME PER COMPONENT =    11.00 MINS
     DEFECT RATE OF THE CELL       =     2.00 %
     (Assuming that bad components cannot be reworked)
```

```
             CAPACITY PLANNING FOR MIXED PRODUCTION SYSTEMS
                                                        Page 4
+------------------------------------------------------------------+

     CALCULATED RESULTS
     ------------------

     RAW PARTS REQUIRED PER PERIOD        =      39.00
     UNADJUSTED EQUIPMENT REQUIREMENT =          1.02 FMS CELLS
     ADJUSTED EQUIPMENT REQUIREMENT   =          1.00 FMS CELLS



     ****************
        FMS CELL 8
     ****************

     INPUT DATA ECHO
     ---------------

     LENGTH OF PRODUCTION PERIOD   =      8.00 HOURS
     FMS CELL EFFICIENCY           =     93.00 %
     PROCESSING TIME PER COMPONENT =      5.00 MINS
     DEFECT RATE OF THE CELL       =      2.00 %
     (Assuming that bad components cannot be reworked)

     CALCULATED RESULTS
     ------------------

     RAW PARTS REQUIRED PER PERIOD        =      38.00
     UNADJUSTED EQUIPMENT REQUIREMENT =          0.43 FMS CELLS
     ADJUSTED EQUIPMENT REQUIREMENT   =          1.00 FMS CELLS



     ****************
        FMS CELL 9
     ****************

     INPUT DATA ECHO
     ---------------

     LENGTH OF PRODUCTION PERIOD   =      8.00 HOURS
     FMS CELL EFFICIENCY           =     98.00 %
     PROCESSING TIME PER COMPONENT =      2.50 MINS
     DEFECT RATE OF THE CELL       =      1.00 %
     (Assuming that bad components cannot be reworked)

     CALCULATED RESULTS
     ------------------

     RAW PARTS REQUIRED PER PERIOD        =      37.00
     UNADJUSTED EQUIPMENT REQUIREMENT =          0.20 FMS CELLS
     ADJUSTED EQUIPMENT REQUIREMENT   =          1.00 FMS CELLS



     ****************
        FMS CELL 10
     ****************

     INPUT DATA ECHO
     ---------------

     LENGTH OF PRODUCTION PERIOD   =      8.00 HOURS
     FMS CELL EFFICIENCY           =     96.00 %
     PROCESSING TIME PER COMPONENT =      8.00 MINS
     DEFECT RATE OF THE CELL       =      2.00 %
     (Assuming that bad components cannot be reworked)
```

CAPACITY PLANNING FOR MIXED PRODUCTION SYSTEMS

```
+------------------------------------------------------------------------+
```

CALCULATED RESULTS

```
RAW PARTS REQUIRED PER PERIOD         =      37.00
UNADJUSTED EQUIPMENT REQUIREMENT  =       0.64 FMS CELLS
ADJUSTED EQUIPMENT REQUIREMENT        =      1.00 FMS CELLS
```

```
****************
    FMS CELL 11
****************
```

INPUT DATA ECHO

```
LENGTH OF PRODUCTION PERIOD    =       8.00 HOURS
AVERAGE DOWNTIME               =     243.00 MINS
AVERAGE SETUP TIME             =     355.00 MINS
ANALYSED PERIOD OF TIME        =      24.00 HOURS

PROCESSING TIME PER COMPONENT =      12.00 MINS
DEFECT RATE OF THE CELL        =       4.00 %
```
(Assuming that bad components cannot be reworked)

CALCULATED RESULTS

```
FMS CELL EFFICIENCY                    =      58.47 %
RAW PARTS REQUIRED PER PERIOD          =      36.00
UNADJUSTED EQUIPMENT REQUIREMENT  =       1.54 FMS CELLS
ADJUSTED EQUIPMENT REQUIREMENT    =       2.00 FMS CELLS
```

```
****************
    FMS CELL 12
****************
```

INPUT DATA ECHO

```
LENGTH OF PRODUCTION PERIOD    =       8.00 HOURS
FMS CELL EFFICIENCY            =      93.00 %
PROCESSING TIME PER COMPONENT =       8.10 MINS
DEFECT RATE OF THE CELL        =       2.00 %
```
(Assuming that bad components cannot be reworked)

CALCULATED RESULTS

```
RAW PARTS REQUIRED PER PERIOD         =      35.00
UNADJUSTED EQUIPMENT REQUIREMENT  =       0.64 FMS CELLS
ADJUSTED EQUIPMENT REQUIREMENT        =      1.00 FMS CELLS
```

```
****************
    FMS CELL 13
****************
```

INPUT DATA ECHO

```
LENGTH OF PRODUCTION PERIOD    =       8.00 HOURS
FMS CELL EFFICIENCY            =      87.00 %
PROCESSING TIME PER COMPONENT =       3.00 MINS
DEFECT RATE OF THE CELL        =       3.00 %
```
(Assuming that bad components cannot be reworked)

```
            CAPACITY PLANNING FOR MIXED PRODUCTION SYSTEMS
                                                        Page 6
  +----------------------------------------------------------------+


        CALCULATED RESULTS
        ------------------

        RAW PARTS REQUIRED PER PERIOD     =     34.00
        UNADJUSTED EQUIPMENT REQUIREMENT =      0.24 FMS CELLS
        ADJUSTED EQUIPMENT REQUIREMENT    =     1.00 FMS CELLS


        ****************
        FMS CELL 14
        ****************

        INPUT DATA ECHO
        ---------------

        LENGTH OF PRODUCTION PERIOD    =      8.00 HOURS

        AVERAGE DOWNTIME               =    480.00 MINS
        AVERAGE SETUP TIME             =    325.00 MINS
        ANALYZED PERIOD OF TIME        =     24.00 HOURS

        PROCESSING TIME PER COMPONENT =      8.60 MINS
        DEFECT RATE OF THE CELL        =      2.00 %
        (Assuming that bad components cannot be reworked)

        CALCULATED RESULTS
        ------------------

        FMS CELL EFFICIENCY               =     44.10 %
        RAW PARTS REQUIRED PER PERIOD     =     33.00
        UNADJUSTED EQUIPMENT REQUIREMENT =      1.34 FMS CELLS
        ADJUSTED EQUIPMENT REQUIREMENT    =     2.00 FMS CELLS


    - - - - - end of output - - - - - - end of output - - - - - - -
```

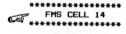

References and further reading

[8.1] P.G.Ranky: The Design and Operation of FMS. (Flexible
      Manufacturing Systems) IFS (Publications) Ltd. and North-
      Holland Publishing Company, 1983. 348 p.

[8.2] P.G.Ránky: The FMS Software Library, CIM software with
      user and system manuals, 1983-84-85. The CAPACITY
      program. (Please contact the author, or Malva, or
      Comporgan System Houses).

[8.3] Mikel P. Groover and Emory W. Zimmers, JR.: CAD/CAM:
      Computer Aided Design and Manufacturing, Prentice/Hall
      International, 1984.

[8.4] O.W. Wight: MRP II: Unlocking America's Productivity
      Potential, Oliver Wight Limited Publications, Inc.,
      Williston, Vt., 1981.

[8.5] K. Hitomi: Manufacturing Systems Engineering, Taylor and Francis Ltd, London, 1979.

[8.6] J.A. Buzacott: Modelling Manufacturing Systems, IFIP/IFAC World Congress, Budapest, 1984. p. 136-141.

[8.7] Paul G. Ránky: A Software Library for Designing and controlling Flexible Manufacturing Systems, IFIP/IFAC World Congress, Budapest, 1984. p. 147-152.

[8.8] M. Horvath and A. Markus: Operation Sequence Planning Using Optimization Concepts and Logic Programming, IFIP/IFAC World Congress, Budapest, 1984. p. 153-156.

[8.9] U. Rembold and R. Dildmann (ed.): Methods and Tools for Computer Integrated Manufacturing, Springer Verlag, 1984.

[8.10] David. A. Bourne and Mark S. Fox: Auonomous Manufacturing: Automating the Job-Shop, Computer, September 1984. p. 76-97.

[8.11] R.K. Sinha and R.H. Hollier: A review of production control problems in cellular manufacture, Int. J. Prod. Res. 1984. Vol. 22, No. 5, p. 773-789.

[8.12] J. Erschler, D. Leveque, F. Roubellat: Periodic Loading of Flexible Manufacturing Systems, Advances in Production Management Systems, Elsevier Science Publ., IFIP 1984.

[8.13] Paul G. Ránky: The FMS Software Library, general Purpose FMS Capacity Planning Program. User and System Manuals, 1983-84.

CHAPTER NINE

Batchsize analysis in FMS

The aim of up-to-date production systems is to design and
manufacture parts requested on demand, rather than on stock in
order to save inventory costs. This concept and production
control is often called JIT, or "just-in-time" production
control and is very much part of FMS and the CIM as a whole,
meaning that one should avoid batches, and rather aim an economic order quantity of one, because:

* The production systems are flexible and thus capable of
  coping equally well economically with single components,
  as with small or medium size batches.

* The setup costs are reduced by design as well as by
  utilizing FMS.

* If something goes wrong it happens to a single part, or
  in the worst case to a few parts only, rather than to
  a whole batch, often representing several days, perhaps
  a week of manufacturing cost, as is often the case in
  conventional batch manufacturing industries.

Unfortunately in even the most flexibly designed manufacturing system, production control may be unable to provide the
most appropriate mix to load the system, at least for a period
of time, because of the volume and mix of received orders.

If, for example, the demand rate for some unforeseen
reason is small compared to the production rate, FMS operation
control should adapt itself to such changes either:

* by reducing the production rate, or

* by changing the product mix, if possible, or

* by carrying on until the problems are resolved and
  manufacture the products periodically in batches.

Also in order to minimize tooling, setup, inventory and
other costs, components can be organized into batches, or
mixed batches.

Small batch type manufacturing with the FMS may also be
advisable, at least for some time, if there is some temporary
shortage in fixtures, or tooling, or if the inventory holding
costs are increasing at a certain date for tax reasons, etc.

In this Chapter the mathematical model of a lotsize analysis program is introduced and a sample run is shown determining the minimum variable cost, or economic batch sizes for
each single batch and the mixed batches. In the mathematical
model as used in this package, the inventory is filled up at a
calculated rate during the production period of the component
and decreases at a calculated rate during the non-productive
period. To determine the minimum cost batch size, all variable costs are minimized.

(Note that the case study in this Chapter represents only
one solution and that there are many different batchsize analysis programs in use, minimizing the number of tool changes
in tool magazines, optimizing for different variables and cost
factors.)

9.1 The mathematical model of the LOTSIZE program

To be able to calculate the economic lot size (i.e. the
minimum cost lot size) all variable costs are minimised. The
variable cost in this case is the sum of the setup cost and
the inventory holding cost for each batch.

The model assumes an inventory management system, in
which the warehouse or the part storage area of the FMS is
filled up at a calculated and close to uniform rate during the
production period and decreases at a calculated rate during
the non-production period. (This model shown in Figure 9.1
follows the shape of a triangle and assumes that the production rate is greater than the demand rate.)

Consider the ABC triangle in Figure 9.1,

$$\tan \gamma = \frac{AB}{BC} = \frac{\text{Actual inventory level}}{\text{Non-productive period}}$$

thus the

Demand rate $= \tan \gamma$

of which $\gamma = \text{Arctan (Demand rate)}$

Similarly from the DBE triangle the production rate can be calculated as follows

$$\tan \alpha = \frac{EB}{DB} = \frac{\text{Max. inventory level}}{\text{Production period}}$$

thus the

Production rate $= \tan \alpha$

of which $\alpha = \text{Arctan (Production rate)}$

and of the EDB and ADB triangles

$$\tan \beta = \frac{\text{Production rate} - \text{Demand rate}}{\text{Production period}}$$

The setup cost must be given for each batch for the total production cycle time.

The total production cycle time is the sum of the production time spent for each batch and the inventory holding (i.e. non-productive) time. (Note that these values must be given

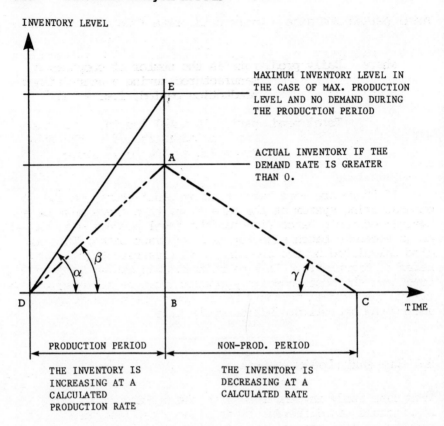

INVENTORY LEVEL

MAXIMUM INVENTORY LEVEL IN
THE CASE OF MAX. PRODUCTION
LEVEL AND NO DEMAND DURING
THE PRODUCTION PERIOD

ACTUAL INVENTORY IF THE
DEMAND RATE IS GREATER
THAN 0.

PRODUCTION PERIOD   NON-PROD. PERIOD

THE INVENTORY IS   THE INVENTORY IS
INCREASING AT A   DECREASING AT A
CALCULATED   CALCULATED RATE
PRODUCTION RATE

Figure 9.1 The inventory model of the LOTSIZE program indicates
          the production rate and the demand rate during the
          total production cycle time. This model assumes that
          the production rate is always higher or equal than
          the rate of demand.

in the same dimension.)

The total variable cost per unit is the sum of the setup
cost and the inventory holding cost per unit. The minimum cost
(i.e. optimum batch size) is determined from these relationships by differentiating the total variable cost equation with
respect to the production quantity and setting it at zero.

The minimum cost production time is then calculated using
this optimum production quantity value.

The total period lot size is calculated as follows:

Total period lot size = Daily prod. rate * Total prod. period

where   Daily prod. rate is the number of components
                         manufactured during a user-defined
                         production period, i.e. a "day"

         Total prod. period is total length of the produc-
                         tion period expressed in the same
                         dimension as the "day" above.

If there are more batches being manufactured on the
manufacturing system at the same time, then there must be an
overall economic batch size for the total product mix, as well
as an economic batch size for each separate batch. This is
also calculated by the program. (To illustrate this point
refer to Figure 9.3.) This means that when mixing batches
together, the most economic production can be achieved for the
mix if the overall economic batch sizes are manufactured
during the optimal production cycle time.

9.2 Case study: Lotsize analysis

This case study is concerned with the problem of how to
manufacture ten different types of batches during different
lengths of time economically as individual batches, or as a
product mix.

The input data collection is shown in Figure 9.2, and the
results are listed in Figure 9.3, showing the calculated economic batch sizes and production cycle times for individual
batches as well as for mixed batch production and the batch
oriented input data and results are shown in the output listing of the LOTSIZE program in Figure 9.4.

```
**************************************
*   TABLE OF CURRENT INPUT DATA   *
*------------------------------------*
*  FMS LOT-SIZE ANALYSIS PROGRAM  *
**************************************
```

PRODUCT TYPE	ANALYZED PROD. PERIOD [DAY]	PRODUCTION RATE [UNIT/DAY]	SETUP COST [BATCTH]	DAILY DEMAND RATE [UNIT/DAY]	INVENTORY COST [UNIT/DAY]
1	30.00	2.00	69.90	1.00	0.43
2	45.00	6.00	102.50	3.00	0.78
3	25.00	22.00	598.00	10.00	0.23
4	52.00	45.00	986.50	27.00	0.99
5	30.00	12.00	342.34	10.00	0.45
6	48.00	132.00	952.00	100.00	0.66
7	20.00	38.00	542.00	23.00	0.78
8	30.00	18.00	769.00	5.00	0.88
9	30.00	21.00	243.66	12.00	0.23
10	52.00	9.00	132.00	2.00	0.16

Figure 9.2 List of current input data of the LOTSIZE analysis program.

```
**************************************
*        TABLE OF RESULTS         *
*------------------------------------*
*  FMS LOT-SIZE ANALYSIS PROGRAM  *
**************************************
```

)))>> NOTE: the given values are not rounded <<(((

**)> The OVERALL ECONOMIC BATCH SIZE is based on the optimal production
cycle time for for all components of = 15.06 DAYS

PRODUCT TYPE	MIN. COST BATCH SIZE [UNIT]	TOTAL PERIOD BATCH SIZE [UNIT]	OVERALL ECON. BATCH SIZE **)> [UNIT]	ECON. PROD. CYCLE TIME [DAY]
1	25.50	60.00	15.06	25.50
2	39.71	270.00	45.18	13.24
3	308.76	550.00	150.61	30.88
4	366.77	2340.00	406.65	13.58
5	302.14	360.00	150.61	30.21
6	1090.87	6336.00	1506.13	10.91
7	284.56	760.00	346.41	12.37
8	110.00	540.00	75.31	22.00
9	243.57	630.00	180.74	20.30
10	65.14	468.00	30.12	32.57

Figure 9.3 List of results of the LOTSIZE analysis program.

Figure 9.4 Sample run of the LOTSIZE program with ten mixed batches.

```
                  LOTSIZE ANALYSIS DEMONSTRATION
                                                        Page 1
+------------------------------------------------------------------+

  ******************************************
  *        THE FMS SOFTWARE LIBRARY        *
  *                                        *
  *    Created by Dr.Paul G Ránky 1983.    *
  *----------------------------------------*
  *    FMS BATCH SIZE ANALYSIS PROGRAM     *
  ******************************************

  INPUT DATA ECHO FOR EACH PRODUCT TYPE
  -------------------------------------

  Please note that a DAY represents a user-specified
  length of time

  PRODUCT TYPE :1  <<<(((*******
  *************

  BATCH CODE = BATCH 1

  THE ANALYZED PRODUCTION PERIOD IN DAYS =      30.00

  DAILY PRODUCTION RATE IN UNITS        =       2.00

  SETUP COST PER BATCH                  =      69.90

  DAILY DEMAND RATE IN UNITS            =       1.00

  THE INVENTORY HOLDING COST PER COMPONENT,
  PER DAY IS =       0.43

  PRODUCT TYPE :2  <<<(((*******
  *************

  BATCH CODE = BATCH 2

  THE ANALYZED PRODUCTION PERIOD IN DAYS =      45.00

  DAILY PRODUCTION RATE IN UNITS        =       6.00

  SETUP COST PER BATCH                  =     102.50

  DAILY DEMAND RATE IN UNITS            =       3.00

  THE INVENTORY HOLDING COST PER COMPONENT,
  PER DAY IS =       0.78

  PRODUCT TYPE :3  <<<(((*******
  *************

  BATCH CODE = BATCH 3

  THE ANALYZED PRODUCTION PERIOD IN DAYS =      25.00

  DAILY PRODUCTION RATE IN UNITS        =      22.00

  SETUP COST PER BATCH                  =     598.00

  DAILY DEMAND RATE IN UNITS            =      10.00

  THE INVENTORY HOLDING COST PER COMPONENT,
  PER DAY IS =       0.23
```

LOTSIZE ANALYSIS DEMONSTRATION

Page 2

+--+

PRODUCT TYPE :4 <<<(((*******
****

BATCH CODE = BATCH 4

THE ANALYZED PRODUCTION PERIOD IN DAYS =   52.00

DAILY PRODUCTION RATE IN UNITS   =   45.00

SETUP COST PER BATCH   =   986.50

DAILY DEMAND RATE IN UNITS   =   27.00

THE INVENTORY HOLDING COST PER COMPONENT,
PER DAY IS =   0.99

PRODUCT TYPE :5 <<<(((*******
****

BATCH CODE = BATCH 5

THE ANALYZED PRODUCTION PERIOD IN DAYS =   30.00

DAILY PRODUCTION RATE IN UNITS   =   12.00

SETUP COST PER BATCH   =   342.34

DAILY DEMAND RATE IN UNITS   =   10.00

THE INVENTORY HOLDING COST PER COMPONENT,
PER DAY IS =   0.45

PRODUCT TYPE :6 <<<(((*******
****

BATCH CODE = BATCH 6

THE ANALYZED PRODUCTION PERIOD IN DAYS =   48.00

DAILY PRODUCTION RATE IN UNITS   =   132.00

SETUP COST PER BATCH   =   952.00

DAILY DEMAND RATE IN UNITS   =   100.00

THE INVENTORY HOLDING COST PER COMPONENT,
PER DAY IS =   0.66

PRODUCT TYPE :7 <<<(((*******
****

BATCH CODE = BATCH 7

THE ANALYZED PRODUCTION PERIOD IN DAYS =   20.00

DAILY PRODUCTION RATE IN UNITS   =   38.00

SETUP COST PER BATCH   =   542.00

DAILY DEMAND RATE IN UNITS   =   23.00

THE INVENTORY HOLDING COST PER COMPONENT,
PER DAY IS =   0.78

LOTSIZE ANALYSIS DEMONSTRATION

+--+

PRODUCT TYPE :8 <<<(((*******
****

BATCH CODE = BATCH 8

THE ANALYZED PRODUCTION PERIOD IN DAYS =   30.00

DAILY PRODUCTION RATE IN UNITS   =   18.00

SETUP COST PER BATCH   =   769.00

DAILY DEMAND RATE IN UNITS   =   5.00

THE INVENTORY HOLDING COST PER COMPONENT,
PER DAY IS =   0.88

PRODUCT TYPE :9 <<<(((*******
****

BATCH CODE = BATCH 9

THE ANALYZED PRODUCTION PERIOD IN DAYS =   30.00

DAILY PRODUCTION RATE IN UNITS   =   21.00

SETUP COST PER BATCH   =   243.66

DAILY DEMAND RATE IN UNITS   =   12.00

THE INVENTORY HOLDING COST PER COMPONENT,
PER DAY IS =   0.23

PRODUCT TYPE :10 <<<(((*******
****

BATCH CODE = BATCH 10

THE ANALYZED PRODUCTION PERIOD IN DAYS =   52.00

DAILY PRODUCTION RATE IN UNITS   =   9.00

SETUP COST PER BATCH   =   132.00

DAILY DEMAND RATE IN UNITS   =   2.00

THE INVENTORY HOLDING COST PER COMPONENT,
PER DAY IS =   0.16

****
*   CALCULATED RESULTS OF THE FMS BATCH SIZE ANALYSIS *
****

**)))>> NOTE: THE OVERALL ECONOMIC BATCH SIZE IS BASED ON THE
              OPTIMAL PRODUCTION CYCLE TIME FOR ALL COMPONENTS
              OF 15.06 DAYS (i.e. period of time)

LOTSIZE ANALYSIS DEMONSTRATION

+--+

PRODUCT TYPE : 1 <<<<(((*****
****

BATCH CODE = BATCH 1

THE MINIMUM COST BATCH SIZE IS	=	25 COMPONENTS
THE TOTAL PERIOD BATCH SIZE IS	=	60 COMPONENTS
THE OVERALL ECONOMIC BATCH SIZE IS	=	15 COMPONENTS
THE ECONOMIC PRODUCTION CYCLE TIME IS	=	25.50 DAYS (i.e. time period)

PRODUCT TYPE : 2 <<<<(((*****
****

BATCH CODE = BATCH 2

THE MINIMUM COST BATCH SIZE IS	=	40 COMPONENTS
THE TOTAL PERIOD BATCH SIZE IS	=	270 COMPONENTS
THE OVERALL ECONOMIC BATCH SIZE IS	=	45 COMPONENTS
THE ECONOMIC PRODUCTION CYCLE TIME IS	=	13.24 DAYS (i.e. time period)

PRODUCT TYPE : 3 <<<<(((*****
****

BATCH CODE = BATCH 3

THE MINIMUM COST BATCH SIZE IS	=	309 COMPONENTS
THE TOTAL PERIOD BATCH SIZE IS	=	550 COMPONENTS
THE OVERALL ECONOMIC BATCH SIZE IS	=	151 COMPONENTS
THE ECONOMIC PRODUCTION CYCLE TIME IS	=	30.88 DAYS (i.e. time period)

PRODUCT TYPE : 4 <<<<(((*****
****

BATCH CODE = BATCH 4

THE MINIMUM COST BATCH SIZE IS	=	367 COMPONENTS
THE TOTAL PERIOD BATCH SIZE IS	=	2340 COMPONENTS
THE OVERALL ECONOMIC BATCH SIZE IS	=	407 COMPONENTS
THE ECONOMIC PRODUCTION CYCLE TIME IS	=	13.58 DAYS (i.e. time period)

+--+

PRODUCT TYPE : 5 <<<(((*****
****

BATCH CODE = BATCH 5

THE MINIMUM COST BATCH SIZE IS   =   302 COMPONENTS

THE TOTAL PERIOD BATCH SIZE IS   =   360 COMPONENTS

THE OVERALL ECONOMIC BATCH SIZE IS   =   151 COMPONENTS

THE ECONOMIC PRODUCTION CYCLE TIME IS   =   30.21 DAYS (i.e. time period)

PRODUCT TYPE : 6 <<<(((*****
****

BATCH CODE = BATCH 6

THE MINIMUM COST BATCH SIZE IS   =   1091 COMPONENTS

THE TOTAL PERIOD BATCH SIZE IS   =   6336 COMPONENTS

THE OVERALL ECONOMIC BATCH SIZE IS   =   1506 COMPONENTS

THE ECONOMIC PRODUCTION CYCLE TIME IS   =   10.91 DAYS (i.e. time period)

PRODUCT TYPE : 7 <<<(((*****
****

BATCH CODE = BATCH 7

THE MINIMUM COST BATCH SIZE IS   =   285 COMPONENTS

THE TOTAL PERIOD BATCH SIZE IS   =   760 COMPONENTS

THE OVERALL ECONOMIC BATCH SIZE IS   =   346 COMPONENTS

THE ECONOMIC PRODUCTION CYCLE TIME IS   =   12.37 DAYS (i.e. time period)

PRODUCT TYPE : 8 <<<(((*****
****

BATCH CODE = BATCH 8

THE MINIMUM COST BATCH SIZE IS   =   110 COMPONENTS

THE TOTAL PERIOD BATCH SIZE IS   =   540 COMPONENTS

THE OVERALL ECONOMIC BATCH SIZE IS   =   75 COMPONENTS

THE ECONOMIC PRODUCTION CYCLE TIME IS   =   22.00 DAYS (i.e. time period)

```
                    LOTSIZE ANALYSIS DEMONSTRATION
                                                        Page 6
+----------------------------------------------------------------+


PRODUCT TYPE : 9 <<<(((*****
**************

BATCH CODE = BATCH 9

THE MINIMUM COST BATCH SIZE IS        =    244 COMPONENTS

THE TOTAL PERIOD BATCH SIZE IS        =    630 COMPONENTS

THE OVERALL ECONOMIC BATCH SIZE IS    =    181 COMPONENTS

THE ECONOMIC PRODUCTION CYCLE TIME IS =  20.30 DAYS (i.e. time period)



PRODUCT TYPE : 10 <<<(((*****
**************

BATCH CODE = BATCH 10

THE MINIMUM COST BATCH SIZE IS        =     65 COMPONENTS

THE TOTAL PERIOD BATCH SIZE IS        =    468 COMPONENTS

THE OVERALL ECONOMIC BATCH SIZE IS    =     30 COMPONENTS

THE ECONOMIC PRODUCTION CYCLE TIME IS =  32.57 DAYS (i.e. time period)
```

References and further reading

[9.1] K. Hitomi: Manufacturing Systems Engineering, Taylor and Francis Ltd., London, 1979.

[9.2] Paul G. Ránky: The FMS Software Library, CIM software with user and system manuals, 1983-84-85. The LOTSIZE program. (Contact the author, Malva, or Comporgan System Houses).

[9.3] D.A. Collier: A comparison of MRP lot sizing methods considering capacity change costs, Journal of Operations Management, Vol.1, No.23.

[9.4] J.J. DeMatteis: An economic lot sizing technique, IBM Systems Journal, Vol.7, No.30.

[9.5] E. Steinberg and H.A. Napier: Optimal, multi-level lot sizing for requirements planning systems, Management Sciences, Vol.26, No.1258.

[9.6] R.K. Sinha and R.H. Hollier: A review of production control problems in cellular manufacture, Int. J. Prod. Res. 1984. Vol.22, No.5, p. 773-789.

[9.7] PERA 1972. The PERA Verimode System of job scheduling. Report No. 252.

[9.8] PERA 1974. iWork Loading and Scheduling in group technology cells. Report No. 282.

[9.9] Harish C. Bahl and Larry P. Ritzman: A cyclical scheduling heuristic for lot sizing with capacity constrains, Int. J. Prod. Res. 1984. Vol.22, No.5, p. 791-800.

[9.10] H.M. Wagner and T.M. Whitin: Dynamic version of the economic lot size model, Management Science, Vol.4, No. 89.

[9.11] E.A. Silver at all.: A simple modification of the EOQ for the case of a varying demand rate, Production and inventory management, 4th Quarter, 1969.

[9.12] R.A. Kaimann: A comperison of EOQ and Dynamic programming inventory models with safety stock and variable lead time considerations, Production and Inventory Management, 1st Quarter, 1974. pp. 1-20.

CHAPTER TEN

Single and mixed product manufacturing and robotized assembly system balancing

In manufacturing, i.e. machining, assembly, test, etc., the sequence of operations is restricted in various ways. As we have seen already in the previous three Chapters, and in particular in Chapter 6, the production control system must store the "knowledge" regarding these restrictions in a rule base, (outlined in Figure 6.39), to enable flexible production despite the restrictions.

For example, in most assembly operations the sequence of assembly steps is restricted too, in terms of the order in which they can be carried out. For example in mechanized assembly, the washer must be placed over the bolt before the nut can be placed, turned and tightened. (These product restrictions are the so called precedence constraints and are illustrated graphically in the form of a precedence diagram in Figure 10.1.)

One can create "variable route" FMS part programs for assembly, and for many other processes too, but in most cases there are certain operations which must be done in an order which satisfies the precedence constraints for the product (as well as the economics of the manufacturing system) for a period of time; meaning that in another period of time perhaps an alternative route could be taken in the manufacturing system for which other precedence constraints would be applicable. As one can see computer programs are used to solve such problems because of the number of different ways in which operations can be grouped.

The aim of this Chapter is to explain and demonstrate:

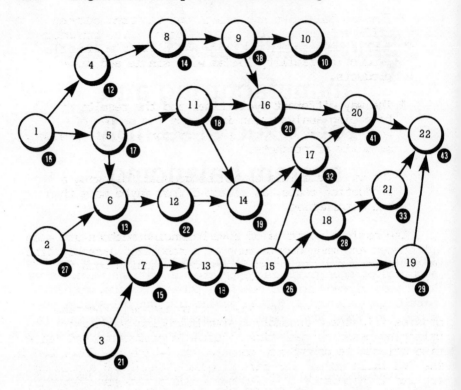

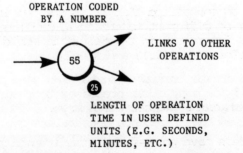

OPERATION CODED
BY A NUMBER

LINKS TO OTHER
OPERATIONS

LENGTH OF OPERATION
TIME IN USER DEFINED
UNITS (E.G. SECONDS,
MINUTES, ETC.)

Figure 10.1 Graphics representation of a precedence diagram.
(Note that the operation number does not represent
the physical order of operations within the assembly or other job. It is simply a code identifying
each operation.)

* The way balancing can smooth out the load between
different workstations of dedicated (i.e. transfer
line type) systems, or between cells of a flexible
production facility loaded with single and mixed
products.

* The way different cycle times and the results of
balancing analysis can influence the number of
workstations, or cells required for a single product,
or a mix of products.

* The way operations should be grouped following a
production route, as selected from maybe more than
one alternative.

The basic algorithms of some balancing methods are also
introduced and many sample runs of a robotized assembly line
balancing program are shown in the case studies both for
single and mixed product assembly.

It must be underlined again, that although the case
studies illustrate assembly system balancing techniques,
balancing based on real-time adjustable production and real-
time selectable precedence constraints (which are described in
the "variable route" FMS part program), could and should be
applied if necessary in all other areas of the manufacturing
process, including machining, test, welding, laser cutting,
sheet metal manufacturing, etc.

It must be emphasized that it is not our intention to
give a survey of different balancing algorithms and programs,
but if one is interested in this area, it would be advisable
to consult references [10.1] and [10.2].

10.1 Application areas of balancing techniques

Balancing models have emerged from the traditional transfer
line, or flow line type of manufacturing methods and systems,
where there are many distinct operations following each other
and processed by rigid (i.e. non-programmable) workstations
in a predefined rigid order, which cannot be altered, meaning
that if any of the workstations breaks down the whole machine
or system must be stopped, thus all processes involved are
disrupted.

As a typical example of such systems Figures 10.2 and
10.3 illustrate a highly efficient, but dedicated assembly
machine consisting of several assembly heads working at a
fixed cycle time. Balancing in the case of designing such
systems is essential.

Balancing however, as we said earlier, can be important
in flexible manufacturing environment too, in particular:

* When products do not arrive entirely in random order,
  but in batches prior to manufacturing.

* When there are overloaded and under-utilized cells in
  the FMS during the analysed period of time.

* If products must be manufactured on the system within
  similar circumstances to those discussed regarding
  economic batch size analysis in the previous Chapter.

Balancing techniques traditionally arrange individual
operations at the workstations so that the total time required
at each workstation is approximately the same, thus the system
works in a balanced way whilst the precedence constraints are
maintained.

In FMS (and at a higher level in CIM) balancing is not
as strict as with flow line systems, because of the built-in
flexibility of FMS, achieved mainly by the cellular design,
the random material handling facility, the possibility of
selecting between alternative production routes in real-time
and dynamic scheduling which is a consequence of all above
listed features.

In most practical situations it is difficult to achieve a
perfect balance. In the case of a transfer line, or a flow
line this will affect the system badly and some workstations
will not work at their maximum capacity, but will wait idle
between the completion time of the last operation on the part
and the arrival of the new part.

FMS by design simplifies and overcomes (within limitations) the "imperfect balance" problem by dynamic rescheduling, or "secondary optimization" of the parts being manufactured. If the imbalance is a major one, resulting in the
exclusion of some cells of the system entirely for the product
mix in question, or the overloading of some cells, dynamic
rescheduling alone might not solve the problem, thus in such

cases lot sizing, balancing and loading sequencing must be evaluated with capacity checks too.

In FMS, balancing can be used both as an off-line simulation tool prior to loading sequencing (and this is very likely the preferred way of using this technique in FMS; refer to Figure 6.37 again), or during the real-time process when components are already under DNC control and waiting inside the system to be processed.

10.2 The mathematical models for balancing

Balancing models generally deal with single product lines, batch model lines and mixed model lines, which are the most difficult to solve.

In our discussion we shall introduce a single product balancing model and we shall apply it also for mixed product manufacturing system balancing for cases where from the single product precedence diagrams involved a single mixed product precedence diagram can be drawn. (Note that the use of the algorithms and the computer programs are demonstrated in the case studies.)

To introduce a very simple, manual assembly line balancing algorithm, (i.e. the largest candidate rule based model) consider the following procedure:

Step 1. List all elements in descending order of the required processing time per operation so, that the largest is at the top of the list.

Step 2. To assign elements to the first assembly head (or robot) start at the top of the list and work down, selecting the first feasible element for placement at the assembly head. (A feasible element is one that satisfies the previously defined precedence requirements and does not cause the sum of the operation time values to exceed the total cycle time.)

Step 3. Continue the process described in Step 2 until no further elements can be added without exceeding the total available cycle time.

Figure 10.2 Mechanized assembly systems need careful balancing at the design stage. (System view 1; Courtesy of Plessey Office Systems Ltd., Beeston, Nottingham.)

Figure 10.3 Mechanized assembly system. View 2. (Courtesy of Plessey Office Systems Ltd., Beeston, Nottingham.)

Step 4. Repeat steps 2 and 3 for the other workstations in the line until all the elements have been assigned.

The above described algorithm is a simplified model, since it does not consider the selection of operations for assignment to stations according to their position in the precedence diagram.

To evaluate all possible solutions in our second algorithm, a heuristic method is used to estimate the best solution. The core of the mathematical model and the developed BALANCE program is based on an extended version of the COMSOAL algorithm as described in [10.3]. The algorithm generates a number of different feasible solutions in a random sequence and selects the answer which has the minimum number of assembly heads, or robots. There is often more than one solution with the minimum number of robots. In this case the program selects the one with the minimum cycle time from the data collection of the minimum number of assembly heads, or work stations.

The program's algorithm employs a method of generating sequences of operations that can be described as follows:

Step 1. For each operation on the manufacturing (or assembly) system form a list of the first number of pre-tasks (i.e. first set of operations performed prior to the operation in question).

Step 2. Form a new list of operations from those which have no immediate pre-tasks and whose duration times are no greater than the remaining time of of the robot station.

Step 3. Form a sub-list of this new list in a random order (i.e. select operations randomly and form a new list).

Step 4. Eliminate the selected operation from the list and deduct 1 from the operations immediately following the selected operation from the list.

Step 5. Reduce the remaining time in the robot cell by the duration of the operation.

Step 6. Repeat Steps 2 to 5 until all tasks have been

allocated. In this procedure evaluation is pro-
gressive, i.e. the available time is diminished
as each operation is generated until the operation becomes too large for the remaining operation.

There are certain restrictions in this algorithm which
effect the rigid transfer line type of systems rather more
than FMS, thus for our purpose and type of applications they
do not represent serious problems.

Here is a summary of the most important conditions one
must ensure when using this algorithm and their effect on
flexible and dedicated or rigid manufacturing systems:

* It is assumed in the algorithm that each operation in
  the manufacturing system, or assembly line could be
  performed by any of the robot cells, as long as the
  precedence diagram is observed.

  By employing flexible robot cells equipped with Automated Hand Changers,this provides no real problem even in
  a mixed product assembly situation. (Further details on
  such cells and Automated Robot Hand Changing systems
  can be found in [10.4]).

  In the case of dedicated mechanized assembly systems,
  or transfer lines, or flow line systems this condition
  must be ensured by design for the selected part, or
  part mix.

* It is also assumed that at each robot a number of operations can be performed and that each robot is allowed
  the same maximum length of time (the user specified
  cycle time) to complete the assigned operations.

* The equal intervals in which an assembled component
  leaves the manufacturing system is the cycle time of
  the line, meaning the sum of the time spent at a
  · single cell plus the time needed to move the component, or sub-assembly between robot cells.

  This is obviously going to be an estimate only for the
  FMS type of systems because of the unpredictable number
  of dynamic rescheduling possibilities and because of
  the different possible alternative routes the part
  could take in the system.

In the case of dedicated lines however, the specified cycle time will be followed in practice.

* It is also assumed that the time spent for material transportation between manufacturing cells is the same between each of them.

  This condition cannot be ensured entirely in practice, but represents no real restriction in flexible systems, where buffer stores will balance the remaining differences. (In the worst case real-time rescheduling will solve this problem if cummulative errors grow too large).

  In conventional systems if one employs a fixed transportation speed material handling system, such as a conveyor, the arrival time differences can be easily calculated, and thus taken into account. In most cases buffer stores will solve the problem unless there are relatively large differences in the access time between different workstations, compared to the specified cycle time.

Finally it must be underlined that this algorithm uses a heuristic method for calculating the minimum number of robots required to perform the specified number of operations in their given order. The program developed generates a number of feasible solutions in a random order and selects the answer which has the minimum number of robots (or assembly heads). Because of the random number generation, different solutions can occur, each time the program is rerun, although the alterations are usually minimal.

10.3 Case study: single product assembly system balancing

The purpose of the first case study is to demonstrate how a general purpose manufacturing system balancing program can be used to simulate a single product assembly system. Before running the program one should prepare the logically correct precedence diagram of the part or parts to be assembled on the line in question ([10.5]). In our example we use a product which has a precedence diagram shown in Figure 10.1.

In most cases the sequence of assembly operations is

restricted in terms of the order in which the operations can
be carried out. Sometimes alternatives are available or can be
generated, but without the support of the manufacturing system
such alternatives cannot be of real use.

In any case, i.e. whether it is a dedicated assembly system, or a flexible assembly system (FAS), it is important to
be able to answer the following questions prior to finalizing
its design:

* What are the necessary operations?

* What are the alternative operations?

* What is the order of the selected operations?

* How many assembly heads, or robots are required to
  perform the selected operations in the designed order?

* What happens if the cycle time increases and what happens if it decreases?

* Should the production rate be higher, or should one
  employ fewer robots? Is there an optimum cycle time,
  if so what is it and which is the most cost effective
  solution?

It is important to remember that in a "real" Flexible
Assembly System robots can accomodate large variety of different jobs in a random order, because they have Automated Robot
Hand Changing facility, the parts are transferred automatically between the robotised cells and because the necessary
sensing, control and data communication tasks are carried out
by the robot cell controller.

In this sense the robot cell is an FMS station capable of
working unmanned, receiving and transferring parts, tools
(i.e. robot hands) and data automatically. A robot in this
sense is really a "computer with hands".

There are two important results which should be observed
from the sample runs shown in Figures 10.4 to 10.7. (Note that
we have provided the full listings of input and output, as
well as the robot cell allocations in the precedence diagrams
in order to be able to follow each case separately, as well as
comparing them with each other).

474   Single and mixed product manufacturing

Figure 10.4/a Single product assembly system balancing. (Target
                 cycle time = 132 units, calculated best = 131
                 units. Note that the "SUB-TOTAL.." values indicate
                 robot cell loading levels for the particular
                 product.)

```
SINGLE PRODUCT ASSEMBLY SYSTEM SIMULATION   (First run)
                                                    Page 1
+---------------------------------------------------------------+


        ************************************************
        *            THE FMS SOFTWARE LIBRARY          *
        *----------------------------------------------*
        *     ROBOTIZED ASSEMBLY SYSTEM BALANCING      *
        ************************************************

INPUT DATA ECHO
***************

BATCH CODE    :Batch A-10-5-3-85
PRODUCT NAME  :Microcomputer
ANALYST       :Paul Ránky
DATE          :5/3/1985

COMMENT       :First run

------------------------------------------------------------------

NUMBER OF OPERATIONS            =     22
MAXIMUM ASSEMBLY CYCLE TIME     =    132.000  <-
NUMBER OF TRIALS IN THIS RUN    =     30


        OPERATION      DURATION         PRECEDENCE CONSTRAINTS
------------------------------------------------------------------

            1          16.000
            2          27.000
            3          21.000
            4          12.000          1,
            5          17.000          1,
            6          13.000          2,   5,
            7          15.000          2,   3,
            8          14.000          4,
            9          38.000          8,
           10          10.000          9,
           11          18.000          5,
           12          22.000          6,
           13          18.000          7,
           14          19.000         11,  12,
           15          26.000         13,
           16          20.000          9,  11,
           17          32.000         14,  15,
           18          28.000         15,
           19          29.000         15,
           20          41.000         16,  17,
           21          33.000         18,
           22          43.000         19,  20,  21,
```

```
SINGLE PRODUCT ASSEMBLY SYSTEM SIMULATION  (First run)
                                               Page 2
+------------------------------------------------------------+
```

CALCULATED RESULTS FOR THE OPTIMUM SOLUTION
****

```
NUMBER OF ROBOT STATIONS REQUIRED         =      4
TOTAL CYCLE TIME REQUIREMENT              =    131.000

CYCLE TIME DIFFERENCE (target-calculated best) =    1.000
```

****** TABLE OF RESULTS ******

ROBOT STATION	OPERATION	DURATION
1	2	27.000
	3	21.000
	1	16.000
	7	15.000
	13	18.000
	5	17.000
	6	13.000
	SUB-TOTAL...	127.000
2	4	12.000
	15	26.000
	11	18.000
	8	14.000
	19	29.000
	18	28.000
	SUB-TOTAL...	127.000
3	12	22.000
	14	19.000
	9	38.000
	17	32.000
	16	20.000
	SUB-TOTAL...	131.000
4	10	10.000
	20	41.000
	21	33.000
	22	43.000
	SUB-TOTAL...	127.000

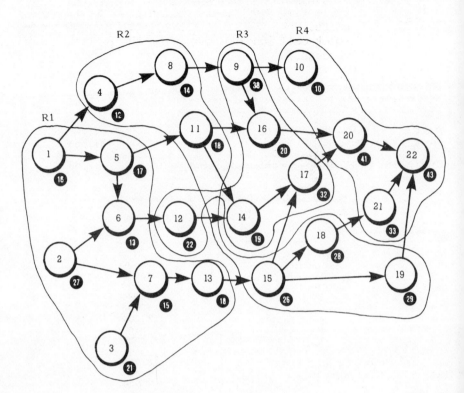

Figure 10.4/b Calculated results indicating graphically the way operations should be grouped for each robotized assembly cell. (Target cycle time = 132 units, number of calculated robots = 4.)

The goal of these runs is:

1. To demonstrate the way different cycle times affect the number of required robot cells, their load and the allocation of different operations on them.

2. To illustrate that such simulation techniques can be used as off-line design tools, as well as real-time software in case any of 'the cells are overloaded and new alternative route selection is required.

Figure 10.5/a Single product assembly system balancing. (Target
cycle time = 90 units, calculated best = 90
units. Note that the "SUB-TOTAL.." values indicate
robot cell loading levels for the particular
product.)

(Second run)

Page 1

+---+

```
**********************************************
*          THE FMS SOFTWARE LIBRARY          *
*--------------------------------------------*
*    ROBOTIZED ASSEMBLY SYSTEM BALANCING     *
**********************************************
```

INPUT DATA ECHO
****

BATCH CODE   :Batch A-10-5-3-85
PRODUCT NAME :Microcomputer
ANALYST   :Paul Ránky
DATE   :5/3/1985

COMMENT   :Second run

NUMBER OF OPERATIONS   =   22
MAXIMUM ASSEMBLY CYCLE TIME   =   90.000 ⬅
NUMBER OF TRIALS IN THIS RUN   =   30

OPERATION	DURATION	PRECEDENCE CONSTRAINTS
1	16.000	
2	27.000	
3	21.000	
4	12.000	1,
5	17.000	1,
6	13.000	2, 5,
7	15.000	2, 3,
8	14.000	4,
9	38.000	8,
10	10.000	9,
11	18.000	5,
12	22.000	6,
13	18.000	7,
14	19.000	11, 12,
15	26.000	13,
16	20.000	9, 11,
17	32.000	14, 15,
18	28.000	15,
19	29.000	15,
20	41.000	16, 17,
21	33.000	18,
22	43.000	19, 20, 21,

```
+-----------------------------------------------------------------------+
```

CALCULATED RESULTS FOR THE OPTIMUM SOLUTION
****

NUMBER OF ROBOT STATIONS REQUIRED   =   6
TOTAL CYCLE TIME REQUIREMENT   =   90.000

CYCLE TIME DIFFERENCE (target-calculated best) =   0.000 ←

****** TABLE OF RESULTS ******

ROBOT STATION	OPERATION	DURATION
1	2	27.000
	3	21.000
	1	16.000
	4	12.000
	8	14.000
	SUB-TOTAL...	90.000
2	9	38.000
	5	17.000
	10	10.000
	11	18.000
	SUB-TOTAL...	83.000
3	6	13.000
	7	15.000
	12	22.000
	13	18.000
	14	19.000
	SUB-TOTAL...	87.000
4	15	26.000
	17	32.000
	19	29.000
	SUB-TOTAL...	87.000
5	16	20.000
	18	28.000
	21	33.000
	SUB-TOTAL...	81.000
6	20	41.000
	22	43.000
	SUB-TOTAL...	84.000

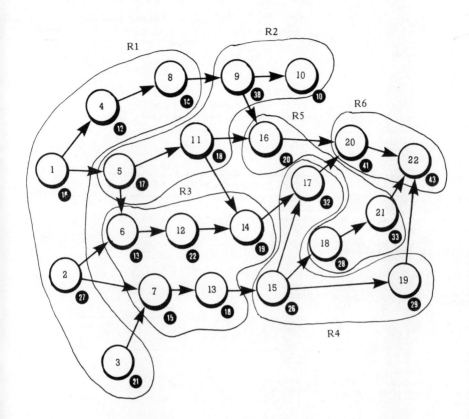

Figure 10.5/b Calculated results indicating graphically the way operations should be grouped for each robotized assembly cell. (Target cycle time = 90 units, number of calculated robots = 6.)

Figure 10.6/a Single product assembly system balancing. (Target
cycle time = 65 units, calculated best = 63
units. Note that the "SUB-TOTAL.." values indicat
robot cell loading levels for the particular
product.)

+--+

```
      ***************************************************
      *              THE FMS SOFTWARE LIBRARY           *
      *-------------------------------------------------*
      *      ROBOTIZED ASSEMBLY SYSTEM BALANCING        *
      ***************************************************

INPUT DATA ECHO
***************

BATCH CODE      :Batch A-10-5-3-85
PRODUCT NAME    :Microcomputer
ANALYST         :Paul Ránky
DATE            :5/3/1985

COMMENT         :Third run
```

```
NUMBER OF OPERATIONS           =     22
MAXIMUM ASSEMBLY CYCLE TIME    =     65.000  ←
NUMBER OF TRIALS IN THIS RUN   =     30
```

OPERATION	DURATION	PRECEDENCE CONSTRAINTS
1	16.000	
2	27.000	
3	21.000	
4	12.000	1,
5	17.000	1,
6	13.000	2, 5,
7	15.000	2, 3,
8	14.000	4,
9	38.000	8,
10	10.000	9,
11	18.000	5,
12	22.000	6,
13	18.000	7,
14	19.000	11, 12,
15	26.000	13,
16	20.000	9, 11,
17	32.000	14, 15,
18	28.000	15,
19	29.000	15,
20	41.000	16, 17,
21	33.000	18,
22	43.000	19, 20, 21,

(Third run)

+---+

CALCULATED RESULTS FOR THE OPTIMUM SOLUTION
****

```
NUMBER OF ROBOT STATIONS REQUIRED          =      9
TOTAL CYCLE TIME REQUIREMENT               =     63.000

CYCLE TIME DIFFERENCE (target-calculated best) =   2.000
```

******   TABLE OF RESULTS   ******

ROBOT STATION	OPERATION	DURATION
1	1	16.000
	5	17.000
	11	18.000
	4	12.000
	SUB-TOTAL...	63.000
2	3	21.000
	2	27.000
	7	15.000
	SUB-TOTAL...	63.000
3	6	13.000
	12	22.000
	14	19.000
	SUB-TOTAL...	54.000
4	13	18.000
	8	14.000
	15	26.000
	SUB-TOTAL...	58.000
5	17	32.000
	19	29.000
	SUB-TOTAL...	61.000
6	18	28.000
	21	33.000
	SUB-TOTAL...	61.000
7	9	38.000
	10	10.000
	SUB-TOTAL...	48.000
8	16	20.000
	20	41.000
	SUB-TOTAL...	61.000
9	22	43.000
	SUB-TOTAL...	43.000

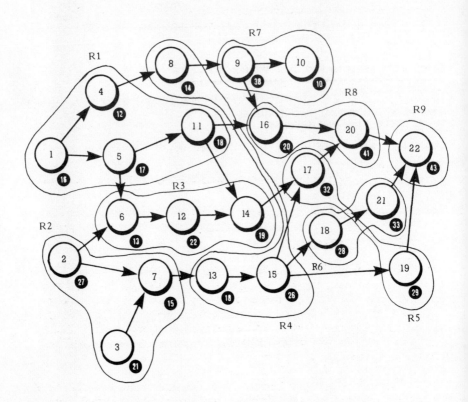

Figure 10.6/b Calculated results indicating graphically the way
operations should be grouped for each robotized
assembly cell. (Target cycle time = 65 units,
number of calculated robots = 9.)

Figure 10.7/a Single product assembly system balancing. (Target
cycle time = 55 units, calculated best = 52
units. Note that the "SUB-TOTAL.." values indicate
robot cell loading levels for the particular
product.)

(Fourth run)

Page 1

+--+

```
*************************************************
*           THE FMS SOFTWARE LIBRARY            *
*-----------------------------------------------*
*    ROBOTIZED ASSEMBLY SYSTEM BALANCING        *
*************************************************
```

INPUT DATA ECHO
****

BATCH CODE   :Batch A-10-5-3-85
PRODUCT NAME :Microcomputer
ANALYST   :Paul Ránky
DATE   :5/3/1985

COMMENT   :Fourth run

NUMBER OF OPERATIONS   =   22
MAXIMUM ASSEMBLY CYCLE TIME   =   55.000 ◄
NUMBER OF TRIALS IN THIS RUN   =   30

OPERATION	DURATION	PRECEDENCE CONSTRAINTS
1	16.000	
2	27.000	
3	21.000	
4	12.000	1,
5	17.000	1,
6	13.000	2, 5,
7	15.000	2, 3,
8	14.000	4,
9	38.000	8,
10	10.000	9,
11	18.000	5,
12	22.000	6,
13	18.000	7,
14	19.000	11, 12,
15	26.000	13,
16	20.000	9, 11,
17	32.000	14, 15,
18	28.000	15,
19	29.000	15,
20	41.000	16, 17,
21	33.000	18,
22	43.000	19, 20, 21,

CIM-Q

(Fourth run)

+---+

CALCULATED RESULTS FOR THE OPTIMUM SOLUTION
****

NUMBER OF ROBOT STATIONS REQUIRED   =   11
TOTAL CYCLE TIME REQUIREMENT   =   52.000

CYCLE TIME DIFFERENCE (target-calculated best) =   3.000

****** TABLE OF RESULTS ******

ROBOT STATION	OPERATION	DURATION
1	2	27.000
	3	21.000
	SUB-TOTAL...	48.000
2	7	15.000
	13	18.000
	1	16.000
	SUB-TOTAL...	49.000
3	4	12.000
	15	26.000
	8	14.000
	SUB-TOTAL...	52.000
4	18	28.000
	5	17.000
	SUB-TOTAL...	45.000
5	9	38.000
	6	13.000
	SUB-TOTAL...	51.000
6	19	29.000
	12	22.000
	SUB-TOTAL...	51.000
7	21	33.000
	10	10.000
	SUB-TOTAL...	43.000
8	11	18.000
	14	19.000
	SUB-TOTAL...	37.000
9	16	20.000
	17	32.000
	SUB-TOTAL...	52.000
10	20	41.000
	SUB-TOTAL...	41.000
11	22	43.000
	SUB-TOTAL...	43.000

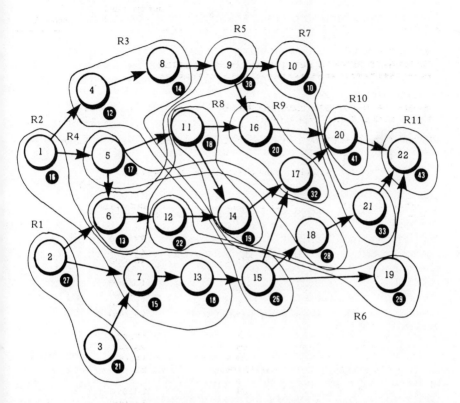

Figure 10.7/b Calculated results indicating graphically the way
operations should be grouped for each robotized
assembly cell. (Target cycle time = 55 units,
number of calculated robots = 11.)

10.4 Case study: mixed product assembly system balancing

The purpose of this case study is to illustrate the way mixed product manufacturing system balancing can be reduced to a single product balancing problem with certain engineering considerations.

Let us assume that the manufacturing system must assemble three different products at the same time, incorporating a few operations which are the same on all three.

Note that this is an important condition in mixed product assembly when more or less dedicated mechanized assembly heads are utilized. (Note also that mixed product assembly is simplified by following a proper "product family design " concept). In the case of "truly flexible" assembly where robotized assembly cells are capable of automated hand changing, part changing, etc. as already explained, it is much easier to provide the necessary facilities for mixed product assembly. The major restriction remaining in such systems is only the number and type of different hands in the magazines of the assembly robots.

For our case the three precedence diagrams of the three different product types are given in Figures 10.8/a-b and c and the combined precedence diagram in Figure 10.8/d. As one can see in this Figure operations marked 9, 10, 13, and 24 are common in two of the designs, and operations 1, 16, 23 in all three. The rest of the operations are individual to one of the three component types.

Because of the common operations on the three different parts, or their batches, one must remember that in the case of operations 9, 10, 13 and 24 double capacity, and in the case of operations 1, 16 and 23 triple capacity is required in order to get three different type of components (or their equal size batches) fully assembled from the line at the end of the specified cycle time values.

From the engineering point of view, there are basically two different ways of solving the problem from here.

1. The first approach suggests a 200% and 300% capacity increase respectively in the case of those workstations where the above indicated "common" operations are going

to be performed. This means that the design will be less flexible, because some heads, or robots will deal with one operation only, but as a bonus the cycle time will be shorter, because the indicated activities will be parallel.

This design is expressed in the multiple product precedence diagram. Note that because of the parallel capacity built into this design model the operation times have not been added together in the case of the multiple operations precedence diagram in Figure 10.8/d.

The sample runs shown in Figures 10.9 to 10.10 give a full list of the precedence rules, the number of required robots and the operation assignment for different cycle time values.

2. In the second solution time is allocated for multiple operations for those needed, thus in the precedence diagram of the product mix (see Figure 10.11/a) the relevant operation times are added together as many times as they occur in the mixed assembly situation.

This solution requires a more flexible manufacturing system than the previous one and will increase the cycle time too, since there will be no parallel cells, or assembly heads, offering parallel capacity, thus the selected cells will have to be able to cope with the increased load and they shall have to be able to perform the operations in a random order.

If the product mix changes often this second approach is the more preferable.

The sample runs shown in Figures 10.12 and 10.13 give a full list of the precedence rules, the number of requi-red robots and the operation assignment for different cycle time values.

To summarize, one can see how different cycle time values affect the number of required robot cells and the allocation of different operations. It must also be mentioned that the illustrated simulation models consider time related activities only, but in practice engineering decisions regarding cost, cell availability, robot hand availability, etc. would influence the design and the selected solution.

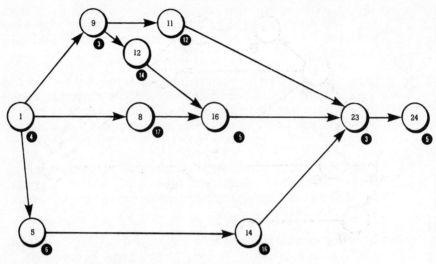

Figure 10.8/a Mixed production assembly system balancing.
Product A1.

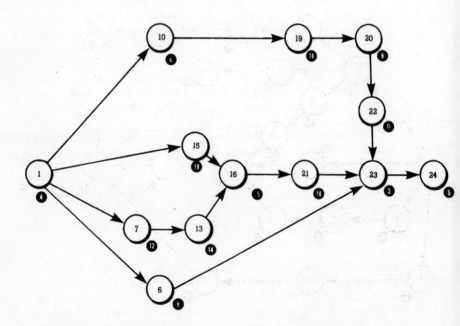

Figure 10.8/b Mixed production assembly system balancing.
Product A2.

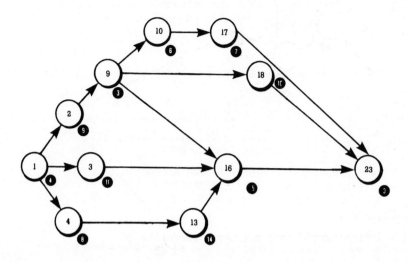

Figure 10.8/c Mixed production assembly system balancing.
Product B12.

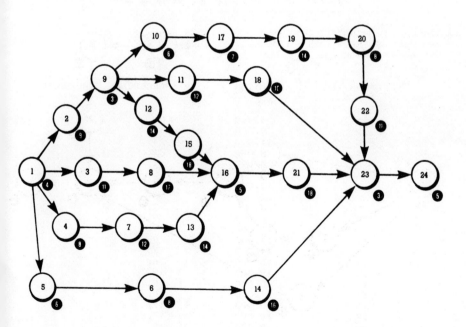

Figure 10.8/d Mixed production assembly system balancing.
Product mix.

Figure 10.9/a Mixed production assembly system balancing.
Product mix. (Target cycle time = 25 units,
calculated best = 24 units.)


```
          MIXED BATCH ASSEMBLY SYSTEM BALANCING   (First run)
                                                      Page 1
+----------------------------------------------------------------+



          ***************************************************
          *            THE FMS SOFTWARE LIBRARY           *
          *-----------------------------------------------*
          *     ROBOTIZED ASSEMBLY SYSTEM BALANCING       *
          ***************************************************

   INPUT DATA ECHO
   ***************

   BATCH CODE    :Batch-A1, Batch-A2, Batch-B12
   PRODUCT NAME  :Display terminal Type A1, A2 and B12
   ANALYST       :Paul Ranky
   DATE          :5/2/1985

   COMMENT       :Mixed batch balancing with parallel
                 capacity for multiple operations.
   -------------------------------------------------------------

   NUMBER OF OPERATIONS              =    24
   MAXIMUM ASSEMBLY CYCLE TIME       =    25.000  <---
   NUMBER OF TRIALS IN THIS RUN      =    30


          OPERATION       DURATION        PRECEDENCE CONSTRAINTS
   -------------------------------------------------------------

              1            4.000
              2            9.000          1,
              3           11.000          1,
              4            8.000          1,
              5            6.000          1,
              6            8.000          5,
              7           12.000          4,
              8           17.000          3,
              9            3.000          2,
             10            6.000          9,
             11           12.000          9,
             12           14.000          9,
             13           14.000          7,
             14           16.000          6,
             15           18.000         12,
             16            5.000          8, 15, 13,
             17            7.000         10,
             18           10.000         11,
             19           14.000         17,
             20            8.000         19,
             21           18.000         16,
             22           11.000         20,
             23            3.000         14, 18, 21, 22,
             24            5.000         23,
```

Case study: mixed product assembly system balancing   491

```
MIXED BATCH ASSEMBLY SYSTEM BALANCING   (First run)
                                                  Page 2
+-----------------------------------------------------------------+

CALCULATED RESULTS FOR THE OPTIMUM SOLUTION
*******************************************
NUMBER OF ROBOT STATIONS REQUIRED          =      11
TOTAL CYCLE TIME REQUIREMENT               =      24.000

CYCLE TIME DIFFERENCE (target-calculated best) =    1.000

              ******   TABLE OF RESULTS   ******

   ROBOT STATION          OPERATION              DURATION
----------------------------------------------------------------

         1                    1                  4.000
                              5                  6.000
                              4                  8.000
                         SUB-TOTAL...           18.000

         2                    6                  8.000
                             14                 16.000
                         SUB-TOTAL...           24.000

         3                    2                  9.000
                              9                  3.000
                             11                 12.000
                         SUB-TOTAL...           24.000

         4                   12                 14.000
                             18                 10.000
                         SUB-TOTAL...           24.000

         5                    3                 11.000
                              7                 12.000
                         SUB-TOTAL...           23.000

         6                   13                 14.000
                             10                  6.000
                         SUB-TOTAL...           20.000

         7                   17                  7.000
                              8                 17.000
                         SUB-TOTAL...           24.000

         8                   19                 14.000
                             20                  8.000
                         SUB-TOTAL...           22.000

         9                   15                 18.000
                             16                  5.000
                         SUB-TOTAL...           23.000

        10                   21                 18.000
                         SUB-TOTAL...           18.000

        11                   22                 11.000
                             23                  3.000
                             24                  5.000
                         SUB-TOTAL...           19.000

CIM-Q*
```

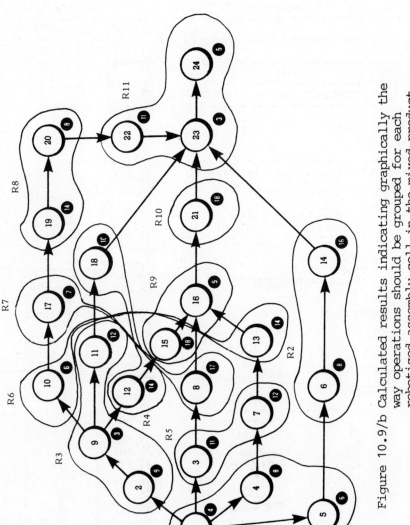

Figure 10.9/b Calculated results indicating graphically the way operations should be grouped for each robotized assembly cell in the mixed product assembly system. (Target cycle time = 25 units, number of calculated robots = 11.)

Figure 10.10/a Mixed production assembly system balancing.
Product mix. (Target cycle time = 25 units,
calculated best = 24 units.)

```
     MIXED BATCH ASSEMBLY SYSTEM BALANCING   (Second run)
                                                     Page 1
+-------------------------------------------------------------------+



          *********************************************
          *          THE FMS SOFTWARE LIBRARY         *
          *-------------------------------------------*
          *     ROBOTIZED ASSEMBLY SYSTEM BALANCING   *
          *********************************************

     INPUT DATA ECHO
     ***************

     BATCH CODE    :Batch-A1, Batch-A2, Batch-B12
     PRODUCT NAME  :Display terminal Type A1, A2 and B12
     ANALYST       :Paul Ránky
     DATE          :5/2/1985

     COMMENT       :Mixed batch balancing with parallel
                    capacity for multiple operations.
     -----------------------------------------------------------

     NUMBER OF OPERATIONS            =    24
     MAXIMUM ASSEMBLY CYCLE TIME     =    42.000  <-
     NUMBER OF TRIALS IN THIS RUN    =    30


          OPERATION       DURATION          PRECEDENCE CONSTRAINTS
     ------------------------------------------------------------------

             1             4.000
             2             9.000          1,
             3            11.000          1,
             4             8.000          1,
             5             6.000          1,
             6             8.000          5,
             7            12.000          4,
             8            17.000          3,
             9             3.000          2,
            10             6.000          9,
            11            12.000          9,
            12            14.000          9,
            13            14.000          7,
            14            16.000          6,
            15            18.000          12,
            16             5.000          8,  15,  13,
            17             7.000          10,
            18            10.000          11,
            19            14.000          17,
            20             8.000          19,
            21            18.000          16,
            22            11.000          20,
            23             3.000          14,  18,  21,  22,
            24             5.000          23,
```

494 Single and mixed product manufacturing

```
MIXED BATCH ASSEMBLY SYSTEM BALANCING   (Second run)
                                                          Page 2
+-------------------------------------------------------------------+

CALCULATED RESULTS FOR THE OPTIMUM SOLUTION
*******************************************


NUMBER OF ROBOT STATIONS REQUIRED          =      6
TOTAL CYCLE TIME REQUIREMENT               =      42.000

CYCLE TIME DIFFERENCE (target-calculated best) =   0.000  <-


         ******   TABLE OF RESULTS   ******


  ROBOT STATION          OPERATION          DURATION
---------------------------------------------------------------

         1                  1                4.000
                            4                8.000
                            3               11.000
                            2                9.000
                            5                6.000
                            9                3.000
                         SUB-TOTAL...       41.000

         2                  7               12.000
                           11               12.000
                           18               10.000
                            6                8.000
                         SUB-TOTAL...       42.000

         3                  8               17.000
                           10                6.000
                           12               14.000
                         SUB-TOTAL...       37.000

         4                 13               14.000
                           15               18.000
                           16                5.000
                         SUB-TOTAL...       37.000

         5                 21               18.000
                           14               16.000
                           17                7.000
                         SUB-TOTAL...       41.000

         6                 19               14.000
                           20                8.000
                           22               11.000
                           23                3.000
                           24                5.000
                         SUB-TOTAL...       41.000
```

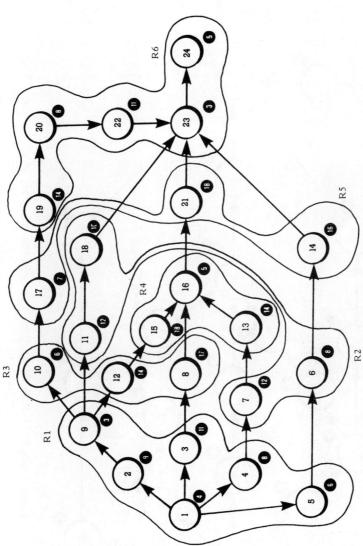

Figure 10.10/b Calculated results indicating graphically the way operations should be grouped for each robotized assembly cell in the mixed product assembly system. (Target cycle time = 25 units, number of calculated robots = 11.)

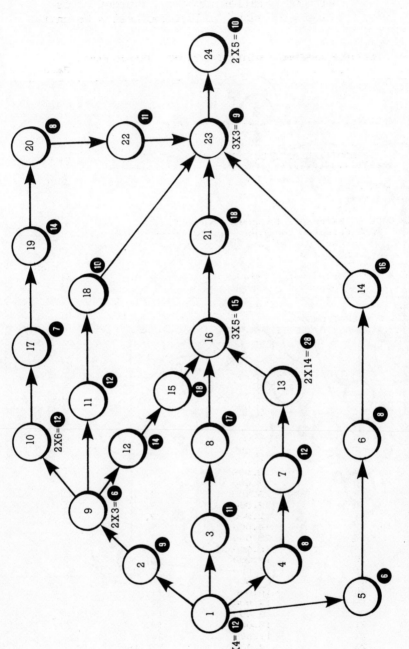

Figure 10.11/a Mixed product precedence diagram without parallel capacity.

Figure 10.11/b Mixed production assembly system balancing
without parallel capacity. (Target cycle
time = 28 units, calculated best = 28 units.)

```
            FLEXIBLE ASSEMBLY SYSTEM BALANCING   (First run)
                                                        Page 1
+------------------------------------------------------------------+


        ***********************************************
        *           THE FMS SOFTWARE LIBRARY          *
        *---------------------------------------------*
        *    ROBOTIZED ASSEMBLY SYSTEM BALANCING      *
        ***********************************************

INPUT DATA ECHO
***************

BATCH CODE    :Batch-A1, Batch-A2, Batch-B12
PRODUCT NAME :Display terminal Type A1, A2 and B12
ANALYST       :Paul Ránky
DATE          :5/2/1985

COMMENT       :Mixed batch balancing of a Flexible
               Robotised Assembly line
------------------------------------------------------------------

NUMBER OF OPERATIONS               =    24
MAXIMUM ASSEMBLY CYCLE TIME        =    28.000
NUMBER OF TRIALS IN THIS RUN       =    30


        OPERATION      DURATION            PRECEDENCE CONSTRAINTS
------------------------------------------------------------------


            1          12.000
            2           9.000          1,
            3          11.000          1,
            4           8.000          1,
            5           6.000          1,
            6           8.000          5,
            7          12.000          4,
            8          17.000          3,
            9           6.000          2,
           10          12.000          9,
           11          12.000          9,
           12          14.000          9,
           13          28.000          7,
           14          16.000          6,
           15          18.000         12,
           16          15.000          8, 13, 15,
           17           7.000         10,
           18          10.000         11,
           19          14.000         17,
           20           8.000         19,
           21          18.000         16,
           22          11.000         20,
           23           9.000         22, 18, 21, 14,
           24          10.000         23,
```

```
         FLEXIBLE ASSEMBLY SYSTEM BALANCING   (First run)
                                                          Page 2
+-----------------------------------------------------------------+

CALCULATED RESULTS FOR THE OPTIMUM SOLUTION
*********************************************


NUMBER OF ROBOT STATIONS REQUIRED              =     12
TOTAL CYCLE TIME REQUIREMENT                   =     28.000

CYCLE TIME DIFFERENCE (target-calculated best) =     0.000  <--

            ******   TABLE OF RESULTS   ******
    ROBOT STATION        OPERATION          DURATION
    -------------------------------------------------------

          1                 1              12.000
                            5               6.000
                            2               9.000
                      SUB-TOTAL...         27.000

          2                 4               8.000
                            7              12.000
                            6               8.000
                      SUB-TOTAL...         28.000

          3                13              28.000
                      SUB-TOTAL...         28.000

          4                 9               6.000
                           11              12.000
                           18              10.000
                      SUB-TOTAL...         28.000

          5                14              16.000
                            3              11.000
                      SUB-TOTAL...         27.000

          6                 8              17.000
                      SUB-TOTAL...         17.000

          7                12              14.000
                           10              12.000
                      SUB-TOTAL...         26.000

          8                17               7.000
                           19              14.000
                      SUB-TOTAL...         21.000

          9                15              18.000
                           20               8.000
                      SUB-TOTAL...         26.000

         10                16              15.000
                           22              11.000
                      SUB-TOTAL...         26.000

         11                21              18.000
                           23               9.000
                      SUB-TOTAL...         27.000

         12                24              10.000
                      SUB-TOTAL...         10.000
```

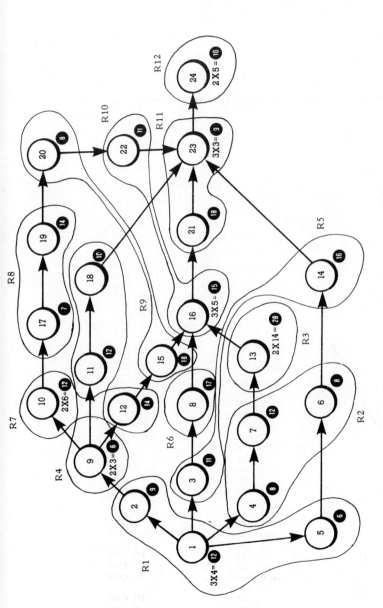

Figure 10.11/c Calculated results indicating graphically the way operations should be grouped for each robotized assembly cell in the mixed product assembly system. (Target cycle time = 28 units, number of calculated robots = 12.)

Figure 10.12/a Mixed production assembly system balancing
without parallel capacity. (Target cycle
time = 35 units, calculated best = 35 units.)

```
          FLEXIBLE ASSEMBLY SYSTEM BALANCING   (Second run)
                                                      Page 1
+----------------------------------------------------------------+



          **************************************************
          *           THE FMS SOFTWARE LIBRARY            *
          *----------------------------------------------*
          *     ROBOTIZED ASSEMBLY SYSTEM BALANCING       *
          **************************************************

INPUT DATA ECHO
***************

BATCH CODE     :Batch-A1, Batch-A2, Batch-B12
PRODUCT NAME   :Display terminal Type A1, A2 and B12
ANALYST        :Paul Ránky
DATE           :5/2/1985

COMMENT        :Mixed batch balancing of a Flexible
                Robotised Assembly line
-----------------------------------------------------------------

NUMBER OF OPERATIONS            =    24
MAXIMUM ASSEMBLY CYCLE TIME     =    35.000  <-
NUMBER OF TRIALS IN THIS RUN    =    30


     OPERATION        DURATION           PRECEDENCE CONSTRAINTS
-----------------------------------------------------------------

          1           12.000
          2            9.000          1,
          3           11.000          1,
          4            8.000          1,
          5            6.000          1,
          6            8.000          5,
          7           12.000          4,
          8           17.000          3,
          9            6.000          2,
         10           12.000          9,
         11           12.000          9,
         12           14.000          9,
         13           28.000          7,
         14           16.000          6,
         15           18.000         12,
         16           15.000          8, 13, 15,
         17            7.000         10,
         18           10.000         11,
         19           14.000         17,
         20            8.000         19,
         21           18.000         16,
         22           11.000         20,
         23            9.000         22, 18, 21, 14,
         24           10.000         23,
```

Case study: mixed product assembly system balancing   501

```
FLEXIBLE ASSEMBLY SYSTEM BALANCING   (Second run)
                                              Page 2
+----------------------------------------------------------------+


CALCULATED RESULTS FOR THE OPTIMUM SOLUTION
*******************************************


NUMBER OF ROBOT STATIONS REQUIRED            =     9
TOTAL CYCLE TIME REQUIREMENT                 =     35.000

CYCLE TIME DIFFERENCE (target-calculated best) =    0.000 <-

          ******   TABLE OF RESULTS   ******
   ROBOT STATION        OPERATION           DURATION
----------------------------------------------------------

         1                 1                12.000
                           2                 9.000
                           5                 6.000
                           4                 8.000
                      SUB-TOTAL...          35.000

         2                 6                 8.000
                           3                11.000
                           7                12.000
                      SUB-TOTAL...          31.000

         3                 8                17.000
                           9                 6.000
                          11                12.000
                      SUB-TOTAL...          35.000

         4                10                12.000
                          12                14.000
                          17                 7.000
                      SUB-TOTAL...          33.000

         5                13                28.000
                      SUB-TOTAL...          28.000

         6                14                16.000
                          15                18.000
                      SUB-TOTAL...          34.000

         7                19                14.000
                          20                 8.000
                          18                10.000
                      SUB-TOTAL...          32.000

         8                16                15.000
                          21                18.000
                      SUB-TOTAL...          33.000

         9                22                11.000
                          23                 9.000
                          24                10.000
                      SUB-TOTAL...          30.000
```

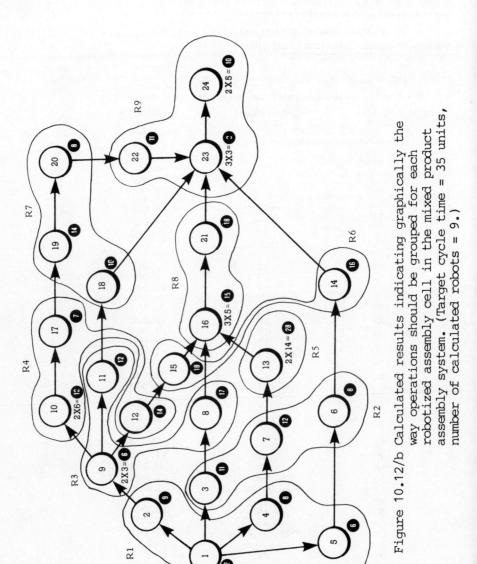

Figure 10.12/b Calculated results indicating graphically the
way operations should be grouped for each
robotized assembly cell in the mixed product
assembly system. (Target cycle time = 35 units,
number of calculated robots = 9.)

Figure 10.13/a Mixed production assembly system balancing
without parallel capacity. (Target cycle
time = 42 units, calculated best = 40 units.)

```
       FLEXIBLE ASSEMBLY SYSTEM BALANCING  (Third run)
                                                  Page 1
+------------------------------------------------------------+


       **********************************************
       *          THE FMS SOFTWARE LIBRARY          *
       *--------------------------------------------*
       *    ROBOTIZED ASSEMBLY SYSTEM BALANCING     *
       **********************************************

INPUT DATA ECHO
***************

BATCH CODE    :Batch-A1, Batch-A2, Batch-B12
PRODUCT NAME  :Display terminal Type A1, A2 and B12
ANALYST       :Paul Ránky
DATE          :5/2/1985

COMMENT       :Mixed batch balancing of a Flexible
               Robotised Assembly line
------------------------------------------------------------

NUMBER OF OPERATIONS             =    24
MAXIMUM ASSEMBLY CYCLE TIME      =    42.000  <---
NUMBER OF TRIALS IN THIS RUN     =    30

       OPERATION       DURATION        PRECEDENCE CONSTRAINTS
-----------------------------------------------------------------

           1           12.000
           2            9.000          1,
           3           11.000          1,
           4            8.000          1,
           5            6.000          1,
           6            8.000          5,
           7           12.000          4,
           8           17.000          3,
           9            6.000          2,
          10           12.000          9,
          11           12.000          9,
          12           14.000          9,
          13           28.000          7,
          14           16.000          6,
          15           18.000         12,
          16           15.000          8,  13,  15,
          17            7.000         10,
          18           10.000         11,
          19           14.000         17,
          20            8.000         19,
          21           18.000         16,
          22           11.000         20,
          23            9.000         22,  18,  21,  14,
          24           10.000         23,
```

```
            FLEXIBLE ASSEMBLY SYSTEM. BALANCING  (Third run)
                                                              Page 2
+-----------------------------------------------------------------+

CALCULATED RESULTS FOR THE OPTIMUM SOLUTION
*****************************************


NUMBER OF ROBOT STATIONS REQUIRED          =     8
TOTAL CYCLE TIME REQUIREMENT               =    40.000

CYCLE TIME DIFFERENCE (target-calculated best) =    2.000

             ******   TABLE OF RESULTS   ******
   ROBOT STATION         OPERATION           DURATION
---------------------------------------------------------------

        1                    1              12.000
                             4               8.000
                             3              11.000
                             2               9.000
                        SUB-TOTAL...        40.000

        2                    7              12.000
                            13              28.000
                        SUB-TOTAL...        40.000

        3                    5               6.000
                             8              17.000
                             9               6.000
                             6               8.000
                        SUB-TOTAL...        37.000

        4                   10              12.000
                            11              12.000
                            12              14.000
                        SUB-TOTAL...        38.000

        5                   14              16.000
                            17               7.000
                            18              10.000
                        SUB-TOTAL...        33.000

        6                   19              14.000
                            15              18.000
                            20               8.000
                        SUB-TOTAL...        40.000

        7                   22              11.000
                            16              15.000
                        SUB-TOTAL...        26.000

        8                   21              18.000
                            23               9.000
                            24              10.000
                        SUB-TOTAL...        37.000
```

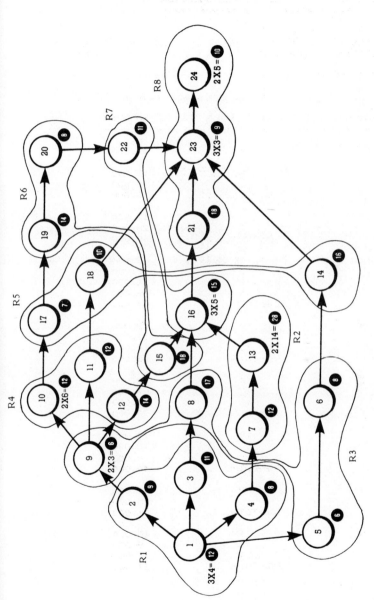

Figure 10.13/b Calculated results indicating graphically the way operations should be grouped for each robotized assembly cell in the mixed product assembly system. (Target cycle time = 42 units, number of calculated robots = 8.)

506   Single and mixed product manufacturing

References and further reading

[10.1] Frederic S. Hillier and Gerald J. Lieberman: Introduction to Operations Research, Holden-Day Inc., San Francisco, 1981.

[10.2] Elwood S. Buffa and William H. Taubert: Production-Inventory Systems, Planning and Control, Richard D. Irwin, Inc. 1974.

[10.3] Albert L. Arcus: COMSOAL, A computer method of sequencing operations for assembly lines The International Journal of Production Research, (1966), Volume 4, No.4, p 259-277.

[10.4] Paul G. Ránky, C.Y. Ho: Robot Modelling, Control and Applications with Software, IFS (Publications) Ltd. and Springer Verlag, 1985.

[10.5] Paul G. Ránky: The FMS Software Library. CIM software with user and system manuals. The BALANCE program. (Contact the author, or Malva, or Comporgan System Houses.)

Closing remarks

Every time I have a chance to visit the Science Museum in London, I am amazed by the enormous speed of development, particularly in the data processing industry.

Just to mention a few areas of interests there: compare a 'powerful' disk drive designed and manufactured only 10 years ago with an up-to-date Winchester drive, or a 'programmed machine tool' of the 18th Century with a current CNC machining center.

It is quite clear that the growth-rate of any product or machine incorporating computers is exponential. It is undoubtedly fascinating as well as sometimes dramatic to live in this century and experience this rapid development of science and technology.

I believe that the main task for all of us in the manufacturing industry is to create machines, robots, test equipment, material handling and storage devices, etc. which demonstrate some level of intelligence, which are reprogrammable and are able to communicate with each other.

Communications is the key issue of this decade and because of this Local Area Networks and wire-less data communication methods are increasingly important design goals for every manufacturer.

One should not forget either that the young engineering generation must understand and apply this increasingly complex technology, often referred to as CIM.

I have written this book because it may help people to learn, and to go on and prosper and create new jobs by not remaining on the periphery of such vital developments.

Paul G. Ránky
May 1985

Index